Fractional Calculus for Hydrology, Soil Science and Geomechanics

Ninghu Su

James Cook University
Cairns, Queensland
Australia

CRC Press
Taylor & Francis Group
Boca Raton   London   New York

CRC Press is an imprint of the
Taylor & Francis Group, an **informa** business

A SCIENCE PUBLISHERS BOOK

T0174905

Cover credit: Cover illustration reproduced by kind courtesy of Dr. Ninghu Su. It is a photograph of a small-scale physical model designed by the author for a hydrological system consisting of soils and different types of aquifers.

CRC Press
Taylor & Francis Group
6000 Broken Sound Parkway NW, Suite 300
Boca Raton, FL 33487-2742

© 2021 by Taylor & Francis Group, LLC
CRC Press is an imprint of Taylor & Francis Group, an Informa business

No claim to original U.S. Government works

Version Date: 20200523

ISBN: 978-1-138-49166-3 (hbk)
ISBN: 978-0-367-51703-8 (pbk)

Visit the Taylor & Francis Web site at
http://www.taylorandfrancis.com

and the CRC Press Web site at
http://www.routledge.com

Foreword

This book offers a complete and self-contained overview of the theoretical aspects and applications of fractional calculus-based models in soil physics and hydrology, as well as poroelastic properties of porous media. It addresses water flow and solute movement in surface water, soils and groundwater systems. Further, this work also discusses some fractional generalizations of the main problems associated with flow and transport with these theories.

With the comprehensive and clear evidence of the practical implications of fractional calculus and the advantages of this theory to the current lore, this book represents a remarkable piece of literature that will certainly be a fundamental reference for those who wish to pursue theoretical studies and applications of fractional calculus in hydrology, soil science and related topics.

May 2019

Francesco Mainardi
University of Bologna
Italy

Preface

This book focuses on the development of fractional calculus-based mathematical models and their applications in hydrology, soil science and mechanics of flow in porous media.

Fractional calculus has been widely applied to numerous fields since its inception in 1695. Hydrology, soil science, mechanics of flow in porous media and other branches of geoscience and environmental science are among the fields where different mathematical models based on fractional calculus are extensively used. In spite of the wide range of applications of fractional calculus, there is no single treatise which presents, in a systematic form, its background, theory, models and applications with examples in hydrology, soil science, mechanics of flow in porous media, and other related topics in geoscience. The reality at the moment is that extensive reports on the development of fractional calculus and its applications in the form of fractional partial differential equations (fPDEs) and fractional integral equations (fIEs) in these fields appear in a wide range of journals and some are scattered in a limited number of books.

In this book, I endeavor to bring together the essential mathematics of fractional calculus, particularly the theory of fPDEs and their applications as models and in related topics, for readers in hydrology, soil science, mechanics of flow in porous media and some related branches of geo-environmental sciences.

Furthermore, I present the majority of these mathematical models as fPDEs in different forms and, in limited cases, fIEs and fractional differential equations (fDEs) reported by others (in literature) and developed by myself (including some unpublished results). I have also cited more than 900 references through the length of the book, listed in the Bibliography, which include both important historical and contemporary contributions in fractional calculus and pertinent topics. I believe that these references are very valuable and can provide useful information to the readers.

I aim to present these materials as a summary of the most relevant reported models, along with my own research findings in these fields. With this objective, I hope the content of this book to be of interest to senior undergraduates and postgraduates for their studies, and useful for scientists, engineers and practitioners with an interest in the background, theory and applications of fractional calculus-based models in related fields.

I first became acquainted with the term Fractional Calculus in my initial days serving as a Senior Research Associate at the School of Mathematical Sciences, Queensland University of Technology (QUT) in Brisbane, Australia. At a School seminar in late 2000, I was sitting next to Professor Fawang Liu, a colleague at the School. In 2012, he was awarded the Mittag-Leffler prize at the *International Conference on Fractional Differentiation and Its Applications* (FDA'12) held in Nanjing, China, for his pioneer contributions to numerical methods for fPDEs and related models. I noticed that he was holding a book titled Fractional Differential Equations by Igor Podlubny. I asked, "Can differential equations be fractional?" From this point, I started learning fractional calculus, particularly, fPDEs and fDEs and their applications.

Having published some papers on fPDEs developed for analyzing water movement in soils and aquifers, since 2009, I have increasingly realized the absence of a compact book which systematically presents the fundamentals and the most relevant materials on fPDEs and their applications in hydrology, soil science and mechanics of flow in porous media. In April 2016, Mr. Raju Primlani at Science Publishers, CRC Press/Taylor & Francis Group contacted me for writing a book on certain issues, but I had not decided what to write then. I am thankful for Mr. Primlani for his communication over the past few years which eventually resulted in this book.

In October 2016, following my visit to Ningxia University, I met my uncle, Mr. Su Haidong, a retired public servant from the county of Longde, Ningxia, and an amateur poet and calligrapher. During one of our conversations, he suggested me to write a book on the topics I am working on. I promised him, and decided then, to write something on fPDEs and their applications. I am very grateful for his interest in my study and career as he has always been encouraging since my childhood.

I am extremely grateful to Professor Francesco Mainardi at the University of Bologna, Italy, who was awarded the Mittag-Leffler prize at the FDA'14 held in Catania, Italy, for his pioneer contributions to the applications of the Mittag-Leffler function in problems of fractional calculus. I communicated with him over six years ago regarding his papers as a part of the materials for my self-education on fPDEs and related topics. In December 2018, I also requested him to comment on the draft book chapters. With the help of Professor Mainardi, who also sent me some of the papers he had published, I was able to apply the continuous-time random walk (CTRW) theory to derive fPDEs and carry out related asymptotic analyses of the solutions of fPDEs, which resulted in my development of the models for water movement in soils (2014) and in aquifers (2015, 2017), all of which have been published in the Journal of Hydrology. Through my communication with Professor Mainardi, following his comment on my draft chapters in April 2019, I was informed of some 'fake' fractional derivatives, in his own words, in literature which are spurious definitions of fractional derivatives proposed by some authors in contrast to the classical definitions identified by Tarasov (doi.org/10.1016/j.cnsns.2018.02.019).

I am also very grateful to late Professor Rudolf Gorenflo, who was originally with the Free University of Berlin, Germany. I contacted him for some papers, particularly on fIEs, that were published by him and/or his colleagues and introduced me to the solutions of fIEs. Originally, I wrote to him in December 2018 to request for his comment on the book's manuscript, in addition to the comments of Professor Mainardi. However, I was sadly informed by his son Harry that Professor R. Gorenflo had passed away in October 2017.

The library staff at James Cook University have been extremely efficient and helpful by locating some very valuable and rare references which could be used and included for my writing. Following the submission of the manuscript of this book, the editorial staff at the CRC Press/Taylor & Francis Group have efficiently provided professional comments and suggestions before the manuscript went into production.

I thank my partner, Rosalind Gilroy, who has always been supportive, positive and interested in what I have been doing. She proofread the whole draft book for errors and suggested changes. Despite her training in economics and marketing, she was remarkable in identifying some typing errors of highly mathematical nature, such as the wrong limits in the definitions of left-hand and right-hand fractional derivatives and special functions, which could ideally be identified only by mathematicians.

I can say that this book is a result of teamwork which originated from an invitation from a publisher, the encouragement from my uncle for its inception, and references and comments from Professor Mainardi and Professor Gorenflo. The corrections and proofreading from Rosalind and the editors are essential for the present form of this book. My connection and cooperation with Professor Fawang Liu at QUT, since 1997, while I was working in New Zealand, has been crucial for me to realign my research directions in hydrological and environmental modelling closer with those of mathematicians. Without this teamwork, the publication of this book would be impossible. I thank you all! However, as the author, I am responsible for everything presented herein.

November 2019

Ninghu Su
James Cook University
Cairns, Queensland, Australia
Ningxia University
Yinchuan, Ningxia, China

Contents

Chapter 1

Application of Fractional Calculus in Water Flow and Related Processes

1. Overview

Water and its movement on land, in soils and aquifers, and in the oceans support terrestrial and marine life on the Earth. Since the very ancient times, human settlements had to either be in the vicinity of viable water sources or build water storage facilities for their survival. In the first instance, the appearance of fresh water sources, in the form of rivers or springs, was essential for humans' survival and, notably, all the earliest civilizations were facilitated by the availability of fresh water such as along the river systems of Nile in Egypt, Tigris and Euphrates in Mesopotamia, Indus in India, and Huanghe (or Dahe, the Yellow River) in China. In the second situation wherein humans lived in dry areas or needed water for special purposes, water engineering works can be traced back to 3,200 BC (Biswas 1967) across the ancient civilizations.

In addition to meeting human survival and basic needs as an essential resource and commodity, water has been the object of spiritual, mythical, mythological, religious and philosophical activities. The earliest Greek philosopher Thales (624 BC–548 BC) (Cartledge 1998) is credited with the hypothesis that water was the underlying factor behind the development of the world. The great Chinese philosopher Laozi (or Laotzu) (~ 571 BC–~ 471 BC) reiterated the virtue of water for humans with the saying, "*the upmost kindness of a man is like water, being the most modest and gracious like water which nourishes all things without conflicts, ends up in the lowest positions and provides services without demand for a reward*".

To understand water and its various properties has always been of constant interest to mankind. Over several millennia, scientific communities have explored water as a central topic for various purposes and utilities with Archimedes' principle (287 BC–212 BC) as one early example. Truesdell (1953) and Darrigol (2005) documented in great detail the evolution of the discipline of hydrodynamics, dealing with the motion of ideal fluids as a highly hypothetical form of water, from Isaac Newton (1643–1727), Daniel Bernoulli (1700–1782) and Leonhard Euler (1707–1783) to George G. Stokes (1819–1903) and Burnett (1935, 1936).

The topic of water waves alone has attracted significant investigations, particularly from the 18th century (Stoker 1958) when many leading scientists and mathematicians joined the race to understand water waves, particularly, Joseph-Louis Lagrange (1736–1813), Seméon D. Poisson (1781–1840), Claude-Louis Navier (1785–1836), Augustin-Louis Cauchy (1789–1857), A.J.C. Barré de Saint-Venant (1797–1886), George G. Stokes (1819–1903), William Thomson (Lord Kelvin, 1824–1907), Joseph Valentin Boussinesq (1842–1929), Horace Lamb (1849–1934) and Jules Henri Poincaré (1854–1912) to list a few.

After centuries of developments by scientists, particularly physicists and mathematicians (Debler 1990), a set of general equations for fluid flow was finalized and named after two important contributors—C.L. Navier and G.G. Stokes. The Navier-Stokes equations (NSEs), discussed in Chapter 3, are fundamental equations that govern the flow of fluids, including that of water in porous media. The NSEs accommodate the viscoelasticity of the porous media, or poroviscoelasticity, and the compressibility of water to the extent of water movement in soils and aquifers.

Just like other fields of human knowledge, the pursuit of better solutions is never-ending. The high-order hydrodynamics of flow, known as Burnett hydrodynamics (Burnett 1935, 1936), is an example capable of explaining more physical mechanisms when the NSEs cease to be valid. The processes in which the high-order hydrodynamics work while the NSEs fail include phenomena such as absorption and dispersion of sound in fluids, dynamics of swarms of particles, structure of different profiles in shock waves at large Mach numbers, Couette flows, in continuum transition flows that appear around space vehicles, and flows in micro-channels (García-Colín et al. 2008).

Classic *hydrodynamics*, evolving since the 17th century, takes largely from mathematics as it deals with imaginary ideal fluids which are frictionless. Its application by experimentalists to real fluids creates the applied field of *hydraulics*. The empirical nature of hydraulics is limited in scope to water only. With the development of and interests in other forms of fluids in aeronautics, petroleum engineering and other areas in civil engineering, a broader field of study was developed—*fluid mechanics*. Fluid mechanics has three branches: *fluid statics* which is concerned with the mechanics of fluids at rest, *kinematics* which deals with velocities and streamlines without considering forces or energy, and *fluid dynamics* for the study of relations between velocities, acceleration and the forces exerted by or upon fluids in motion (Daugherty et al. 1989).

The quest for knowledge about water has created many applied fields in the modern classification of scientific and engineering disciplines. In sciences, water-related fields include fluid mechanics, hydraulics, hydrology, hydrodynamics, meteorology, oceanography, marine science, agricultural science and soil science with water as a key element. On the other hand, water-related fields in engineering comprise hydraulic engineering, irrigation and drainage engineering, marine and coastal engineering, etc.

Water has been a topic for extensive publications in various formats, and the myriad properties and aspects of water find mention in many monographs of hydrology and hydraulics and their sub-disciplines such as groundwater hydrology/hydraulics, surface water hydrology and soil hydrology.

Deformation is another aspect of soils and aquifers, for their physical properties have significant impact on civil engineering infrastructures and geological materials as well as on the environment. In particular, soil mechanics or geomechanics deals with the swelling properties of soil when its water content changes and the reciprocal changes in water pressure as a result of deformation in soil. Many reports in soil mechanics (geomechanics) can be found in parallel with publications in hydrology and hydraulics. Cauchy in 1822 and 1827 laid the foundation of the general theory of elasticity and its extension to mathematical physics (Love 1892). Cauchy's work was followed by Green in 1837 and de Saint-Venant in 1844, with six components of stress and strains investigated (Love 1892). The reports on the subsidence of geological strata first conceptualized by Pratt and Johnson (1926) in oilfields, and by Geertsma (1966) and Verruijt (1969) in aquifers resulting from the extraction of groundwater are regarded as the early examples of investigation in applied areas of geoscience. The term poroelasticity first used by Geertsma (1966) is also important for addressing this specific property of porous media (Wang 2000) in civil and petroleum engineering.

Many reports in hydrology, hydraulics and soil science dealing with soils and aquifers generally ignore the key issues of deformation and stress-strain relations, leaving such discourse for soil mechanics. The separation of soil mechanics and soil physics since the 1930s (Philip 1974) discouraged hydrology and soil science to integrate with geomechanics, making these fields apparently disconnected even though groundwater hydrologists and soil scientists deal with poroviscoelastic soils and aquifers—central topics in soil mechanics.

A weak bridge between hydrology and geomechanics was attempted in limited literature such as the works of Wang (2000), which addressed the linear poroelasticity of porous media, covering a range of issues—geomechanics (soil mechanics), hydrogeology (groundwater hydrology), petroleum engineering, the poroelasticity theory and applications based on the works of Terzaghi (1923) and Biot (1935, 1939, 1941, 1956a, b), and Biot's thermoelasticity (Biot 1941, 1956c). However, Wang's work does not discuss any aspect of the NSEs which govern the flow of water in poroelastic and thermoelastic media, thus eliminating key hydrological elements.

In terms of quantitative methods, integer partial differential equations (PDEs) have enjoyed success as the central mathematical models in hydrology, soil science, hydraulics and geomechanics, etc., for over a century since Darcy (1856) embarked on the use of differential equations (DEs) for describing water flow in porous media and Boussinesq (1904) presented PDEs for groundwater flow in unconfined aquifers.

The decade around 1990 was a turning point when fractional PDEs (fPDEs) appeared as better models, with more information about environmental processes (Lenormand 1992, Zaslavsky 1992, Compte 1997). Environmental processes such as solute transport, sediment transport and groundwater flow, etc., have been shown to be better modelled with fPDEs by Compte (1997); groundwater flow/seepage by He (1998); and solute transport in groundwater by Lenormand (1992) and Benson (1998). These developments were part of the evolution since 1974 when fractional calculus was re-launched, and monographs appeared in applied mathematics and other fields of science (Oldham and Spanier 1974, Samko et al. 1987, 1993, Miller and Ross 1993, Podlubny 1999, Kilbas et al. 2006, Hilfer 2000, Mainardi 2010, Herrmann 2011, Atanacković et al. 2014, Atangana 2018).

2. Objectives of This Book

In order to eliminate the invisible boundaries between hydrology, hydraulics, soil science and geomechanics (soil mechanics), and to address the inconsistent spectrum of mathematical models based on fPDEs in these fields, this book aims to systematically present key concepts, theories, quantitative methods and ideas centered on the application of fractional calculus in hydrology, soil science, flow in porous media and geomechanics (soil mechanics). This book aims to establish frameworks of mathematical models with concepts in fractional calculus, particularly fPDEs and fractional integral equations (fIEs), and stochastic methods such as the continuous-time random walks (CTRW) theory to water flow, solute transport and related processes.

Water flow on land is categorized as *overland flow*, water movement in unsaturated soils as *flow in unsaturated soils*, and that in saturated aquifers as *groundwater flow* (generally classified into two types—confined and unconfined aquifers). With the aforementioned goals and classification in mind, this book presents the following materials:

1. Fundamentals of mathematics in Chapter 2, dealing with concepts commonly used in fractional calculus for models and quantitative methods in hydrology and hydraulics of water flow, solute transport on land, in soils and aquifers, and soil mechanics;
2. Essential properties of soils and aquifers in the context of porous media, in Chapter 3;
3. An overview of the historical transition from quantitative methods based on integer PDEs to fractional calculus-based approach, in Chapter 4, and
4. The remaining Chapters present topics related to water flow and solute transport in *unsaturated soils* (Chapters 5, 6 and 7), *overland flow* (Chapter 8), in *saturated aquifers* (Chapters 9 and 10), and *geomechanics* (Chapter 11).

The topics in this book are central issues in *hydrology*, *soil science* and *geomechanics*, and the fundamentals, models and methodologies used for investigating water-related processes can be categorized in three parts, namely, *fundamentals*, *traditional methods* for quantification and *evolving approaches with fractional calculus*.

3. A Brief Description of Key Concepts

Soils and aquifers are porous media, the former generally unsaturated while the latter is saturated. Unsaturated soils and saturated aquifers are domains of *soil science* and *groundwater hydrology*, respectively. The land surface, both floodplains and undulating terrains, falls within the purview of *surface water hydrology*. *Porous media* of geological origin, such as soils and aquifers, are studied in geomechanics (soil mechanics), centered around the physical phenomena of deformation with viscoelastic or poroelastic properties measured by stress-strain relationships or similar terms.

Soils are either constantly unsaturated by water or variably saturated depending on the local climatic conditions such as rainfall and management options like irrigation. In most cases as reported in literature, there has been a tendency to regard soils as unsaturated porous media even though soils can be saturated. Aquifers, adjacent to the soils in the deeper strata, are physically connected to soils with varying levels of permeability for the exchange of water, solutes (a term representing chemicals, fertilizers, nutrients, microbes and other particles), gases and energy.

Porous media is the largest category of materials on the Earth. It includes a very wide range of soils and other geological strata, biological materials, and extensive types of artificial products. Soils and aquifers are major porous media of geological origin. As the biologically active and productive parts of land, soils are the most important porous media on the Earth, for they support terrestrial life. Soils are also the medium for material and energy exchanges with the atmosphere and the subsurface through a number of physical, chemical and biological processes. These processes include infiltration of water into the soil, evaporation of water from soil, percolation of infiltrated water into aquifers, and runoff on the land (hence forming streams of varying sizes, eventually leading into the oceans).

3.1 Evolution of Mathematical Models in Hydrology, Hydraulics, Soil Science, and Geomechanics Based on Fractional Calculus

Mathematical models in applied fields related to water in porous media were incepted by Darcy (1856), with a differential equation as the flux of water or its velocity in porous media, and by Boussinesq (1904), who used an entire set of PDEs for water movement in unconfined aquifers. The concept of poroelasticity since Terzaghi (1923) and Biot (1935, 1941) entered the investigations of porous media of geological origin in the 20th century. However, fractional calculus was only specifically applied to porous media after 1974 when fractional calculus was re-launched, marked by three events: publication of the first monograph on fractional calculus by Oldham and Spanier (1974); the first PhD conferred to Bertram Ross on the topic of fractional calculus; and the first *International Conference on Fractional Calculus and Its Applications* organized by Bertram Ross and held at the University of Newhaven, Connecticut in June 1974 (Miller et al. 1994).

The terminology of fractional calculus, as recorded in history, was the result of a question raised by the mathematician Marquis de L'Hôpital (1661–1704) to Gottfried Wilhelm Leibniz (1646–1716) in 1695, regarding the meaning of the differentiation of $\frac{d^n y}{dx^n}$ when $n = \frac{1}{2}$ (Kilbas et al. 2006, vii). Leibniz was unable to resolve this query of L'Hôpital's until 1819 when S.F. Lacroix proved that $\frac{d^{1/2} y(x)}{dx^{1/2}} = 2\sqrt{\frac{x}{\pi}}$ for $y(x) = x$ (Miller and Ross 1993). In 1823, Niels Henrik Abel (1802–1829) showed the result $\frac{d^{1/2} k}{dx^{1/2}} = \sqrt{\pi} f(x)$, with the integral equation $k = \int_0^x (x-t)^{-1/2} f(t) dt$ (named after him) to determine $f(x)$ for constant k (Miller and Ross 1993).

The slow evolution of fractional calculus from 1695 to 1974 saw many mathematicians and scientists in different fields applying fractional calculus to investigate various processes and phenomena. However, a complete set of fPDEs readily applicable to water flow and solute transport in soils, aquifers and closely related processes is attributed to the works of Compte (1997). This development will be detailed in Chapter 4 along with other issues.

As will be evident from Chapter 5 onwards, the order of a fractional derivative or fractional integration can be any function rather than being restricted to a fraction. It is, therefore, logical to say that fractional calculus is a generalized form of calculus, with classic integer calculus as a special case.

3.2 *Developments in the Theory and Methodologies for Poroelasticity*

With soil mechanics as an independent field since the 1930s (Philip 1974), the 'standard' text books and research reports on soils and aquifers in hydrology and soil science today almost ignore the deformation and stress-strain relations of porous media. An example of this fact is the Richards equation (1931) for the movement of water in soils, and many procedures dealing with various model parameter estimations.

However, with the property of porous media possessing elasticity termed as poroelasticity (Geertsma 1966, Detournay and Cheng 1993), the historical developments in elasticity, plasticity and viscoelasticity, marked by the contributions of Terzaghi (1923) and Biot (1935, 1939, 1941), signified the most important foundations relevant to flows in poroelastic media. Poroelasticity, a term first used by Geertsma (1966), according to Wang (2000), is also further named poroviscoelasticity in some cases by considering the flow of a viscous liquid in elastic porous media (Bemer et al. 2001). Both poroelasticity and poroviscoelasticity are the central issues in Chapter 11. Besides, the time-dependent deformation process described by Nutting (1921a, b) can be fundamentally treated as a creep-relaxation process that can be interpreted as a property with memory initiated by Caputo and Mainardi (1971a). The general property of thermoelastic relationships as studied by Newmann (1833) (Love 1892) and Zener (1938) constitutes more complex, yet realistic, processes of deformation and related mechanisms when poroviscoelastic materials are subjected to variable conditions.

With the introduction of the concepts of strain and stress, different methods have been developed over time to quantify the relationships of the two basic phenomena of poroelastic media (Wang 2000): *solid-to-fluid coupling* which occurs when a change in applied stress produces a change in fluid pressure or fluid mass, and *fluid-to-solid coupling* as a result of a change in the volume of the porous media due to a change in fluid pressure or fluid mass.

There are extensive literature reviews on poroviscoelastic media, some of which involve water flow and solute transport in porous media (Love 1892, Bagley and Torvik 1983, Koeller 1984, Bemer et al. 2001, Liingaard et al. 2004, Kausel 2010, Koeller 2010, Mainardi 2012, Lai et al. 2016, Sun et al. 2016).

Soil physics, since its separation from soil mechanics in the 1930s, developed its own methodologies to deal with the actual properties of soils. As outlined in section 2.8 of Chapter 3, an integral transform is used to convert the standard Cartesian coordinate in the vertical dimension to a material coordinate (represented by m) for swelling soils (Raats 1965, Raats and Klute 1968, Smiles and Rosenthal 1968, Philip 1969b, Philip and Smiles 1969). With this approach for swelling-shrinking soils, the material coordinate is valid for vertically deforming soils by neglecting the horizontal expansion of the soil resulting from an increase in the water content of swelling soils. This simplification allows deformation to be analyzed with void ratio as the variable which depends on the moisture content in the soil. Similar to the development of ideas for elasticity and deformation, more complicated factors like the thermal effect and time-dependent processes or processes with memory, etc., can also be introduced.

As Mainardi (2010, 26) and Mainardi and Spada (2011) showed, linear elasticity can be quantitatively related as reciprocal relations which have been observed in the field (Hsieh 1996) and are used for estimating aquifer parameters (Burbey 2001). fPDEs have also been developed to investigate water flow in soils by incorporating swelling-shrinking properties of soils (Su 2010, 2012, 2014). These topics are discussed in Chapters 5, 6 and 7.

3.3 Hydrology, Hydraulics, Soil Science, Geomechanics (Soil Mechanics) and Their Relevance to Environmental Issues

Despite its irreplaceable role in all the sectors of society and for sustaining life on Earth, water is, however, often taken for granted. The continual worldwide pollution of soils and land, water bodies and oceans is an example of this ignorance and has been highlighted in numerous reports by FAO (2015), UNEP (2017) and UNESCO (2018).

Solutes in water and in porous media, collectively referring to a number of materials like ions, microbes, natural or artificial particles and pollutants or contaminants of various forms, are central to the issues of environmental and health concerns. The integral analysis of water movement, solute transport and related processes in deforming porous media, with fPDEs and fIEs, is a major focus of this book. To this end, the objective here is to achieve an improved understanding and quantification of water-related processes in soils and aquifers as well as on the surface to sustainably manage the limited resources of water and land.

4. Notation in the Book

Throughout the book, mathematical symbols have been defined independently in each chapter, thus a small number of symbols having different definitions in different chapters. I have tried to ensure consistency in the definition of symbols within each chapter, except in a few cases wherein the original equations have been preserved with the corresponding specific symbols.

Chapter 2

Mathematical Preliminaries

This chapter covers the fundamentals of mathematics used throughout this book in order to ensure the reader's understanding of the essential concepts that are applied to water flow and related processes. The *first part* (sections 1, 2 and 3) provides mathematical preliminaries essential for the book, including the basic concepts of ordinary differential equations (ODEs), partial differential equations (PDEs), integral transforms and a brief outline of the asymptotic analysis of functions. The *second part* (sections 4 and 5) is an introduction to special functions which appear extensively in solutions of fractional differential equations (fDEs) and fractional partial differential equations (fPDEs), the fundamental solution and related concepts. The *third part* (section 6) introduces the definitions of fractional integrals, fractional derivatives and solutions of fDEs and fPDEs. The related materials are presented in the appendices.

1. Introduction

1.1 Basic Concepts, Definitions and Classification of Differential Equations

A mathematical equation which contains certain known and unknown variables and their derivatives is called a differential equation or ordinary differential equation (ODE) (Derrick and Grossman 1987). The simplest example of the ODE is the following model for radioactive decay or material decomposition:

$$\frac{dC}{dt} = -kC \tag{2.1}$$

where C is either the molecular weight or the concentration of a substance, t is the time, and k is the decay coefficient often treated as a constant. The differentiation of concentration with respect to time, $\frac{dC}{dt}$, can often be written as C' for the first-order derivative, as $\frac{d^2C}{dt^2}$ or C'' for the second-order derivative and so on.

A mathematical equation that contains a dependent variable, the independent variable or variables, and one or more partial derivatives of the dependent variable(s) is called a partial differential equation (PDE). For two independent variables such as

x and y for space, t for time, and an independent variable θ, the general form of the PDE can be written as (DuChateau and Zachmann 1986):

$$\theta(x, y, t) = f(x, y, t, \theta, \theta_x, \theta_y, \theta_{xx}, \theta_{yy}, \theta_{xy}...,) \tag{2.2}$$

where θ can be the moisture content or moisture ratio, f denotes the function, $\theta_x = \dfrac{\partial \theta}{\partial x}$ and $\theta_y = \dfrac{\partial \theta}{\partial y}$ are the first-order partial derivatives with respect to x and y respectively, θ_{xx} and θ_{yy} are the second-order partial derivatives, and $\theta_{xy} = \dfrac{\partial^2 \theta}{\partial x \partial y}$ is the mixed partial derivatives.

The *order* of a PDE is the number or the highest-order derivative in a PDE. For example, $\dfrac{\partial \theta}{\partial x}$ and $\dfrac{\partial^2 \theta}{\partial x^2}$ are the first-order and the second-order derivatives of θ with respect to x respectively. When the order of derivatives is not an integer, the derivative is called a fractional derivative and is expressed as $\dfrac{\partial^\lambda \theta}{\partial x^\lambda}$ or other forms. This book focuses on PDEs using fractional derivatives or fPDEs and their applications. Detailed formulation and analysis will be discussed in subsequent chapters.

An example of a PDE is the diffusion equation or the heat equation:

$$\frac{\partial \theta}{\partial t} = \frac{\partial}{\partial x}\left(D \frac{\partial \theta}{\partial x} \right) \tag{2.3}$$

where $\dfrac{\partial \theta}{\partial t} = \theta_t'$ is the first-order partial derivative with respect to time, representing the rate of change in the moisture content, and D is the diffusion constant called diffusivity or the diffusion coefficient.

PDEs fundamentally emerged in an era when ODEs could not solve certain problems in physics, such as the vibration of a string (Evans et al. 1999). With such a background, PDEs are physically-based mathematical models with the time variable specified over some intervals such as $t_1 < t < t_2$, and the space variables defined in some region or domain **R**. The solution of a PDE is a function which satisfies the PDE under a given initial condition (IC) and boundary conditions (BCs).

The initial condition along with the PDE forms the *initial value problem* or *Cauchy problem*. For such a problem, the initial data need not necessarily be time as in the case of steady-state flow of moisture when the time component is absent, the initial data could be that specified for any of x, y, θ_x or θ_{xx}, etc. (Myint-U and Debnath 1987).

Boundary conditions with the PDE give rise to the *boundary value problem*. There are four types of BCs and specific BCs can be stipulated for a PDE for a particular problem (DuChateau and Zachmann 1986, Myint-U and Debnath 1987, Zwillinger 1998):

1. *Dirichlet condition*: In hydrology and soil physics, this is often called the concentration boundary condition. With this type of BC, a given value or a function is specified at the boundary, such as $\theta(x, t) = \theta(0, t)$ for $x = 0$, $t > 0$;

2. *Neumann condition*: This condition is often referred to as the flux boundary condition in hydrology and soil physics. With this type of BC, the derivative of $\theta(x, t)$ is given at the boundary, i.e., $\dfrac{\partial \theta}{\partial x} = i(t)$ for $x = 0$, $t > 0$, where $i(t)$ could be the rainfall intensity or another form of flux.

3. *Mixed boundary condition*: This condition can appear as $\dfrac{\partial \theta}{\partial x} + K\theta = q$, $x = 0$, $t > 0$, with K as a constant or a function and q as a quantity such as flux. This type of BC models the flux at the boundary due to either infiltration or evaporation and is also called *the flux BC*.

4. *Robin problem*: In addition to the above three types of BCs, different BCs can also be specified for different portions of a boundary (Myint-U and Debnath 1987), hence forming this fourth type.

The above BCs can be constants, functions or a moving boundary. A moving BC can be specified for a problem when the location of the boundary is an unknown to be determined, the infiltration front or melting front for instance.

1.2 Major Types of PDEs Used in Flow in Porous Media, Hydrology, Soil Science and Geomechanics

PDEs can be classified into various types. Four forms of classifications depending on the following criteria have been identified in literature:

1. Linearity of the independent variable in a PDE: The terminologies *linear* and *non-linear* often appear in literature. A PDE is called a *linear* PDE if the dependent variable such as θ appears only with an exponent of 0 or 1 (Zwillinger 1998) and if the coefficients of θ and its derivatives are functions only of the independent variables such as x and/or t (DeChateau and Zachmann 1986). Alternatively, a PDE is non-linear if it does not satisfy the conditions for a linear PDE. In some cases, a *quasilinear* PDE appears when the PDE is linear in the highest-order derivative but non-linear in other order derivatives (DeChateau and Zachmann 1986).

2. Types of PDEs: The values of the coefficients of a PDE determine if a PDE is elliptic, hyperbolic, parabolic or degenerate (DuChateau and Zachmann 1986).

3. Homogeneity: The presence or absence of a source term in a PDE also forms another basis of classifying PDEs. A PDE without a source term is called a *homogeneous* PDE, whereas one with a source term is *non-homogeneous*.

4. Types of derivatives: A PDE with integer partial derivatives is simply called a PDE. On the other hand, a PDE qualifies as a fractional PDE (fPDE) when its order of derivatives is not an integer. The fractional order may be a constant or a function.

This book presents mathematical models based on fPDEs for flow and transport in porous media and related topics. For further reading on ODEs and PDEs and their applications in physical sciences, the reader is advised to refer to Carslaw and Jaeger (1959), DuChateau and Zachmann (1986), Derrick and Grossman (1987), Myint-U and Debnath (1987), Kevorkian (1990), Boyce and DiPrima (1997), Zwillinger (1998) and Evans et al. (1999).

2. Integral Transforms

This section briefly outlines integral transforms—Fourier transform and Laplace transform. These transforms are powerful tools for solving the ODEs, PDEs and fPDEs which are demonstrated in the book. Furthermore, Fourier and Laplace transforms and their inverse transforms are encountered in the derivation of fPDEs based on continuous-time random walk (CTRW) concept.

2.1 Fourier Transform and Inverse Fourier Transform

The Fourier transform of a function $f(x)$ is expressed as the following integral (Debnath and Bhatta 2007):

$$\mathcal{F}f(x) = F(k) = \frac{1}{\sqrt{2\pi}} \int_{-\infty}^{\infty} f(x)e^{-ikx}\,dx \tag{2.4}$$

where $F(k)$ is the Fourier transform of the function $f(x)$, $\mathcal{F}$ signifies Fourier transform and k is the Fourier transform variable.

Inverse Fourier transform is defined as:

$$\mathcal{F}^{-1}\{F(k)\} = f(x) = \frac{1}{\sqrt{2\pi}} \int_{-\infty}^{\infty} F(k)e^{ikx}\,dk \tag{2.5}$$

Fourier transform is one of the greatest achievements of Fourier (1822) and is the foundation for other integral transformations. It is a powerful method for solving many problems such as finding solutions of ODEs, PDEs, and their fractional counterparts, fODEs and fPDEs (Debnath and Bhatta 2007).

2.2 Laplace Transform and Inverse Laplace Transform

Laplace transform is a special form of Fourier transform. It is also a powerful method for solving ODEs, PDEs, fODEs and fPDEs. In this section, we only briefly discuss the definitions and major properties of Laplace transform and inverse Laplace transform, which have been most widely used for analyzing the problems discussed in this book. A list of Laplace transforms and inverse Laplace transforms of special functions is summarized in Appendix 2.1.

2.2.1 Definitions

The Laplace transform of a function $f(t)$ is defined by the following integral (Spiegel 1965):

$$\mathcal{L}\{f(t)\} = \tilde{f}(s) = \int_{0}^{\infty} e^{-st} f(t)\,dt \qquad \mathrm{Re}\, s > 0 \tag{2.6}$$

where $\tilde{f}(s)$ is the Laplace transform of $f(t)$, with s being the Laplace transform variable, $\mathcal{L}$ denotes Laplace transform, and Re refers to real numbers that are used to measure continuous quantities as opposed to natural numbers such as 1, 2, 3, ... (Encyclopaedia Britannica 2019).

Inverse Laplace transform is expressed as (Debnath and Bhatta 2007):

$$\mathcal{L}^{-1}\{\tilde{f}(s)\} = f(t) = \frac{1}{2\pi i} \int\limits_{c-i\infty}^{c+i\infty} e^{st} \tilde{f}(s)ds, \qquad c > 0 \tag{2.7}$$

In addition to Laplace transform, defined by Eq. (2.6), the finite Laplace transform of the function $f(t)$ is defined as (Debnath and Bhatta 2007):

$$\mathcal{L}_T\{f(t)\} = \tilde{f}(s,T) = \int\limits_0^T e^{-st} f(t)dt \tag{2.8}$$

2.2.2 Important Properties of Laplace Transform

Laplace transform, defined by Eq. (2.6) has the following salient properties (Spiegel 1965):

1. Linearity

   If $f_1(t)$ and $f_2(t)$ are functions with Laplace transforms $\tilde{f}_1(t)$ and $\tilde{f}_2(t)$ respectively, and c_1 and c_2 are constants, then the following property holds:

   $$\mathcal{L}\{c_1 f_1(t) + c_2 f_2(t)\} = c_1 \tilde{f}_1(s) + c_2 \tilde{f}_2(s) \tag{2.9}$$

2. Translation

   The Laplace transform of the product of the exponential function $e^{at} = \exp(at)$ and an arbitrary function $f(t)$ is given as:

   $$\mathcal{L}\{e^{at} f(t)\} = \tilde{f}(s-a) \tag{2.10}$$

   where a is a constant.

3. Shifting

   For the function

   $$G(t) = \begin{cases} f(t-a), & t > a \\ 0, & t < a \end{cases} \tag{2.11}$$

   where a is a constant, its Laplace transform is:

   $$\mathcal{L}\{G(t)\} = e^{-as} \tilde{f}(s) \tag{2.12}$$

4. Change of Scale

   For a constant a, the Laplace transform of $f(at)$ can be expressed as:

   $$\mathcal{L}\{f(at)\} = \frac{1}{a} \tilde{f}\left(\frac{s}{a}\right) \tag{2.13}$$

The Laplace transform has many other properties which can be found in related literature (Spiegel 1965, Oberhettinger and Badii 1973). Appendix 2.1 provides a summary of Laplace transforms of some special functions and their inverse Laplace transforms which are frequently used for operations in fractional calculus, particularly procedures for solving fPDEs.

3. Asymptotic Analysis

There has been rigorous research recorded in mathematical literature in asymptotic analysis, Copson (1965) for instance. For non-mathematical readers such as hydrologists and soil scientists, a less mathematical explanation of this topic is needed, and this section serves this purpose.

In plain and simple non-mathematical language, the asymptotic analysis is about finding the value of a function when its variable approaches the end values. For example, for a function $f(x)$, the asymptotic analysis may mean determination of the form of the function when $x \to 0$ and/or $x \to \infty$.

The notation $x \to 0$ is often written in two forms: when the variable approaches the origin from the positive side of the axis, it is expressed as $x \to +0$, and it is $x \to -0$ when the variable approaches from the negative direction. Similarly, two forms of expression are used for the variable when it approaches infinity, i.e., $x \to +\infty$ and $x \to -\infty$ when it approaches infinity from the positive side and from the negative side of the axis respectively. In many cases, $x \to +0$ is simply written as $x \to 0$, and $x \to +\infty$ is expressed as $x \to \infty$.

The asymptotic approximation of a function is expressed by different signs such as the approximation signs "~" and "≈", or the equality sign "=" if the full approximation is given (Copson 1965, Spiegel 1965, Podlubny 1999, Mainardi and Garrappa 2015), and, in some cases, by the proportionality sign "∝" (Friedrich 1991). If $f(t) \to a$ as $x \to +0$, we say that for values of t near $t = 0$ from the positive axis, $f(t)$ is close to a (Spiegel 1965).

For general information on asymptotic analysis, the reader may refer to Copson (1965). However, for asymptotic analyses of special functions related to fractional calculus and fPDEs, further reading of Mainardi and Garrappa (2015), Friedrich (1991), Podlubny (1999), Rangarajan and Ding (2000), Chechkin et al. (2002), Sandev et al. (2015) for fPDEs of constant order, and Fedotov et al. (2019) for fPDEs of functional order is suggested.

4. Special Functions

Special functions are very important in operations in fractional calculus, particularly in the derivation of fPDEs and their solutions. Hence, special functions and their properties constitute the majority of this chapter's content.

In solutions of fDEs and fPDEs, a number of functions are related to one particular series in different forms and this series is widely known as the Mittag-Leffler function (MLF) and its extensions such as the Wiman function or two-parameter MLF, generalized Mittag-Leffler function (GML), and more complicated functions like the Prabhakar function (PF) and generalized Prabhakar function (GPF) (Kilbas et al. 2004, 2013, Haubold et al. 2011, Mainardi and Garrappa, 2015, Saxena et al. 2015). In order to reduce repetition and confusion, this section summarizes their major definitions, related functions and major properties.

4.1 Gamma Function and Related Functions

The gamma function and its various combinations with other functions are central to the operations in fractional calculus and fractional calculus-based models. The gamma function is also an important mathematical model for many processes in hydrology and soil science, such as the solute breakthrough curves, the recession curves of groundwater, and storm hydrographs as represented by the well-known Nash model (1957).

In this section, some major properties of the gamma function and other related functions are outlined. The key properties of these functions can be found in the works of Abramowitz and Stegun (1965).

4.1.1 Definitions of the Gamma Function

The gamma function, $\Gamma(x)$, is defined in different forms (Abramowitz and Stegun 1965, 255) which include the following:

1. Euler's integral

$$\Gamma(x) = \int_0^\infty t^{x-1} e^{-t}\, dt \tag{2.14}$$

which is defined for the real number $x > 0$ or Re x. It is also equivalent to

$$\Gamma(x) = k^x \int_0^\infty t^{x-1} e^{-kt}\, dt \tag{2.15}$$

which is defined for $x > 0$ and $k > 0$.

2. Euler's infinite product

$$\frac{1}{\Gamma(x)} = x e^{\gamma x} \prod_{n=1}^\infty \left[\left(1 + \frac{x}{n}\right) e^{-x/n} \right] \tag{2.16}$$

which is defined for $|x| < \infty$, with $\prod_{n=1}^\infty$ representing the product of the n terms, and γ being the Euler number or Euler constant defined as

$$\gamma = \lim_{m \to \infty} \left[1 + \frac{1}{2} + \frac{1}{3} + \ldots \frac{1}{m} - \ln m \right] \approx 0.57721 \tag{2.17}$$

where ln is the natural logarithm of m.

4.1.2 Special Forms of the Gamma Function and Related Formulae

4.1.2.1 Factorials for Integers

When the argument of the gamma function is an integer, the following applies:

$$\Gamma(n+1) = 1 \cdot 2 \cdot 3 \ldots (n-1)n = n! \tag{2.18}$$

When the argument of the gamma function is a fraction, tabulated values or a computer code can be used to find the values of the gamma function.

4.1.2.2 Recurrence Formula and Reflection Formula

The gamma function of an integer can also be written using the *recurrence formula* (Abramowitz and Stegun 1965) as:

$$\Gamma(x+1) = x\Gamma(x) = x! = x(x-1)!$$ (2.19)

While deriving and operating on fPDEs, a negative argument in the gamma function occasionally appears, which requires the use of the *reflection formula*:

$$\Gamma(x)\Gamma(1-x) = -x\Gamma(-x)\Gamma(x) = \pi\csc(\pi x)$$

$$= \int_0^\infty \frac{t^{x-1}}{1+t}\,dt$$ (2.20)

where $\csc(\pi x)$ is the trigonometric function. This formula is applicable in the range $0 < x < 1$. Rearranging the *reflection formula* in Eq. (2.20) gives the gamma function with a negative argument which is used in some operations (Gradshteyn and Ryzhik 1994, 256, Eq. (6.1.17)) as:

$$\Gamma(-x) = -\frac{\Gamma(1-x)}{x}$$ (2.21)

4.1.2.3 Binomial Coefficient

$$\binom{x}{y} = \begin{cases} \dfrac{x!}{y!(x-y)!} & \text{for } 0 \le y \le x \\ 0 & \text{for } 0 \le x < y \end{cases}$$ (2.22)

In the case of $0 \le y \le x$, this can be written as a combination of gamma functions:

$$\binom{x}{y} = \frac{\Gamma(x+1)}{\Gamma(y+1)\Gamma(x-y+1)}$$ (2.23)

4.2 Incomplete Gamma Function

Several forms of the incomplete gamma function have been defined in literature, which include the following (Abramowitz and Stegun 1965):

$$\gamma(a,x) = \int_0^x e^{-t} t^{a-1}\,dt$$ (2.24)

and

$$\gamma^*(a,x) = \frac{x^{-a}}{\Gamma(a)}\gamma(a,x) = \frac{x^{-a}}{\Gamma(a)}\int_0^x e^{-t} t^{a-1}\,dt$$ (2.25)

which has a series representation as:

$$\gamma^*(a,x) = e^{-x}\sum_{n=0}^\infty \frac{x^n}{\Gamma(a+n+1)} = \frac{1}{\Gamma(a)}\sum_{n=0}^\infty \frac{(-x)^n}{(a+n)n!}$$ (2.26)

Another incomplete gamma function is the following identity which relates the gamma function and the incomplete gamma function:

$$\Gamma(a,x) = \Gamma(a) - \gamma(a,x)$$

$$= \int_x^\infty e^{-t} t^{a-1} dt \qquad (2.27)$$

4.3 Beta Function

The beta function is defined as:

$$B(z,w) = \int_0^1 t^{z-1} (1-t)^{w-1} dt = \int_0^\infty \frac{t^{z-1}}{(1+t)^{z+w}} dt$$

$$= \int_0^{\pi/2} (\sin t)^{2z-1} (\cos t)^{2w-1} dt \qquad (2.28)$$

with $z > 0$ and $w > 0$. The beta function is related to the gamma function through the following identity (Abramowitz and Stegun 1965):

$$B(z,w) = B(w,z) = \frac{\Gamma(z)\Gamma(w)}{\Gamma(z+w)} \qquad (2.29)$$

where $\Gamma(z)$ and $\Gamma(w)$ are gamma functions of z and w respectively.

4.4 Mittag-Leffler Function and Related Functions

In fractional calculus and solutions of fDEs and fPDEs, the Mittag-Leffler function (MLF) and its variants are very frequently used special functions, and the mathematical structures of which are briefly outlined in this section. Their connections with some more complicated functions are listed in Appendix 2.3 at the end of this section.

In addition to the MLF and other simpler functions outlined above, there are many forms of complicated functions associated with the MLF in solutions of different fDEs and fPDEs. This section only briefly discusses the more relevant forms of these functions which appear in the operations of fDEs and fPDEs in this book.

4.4.1 Mittag-Leffler Function

The Mittag-Leffler function was named after Gösta Mittag-Leffler, who introduced it in 1903 as an infinite series of the following form:

$$E_\alpha(x) = \sum_{k=0}^\infty \frac{x^k}{\Gamma(\alpha k + 1)}, \qquad (\text{Re } \alpha > 0, x \in C) \qquad (2.30)$$

In addition to the above, for an integer α, the following special identities hold (Kilbas et al. 2006, Valerio et al. 2013):

$$E_0(x) = \frac{1}{1-x}, \qquad |x| < 1 \qquad (2.31)$$

$$E_1(x) = e^x \tag{2.32}$$

$$E_2(x) = \cosh\left(\sqrt{x}\right) \tag{2.33}$$

4.4.1.1 Differentiation

For $\alpha = n$ as integers, the following properties hold (Kilbas et al. 2006):

$$\frac{d^n}{dx^n} E_n\left(\lambda x^n\right) = \lambda E_n\left(\lambda x^n\right) \qquad (n \in N; \lambda \in C) \tag{2.34}$$

$$\frac{d^n}{dx^n}\left[x^{n-1} E_n\left(\frac{\lambda}{x^n}\right)\right] = \frac{(-1)^n \lambda}{x^{n+1}} E_n\left(\frac{\lambda}{x^n}\right) \qquad (z \neq 0; n \in N; \lambda \in C) \tag{2.35}$$

and

$$\frac{d^n}{dx^n} E_\alpha(x) = n! E_{\alpha,1+\alpha n}^{n+1}(x) \qquad (n \in N) \tag{2.36}$$

where $E_{\alpha,1+\alpha n}^{n+1}(x)$ is the generalized MLF or the Prabhakar function discussed in later sections.

The partial differentiation of the MLF with respect to x is given as (Kilbas et al. 2006):

$$\frac{\partial^n}{\partial x^n} E_\alpha\left(\lambda x^\alpha\right) = x^{-n} E_{\alpha,1-n}\left(\lambda x^\alpha\right) \tag{2.37}$$

and the partial differentiation of the MLF with respect to λ is given as:

$$\frac{\partial^n}{\partial \lambda^n} E_\alpha\left(\lambda x^\alpha\right) = n! x^{n\alpha} E_{\alpha,n\alpha+1}^{n+1}\left(\lambda x^\alpha\right) \tag{2.38}$$

where $E_{\alpha,n\alpha+1}^{n+1}\left(\lambda x^\alpha\right)$ is the generalized MLF or the Prabhakar function.

The fractional differentiation of the MLF was given by Kilbas et al. (2004) as:

$$D_{0+}^v\left[E_\alpha\left(\pm\lambda t^\alpha\right)\right](x) = x^{-v} E_{\alpha,1-v}\left(\pm\lambda x^\alpha\right) \tag{2.39}$$

where $E_{\alpha,1-v}\left(\pm\lambda x^\alpha\right)$ is the two-parameter MLF or the Wiman function discussed in subsequent sections.

4.4.1.2 Laplace Transform of the MLF

The Laplace transform of the MLF is expressed as (Valério et al. 2013):

$$\mathcal{L}\left[E_\alpha\left(\pm\lambda x^\alpha\right)\right] = \frac{s^{\alpha-1}}{s^\alpha \mp \lambda} \tag{2.40}$$

In fractional calculus operations, the MLF is often combined with other algebraic functions to form more complex functions and their Laplace transforms are more complicated (Valério et al. 2013). Some frequently used special functions and their Laplace transforms are hence listed in Appendix 2.1.

4.4.2 Two-parameter Mittag-Leffler Function or the Wiman Function

Often called the two-parameter Mittag-Leffler function, the Wiman function was introduced by and named after Wiman (1905):

$$E_{\alpha,\beta}(x) = \sum_{k=0}^{\infty} \frac{x^k}{\Gamma(\alpha k + \beta)} \tag{2.41}$$

Some special cases of the Wiman function have been described hereunder (Kilbas et al. 2006).

$$E_{\alpha,1}(x) = E_{\alpha}(x) \tag{2.42}$$

The two parameters in the Wiman function with special values such as $\alpha < 0$ or $\alpha = 0$ can be written as simple algebraic functions as (Podlubny 1999, Hanneken et al. 2009, Kilbas et al. 2006, 2013, Tomovski et al. 2014):

$$E_{1,1}(x) = e^x \tag{2.43}$$

$$E_{1,2}(x) = \frac{e^x - 1}{x} \tag{2.44}$$

$$E_{1,m}(x) = \frac{1}{x^{m-1}} \left(x^m - \sum_{k=0}^{m-2} \frac{x^k}{k!} \right) \tag{2.45}$$

$$E_{0,\beta}(x) = \frac{1}{\Gamma(\beta)(1-x)}, \qquad\qquad |x| < 1 \tag{2.46}$$

$$E_{-\alpha,\beta}(x) = -\sum_{k=0}^{\infty} \frac{1}{\Gamma(\alpha k + \beta)} \left(\frac{1}{x} \right)^k \tag{2.47}$$

$$E_{-\alpha,\beta}(x) = -\frac{1}{x} E_{\alpha,\alpha+\beta} \left(\frac{1}{x} \right) \tag{2.48}$$

$$E_{-\alpha,\beta}(x) = \frac{1}{\Gamma(\beta)} - E_{\alpha,\beta} \left(\frac{1}{x} \right) \approx \frac{1}{\Gamma(\beta)} - \frac{x}{\Gamma(\beta-\alpha)}, \qquad x \ll 1 \tag{2.49}$$

with the identity $E_{\alpha,1}(x) = E_{\alpha}(x)$, for $\beta = 1$, Eq. (2.49) can be rewritten as:

$$E_{-\alpha,1}(x) = E_{-\alpha}(x)$$

$$= 1 - E_{\alpha} \left(\frac{1}{x} \right) \approx 1 - \frac{x}{\Gamma(1-\alpha)}, \qquad x \ll 1 \tag{2.50}$$

Valério et al. (2013) further listed additional special properties of the Wiman function, and Hanneken et al. (2009) presented other special properties, including that of $E_{-\alpha,\beta}(x)$. With some of the above identities, the Wiman function or the MLF with a negative parameter can be expressed as a different MLF with positive parameters.

4.4.2.1 Differentiation

The integer differentiations of the Wiman function are defined as follows (Kilbas et al. 2004, 2006):

$$\frac{d^n}{dx^n} E_{\alpha,\beta}(x) = n! E_{\alpha,\beta+\alpha n}^{n+1}(x) \qquad (n \in N) \tag{2.51}$$

$$\frac{d^n}{dx^n} x^{\beta-1} E_{\alpha,\beta}\left(\lambda x^n\right) = x^{\beta-n+1} E_{\alpha,\beta-n}\left(\lambda x^n\right) \tag{2.52}$$

$$\frac{d^n}{dx^n} x^{n-\beta} E_{n,\beta}\left(\frac{\lambda}{x^n}\right) = \frac{(-1)^n \lambda}{x^{n+\beta}} E_{n,\beta}\left(\frac{\lambda}{x^n}\right) \tag{2.53}$$

The fractional differentiation of the Wiman function can be achieved (Kilbas et al. 2004) as:

$$D_{0+}^{v}\left[x^{\beta-1} E_{\alpha,\beta}\left(\pm\lambda x^{\alpha}\right)\right](x) = x^{\beta-v-1} E_{\alpha,\beta-v}\left(\pm\lambda x^{\alpha}\right) \tag{2.54}$$

4.4.2.2 Integration

The integration of the Wiman function is given as (Kilbas et al. 2004):

$$\int_0^x t^{\beta-1} E_{\alpha,\beta}\left(\lambda t^\rho\right) dt = x^\beta E_{\alpha,\beta+1}\left(\lambda x^\rho\right) \tag{2.55}$$

In particular, for $\beta = 1$ in Eq. (2.55), the integration of the MLF can be simplified as:

$$\int_0^x E_\alpha\left(\lambda t^\rho\right) dt = x E_{\alpha,2}\left(\lambda x^\rho\right) \tag{2.56}$$

4.4.2.3 Laplace Transform

The Laplace transform of the Wiman function results in the following expression (Valério et al. 2013, Tomovski et al. 2014):

$$\mathscr{L}\left[x^{\beta-1} E_{\alpha,\beta}\left(\pm\lambda x^{\alpha}\right)\right] = \frac{s^{\alpha-\beta}}{s^{\alpha} \mp \lambda} \tag{2.57}$$

For $\beta = 1$, Eq. (2.57) is simplified as Eq. (2.40).

4.4.3 Prabhakar Function (Generalized Mittag-Leffler Function), Generalized Prabhakar Function

A further generalization of the Wiman function has been accredited to Prabhakar (1971) for extending the Wiman function to a series with three parameters. This has been referred to as the Prabhakar function (PF) in this book, although, in literature, it is also known by other names such as the generalized MLF, etc. (Kilbas et al. 2006). This function is expressed as:

$$E_{\alpha,\beta}^{\gamma}(x) = \sum_{k=0}^{\infty} \frac{(\gamma)_k x^k}{k! \Gamma(\alpha k + \beta)} \tag{2.58}$$

where $\Gamma(\alpha k + \beta)$ is the gamma function of $\alpha k + \beta$, and $(\gamma)_k$ is the Pochhammer symbol or the Pochhammer function defined as:

$$(\gamma)_k = \gamma(\gamma+1)(\gamma+2)\dots(\gamma+k-1) = \frac{\Gamma(\gamma+k)}{\Gamma(\gamma)} \tag{2.59}$$

In addition, with a negative subscript, $-k$, this function becomes the shifted factorial written as follows (Gasper and Rahman 1990):

$$(\gamma)_{-k} = \frac{1}{(\gamma-1)(\gamma-2)\dots(\gamma-k)} = \frac{1}{(\gamma-k)_k}$$
$$= \frac{(-1)^k}{(1-\gamma)_k} \tag{2.60}$$

As can be seen below, the PF can also be expressed as a special form of Fox's H-function as (Kilbas et al. 2002a, Haubold et al. 2011):

$$E_{\alpha,\beta}^{\gamma}(x) = \frac{1}{\Gamma(\gamma)} H_{1,2}^{1,1}\left[-x\left|\begin{matrix}(1-\gamma,1)\\(0,1,)(1-\beta,\alpha)\end{matrix}\right.\right] \tag{2.61}$$

In particular, $(\gamma)_0 = 1$ for $k = 0$ and $(1)_n = n!$ (n is an integer) (Srivastava and Tomovski 2009).

For $\gamma = 0$, the PF becomes the inverse gamma function (Kilbas et al. 2002a):

$$E_{\alpha,\beta}^{0}(x) = \frac{1}{\Gamma(\beta)} \tag{2.62}$$

For $\gamma = 1$, the PF reduces to the Wiman function, which further reduces to the MLF when $\beta = 1$ (Kilbas et al. 2004):

$$E_{\alpha,\beta}^{1}(x) = E_{\alpha,\beta}(x) \tag{2.63}$$

$$E_{\alpha,1}^{1}(x) = E_{\alpha}(x) \tag{2.64}$$

The Prabhakar function can be further extended to yield the generalized Prabhakar function (GPF) (Prabhakar 1971, Kilbas et al. 2004, Tomovski et al. 2014),

$$e_{\alpha,\beta}^{\gamma}(t;\lambda) = t^{\beta-1}E_{\alpha,\beta}^{\gamma}\left(-\lambda t^{\alpha}\right) \tag{2.65}$$

where $E_{\alpha,\beta}^{\gamma}\left(-\lambda t^{\alpha}\right)$ is the PF given by Eq. (2.58).

For integer $\gamma < 0$ and $\gamma = m < 0$, i.e., $E_{k,c+1}^{-n}(x^k)$, the PF is related to the Konhauser polynomial or the generalized Laguerre polynomial, $Z_n^c(x; k)$ (Konhauser 1967, Prabhakar 1971, Kilbas et al. 2004), which is further related to the Laguerre polynomial, $L_n^c(x)$, as:

$$E_{k,c+1}^{-n}(x^k) = \frac{\Gamma(n+1)}{\Gamma(kn+c+1)}Z_n^c(x;k) \tag{2.66}$$

where

$$Z_n^c(x;k) = \frac{\Gamma(kn+c+1)}{\Gamma(n+1)}\sum_{j=0}^{n}(-1)^j\binom{n}{j}\frac{x^{kj}}{\Gamma(kj+c+1)} \tag{2.67}$$

is the generalized Laguerre polynomial with k and n as integers, and $\Gamma(n + 1) = n!$ used in some cases.

From the generalization of Konhauser (1967), it can be seen that

$$L_n^c(x) = Z_n^c(x; 1) \tag{2.68}$$

4.4.3.1 Differentiation of the Generalized Prabhakar Function

As the GPF generalizes the MLF, the Wiman function, and the PF, its major properties can also be applicable to these other functions when the three parameters in the GPF take specific values.

Kilbas et al. (2004) documented the major properties of the GPF and Srivastava and Tomovski (2009) demonstrated that with the Riemann-Liouville fractional derivative, $\dfrac{d^\gamma}{dx^\gamma} = D_x^\gamma$, the following identity holds:

$$E_{\alpha,\beta}^\gamma(x) = \frac{1}{\Gamma(\gamma)} \frac{d^\gamma}{dx^\gamma}\left[x^{\gamma-1} E_{\alpha,\beta}^\gamma(x)\right] \tag{2.69}$$

which connects the PF and the MLF. Furthermore,

$$\frac{d^n}{dt^n}\left[e_{\alpha,\beta}\left(t;\lambda\right)\right] = t^{\beta-n-1} E_{\alpha,\beta-n}^\gamma\left(-\lambda t^\alpha\right) \tag{2.70}$$

For $\gamma = 1$, Eq. (2.70) becomes

$$\frac{d^n}{dt^n}\left[t^{\beta-1} E_{\alpha,\beta}\left(-\lambda t^\alpha\right)\right] = t^{\beta-n-1} E_{\alpha,\beta-n}\left(-\lambda t^\alpha\right) \tag{2.71}$$

4.4.3.2 Partial Differentiation

The partial differentiation of the GPF with respect to x is given as (Kilbas et al. 2006):

$$\frac{\partial^n}{\partial x^n}\left[x^{\beta-1} E_{\alpha,\beta}^\gamma\left(\lambda x^\alpha\right)\right] = x^{\beta-n-1} E_{\alpha,\beta-n}^\gamma\left(\lambda x^\alpha\right) \tag{2.72}$$

and that with respect to λ can be expressed as:

$$\frac{\partial^n}{\partial \lambda^n}\left[x^{\beta-1} E_{\alpha,\beta}^\gamma\left(\lambda x^\alpha\right)\right] = n! x^{\alpha n+\beta-1} E_{\alpha,\alpha n+\beta}^{n+1}\left(\lambda x^\alpha\right) \tag{2.73}$$

4.4.3.3 Integration

The fractional integration of the GPF is written as (Kilbas et al. 2004):

$$\int_0^x t^{\beta-1} E_{\alpha,\beta}^\gamma(-\lambda t^\alpha)dt = x^\beta E_{\alpha,\beta+1}^\gamma(-\lambda x^\alpha) \tag{2.74}$$

In some applications, double GPFs such as the following are also encountered (Kilbas et al. 2004):

$$\int_0^x (x-t)^{\mu-1} E_{\alpha,\mu}^\gamma[\lambda(x-t)^\alpha] t^{\nu-1} E_{\alpha,\nu}^\delta\left(\lambda t^\alpha\right)dt = x^{\mu+\nu+1} E_{\alpha,\mu+\nu}^{\gamma+\delta}\left(\lambda x^\alpha\right) \tag{2.75}$$

4.4.3.4 Laplace Transform of the Generalized Prabhakar Function

The Laplace transform of the GPF is given as follows (Prabhakar 1971, Valério et al. 2013):

$$\mathscr{L}\left[x^{\beta-1}E_{\alpha,\beta}^{\gamma}\left(\lambda x^{\alpha}\right)\right]=s^{-\beta}\left(1-\lambda s^{-\alpha}\right)^{-\gamma}=\frac{s^{\alpha\gamma-\beta}}{\left(s^{\alpha}-\lambda\right)^{\gamma}},\ \left(s>\left|\lambda\right|^{1/\alpha}\right) \tag{2.76}$$

4.4.3.5 Asymptotic Properties of the Prabhakar Function and the Generalized Prabhakar Function

1. Asymptotic properties of the Prabhakar function

Sandev et al. (2015) presented two forms of asymptotic results for the PF:

$$E_{\alpha,\beta}^{\gamma}\left(-t^{\alpha}\right)\sim\frac{1}{\Gamma(\beta)}-\frac{t^{\alpha}}{\Gamma(\alpha+\beta)}$$
$$\sim\frac{1}{\Gamma(\beta)}\exp\left[-\frac{\gamma\,\Gamma(\beta)}{\Gamma(\alpha+\beta)}t^{\alpha}\right], \qquad t\to 0 \tag{2.77}$$

and

$$E_{\alpha,\beta}^{\gamma}\left(-t^{\alpha}\right)\sim\frac{t^{-\alpha\gamma}}{\Gamma(\beta-\alpha\gamma)}, \qquad\qquad t\to\infty \tag{2.78}$$

For $\gamma=1$, the above results can be applied to the Wiman function or the two-parameter MLF.

2. Asymptotic properties of the generalized Prabhakar function

The GPF in Eq. (2.64) (Prabhakar 1971, Friedrich 1991, Tomovski et al. 2014, Mainardi and Garrappa 2015) has two forms of asymptotic solutions under different conditions as $t\to+\infty$ (Mainardi and Garrappa 2015):

Case 1: $0<\alpha\gamma<\beta\leq 2.0$

$$e_{\alpha,\beta}^{\gamma}\left(t;\lambda\right)\sim\frac{(\lambda t)^{\beta-\alpha\gamma-1}}{\Gamma(\beta-\alpha\gamma)}, \qquad\qquad t\to+\infty \tag{2.79}$$

Case 2: $0<\alpha\gamma=\beta\leq 2.0$

$$e_{\alpha,\beta}^{\gamma}\left(t;\lambda\right)\sim\gamma\frac{(\lambda t)^{-\alpha-1}}{\Gamma(-\alpha)}, \qquad\qquad t\to+\infty \tag{2.80}$$

As the GPF becomes the Wiman function for $\gamma=1$, the above asymptotic results can be applied as:

$$t^{\beta-1}E_{\alpha,\beta}\left(-\lambda t^{\alpha}\right)\sim\begin{cases}\dfrac{(\lambda t)^{\beta-\alpha-1}}{\Gamma(\beta-\alpha)}, & 0<\alpha<\beta\leq 2.0\\[2ex]\dfrac{(\lambda t)^{-\alpha-1}}{\Gamma(-\alpha)}, & 0<\alpha=\beta\leq 2.0\end{cases}, \qquad t\to+\infty \tag{2.81}$$

For $\gamma = 1$, these asymptotic results apply to the Wiman function, and for $\gamma = 1$ and $\beta = 1$, they are applicable to the MLF. In addition to the case $t \to +\infty$, Friedrich (1991) showed that the following asymptotes result as $t \to 0$ for a special form of the GPF:

$$t^{\alpha,\beta} E_{\alpha,k}\left(-\lambda t^{\alpha}\right) \sim \frac{(\lambda t)^{\alpha-\beta}}{\Gamma(k)}, \qquad t \to 0 \qquad (2.82)$$

4.5 Wright Function and Generalized Wright Function

Both the Wright function and the generalized Wright function (GWF) can be expressed as special cases of the H-function (Kilbas et al. 2002a). The Wright function is written as (Gorenflo et al. 1999):

$$\phi(\alpha,\beta;x) = \sum_{k=0}^{\infty} \frac{x^k}{k!\Gamma(\alpha k+\beta)}, \qquad \alpha > -1 \qquad (2.83)$$

The Wright function has a special role in fractional calculus, its Laplace transform being the Wiman function (Podlubny 1999):

$$\mathscr{L}\left(\sum_{k=0}^{\infty} \frac{x^k}{k!\Gamma(\alpha k+\beta)}\right) = \sum_{k=0}^{\infty} \frac{1}{\Gamma(\alpha k+\beta)} \frac{1}{s^{k+1}} = s^{-1} E_{\alpha,\beta}\left(s^{-1}\right) \qquad (2.84)$$

The left-sided and right-sided Riemann-Liouville fractional integration and fractional differentiation of Eq. (2.83) were investigated by Kilbas (2005). Other properties of the Wright function, such as the Laplace transforms and the asymptotic properties of functions related to the Wright function, have been discussed by Stanković (1970) and Gorenflo et al. (1999).

The Wright function is a special case of the generalized hypergeometric series (GHS) which can be expressed as follows (Kilbas et al. 2002a, Kilbas 2005):

$$_0\Psi_1\left[{(\beta,\alpha)}\middle|x\right] = \phi(\alpha,\beta;x) = \sum_{k=0}^{\infty} \frac{x^k}{k!\Gamma(\alpha k+\beta)} \qquad (2.85)$$

A more generic form of the above function is the GWF (Kilbas, 2005) or Wright's generalized hypergeometric function (WGHF) (Haubold et al. 2011) and is expressed as under:

$$_p\Psi_q\left[{(a_i,\alpha_i)_{1,p} \atop (b_j,\beta_j)_{1,q}}\middle|x\right] = \sum_{k=0}^{\infty} \frac{\prod\limits_{i=1}^{p}\Gamma(a_i+\alpha_i k)}{\prod\limits_{j=1}^{q}\Gamma(b_j+\beta_j k)}\frac{x^k}{k!} \qquad (2.86)$$

where $\prod\limits_{i=1}^{p}\Gamma(a_i+\alpha_i k)$ and $\prod\limits_{j=1}^{q}\Gamma(b_j+\beta_j k)$ are the products of the gamma functions $\Gamma(a_i+\alpha_i k)$ and $\Gamma(b_j+\beta_j k)$ respectively.

The GWF can also be expressed as a special case of the H-function (Kilbas et al. 2002a) in the following manner:

$$_p\Psi_q\left[{(a_i,\alpha_i)_{1,p} \atop (b_j,\beta_j)_{1,q}}\middle|x\right] = H_{p,q+1}^{1,p}\left(-x\middle|{(1-a_i,\alpha_i)_{1,p} \atop (0,1),(1-b_j,\beta_j)_{1,q}}\right) \qquad (2.87)$$

which can be reduced to the Wright function as:

$$\phi(\alpha,\beta;x) = H_{0,2}^{1,0}\left(-x \left|\begin{array}{c}- - - - - - - - \\ (0,1),(1-\beta,\alpha)\end{array}\right.\right) \tag{2.88}$$

Further discussion on the Wright function and the GWF, including investigations on the asymptotic behaviors of the GWF, can be found in Kilbas et al. (2002a), Kilbas (2005) and Stanković (1970).

Further forms of special functions associated with those discussed in this section have been outlined in Appendix 2.3 at the end of this chapter.

5. Fundamental Solution, Green Function, Delta Functions and Generalized Functions

5.1 Historical Development of the Ideas

The delta function is a very important concept and tool for the derivation of the fundamental solution or Green's function of a PDE or an fPDE subject to an instantaneous input or a pulse input in physical terms. It is also associated with the derivation of an fPDE from the CTRW theory. The methods and applications of the special delta function, known as the Dirac delta function, will be detailed in Chapters 3, 4 and 5 in the context of soil water movement, solute transport in soils, and groundwater flow respectively.

Both the terminologies, *fundamental solution* and *Green's function*, appear in literature. The difference between them is that the fundamental solution is derived for the entire space of concern (mathematically, it is expressed as R^n and is in the range of $-\infty < x < +\infty$ in one dimension), whereas Green's function is a response to a source with a homogeneous (or zero) boundary condition, derived only for a limited domain Ω of the entire space ($\Omega < R^n$) (Kevorkian 1990), $0 \le x < +\infty$ in one dimension for instance. These concepts also apply to fPDEs and fDEs (Podlubny 1999). For both Green's function and the fundamental solution, the delta function can appear either as a part of the differential equation as a source or in the boundary condition (Zwillinger 1998).

Depending on the type of BCs from which a Green's function is derived, it can be grouped into the first kind and the second kind (Kevorkian 1990). The Green's function of the first kind is derived from the BC of the first kind while the Green's function of the second kind is derived from the BC of the second kind (Kevorkian 1990).

It is worthwhile to briefly recount the development of the delta function and its properties, particularly the Dirac delta function, which popularized the concept of the delta function and its application beyond the field of quantum mechanics (Dirac 1934, 1947) and found widespread applications in other fields, including hydrology and soil physics.

Wagner (2004) and Katz and Tall (2013) briefly traced the historical developments of the concept of the delta function, its applications in finding the fundamental solution of a PDE, and other roles in mathematics and mathematical physics. The essential concept associated with the delta function is the fundamental solution of a DE, a PDE or an fPDE, which is the solution which is derived from an instantaneous initial condition or, as it was called later, the Dirac delta function as the initial condition.

The following timeline can be observed to have been recorded in literature (Wagner 2004, Katz and Tall 2013):

1747: d'Alembert was probably the first mathematician to develop a fundamental solution of the wave equation, a second order PDE, based on the concept of an initial condition—an instantaneous input equivalent to the modern Dirac delta function.

1789 and 1809: Laplace investigated the fundamental solutions of elliptic and parabolic operators.

1818: Fourier worked out the fundamental solution of the dynamic deflections of beams; Poisson generalized d'Alembert's fundamental solution to three spatial dimensions by representing the solutions of the wave equation as a convolution of the fundamental solution. Poisson's approach is being applied in many fields today (Evans et al. 1999).

1828: Green determined the fundamental solutions of the Laplace equation and the Poisson equation; his method of solutions bears his name as Green's function in mathematical literature.

1829: Poisson further analyzed waves in elastic media; in 1849, Stokes made similar investigations on this topic. From this time, the modern-day terminology, kernel, in convolution integrals appeared to be formalized.

1894: Volterra developed a fundamental solution of the wave equation in two dimensions.

1882: According to Wagner (2004), the notation δ for the delta function was first used by Kirchhoff. It bears Dirac's name due to Dirac's extensive investigation and analysis of this function in quantum mechanics and extension to other fields.

1947: Precisely 200 years after d'Alembert's inception of the idea of the fundamental solution, Paul Dirac (1947) investigated the Dirac delta function and its related properties in physics.

1950: It was Schwartz and Lighthill's theory of distributions in the 1950s which laid a firm foundation for a formal and mathematically rigorous interpretation of the delta function (Evans et al. 1999) and the fundamental solution (Wagner 2004). The scientific community took more than 200 years to reach a point where these two scientists developed the theory of distributions for the delta function to be formalized in mathematics.

1974: According to Podlubny (1999), Green's function of fractional differential equations was first investigated by Meshkov (1974).

Green's function or the fundamental solution can be used as a kernel to construct a solution of a PDE or an fPDE, subject to an arbitrary input, with the aid of the convolution integral. Further applications of these concepts associated with the Dirac delta function and the generalized function, and the connections between the derivation of fPDEs and the parameters in the CTRW theory, etc., have been detailed in subsequent chapters.

Miller and Ross (1993) showed that, similar to the procedures for solving PDEs with integer derivatives, the solutions of a non-homogeneous fPDE in terms of probability $p(x, t)$ (water content, solute concentration or any other variable) can be derived by the convolution integral of the fractional Green's function and the source term $f(t)$ as:

$$p(x,t) = \int_0^t G_0(x,t-\tau)f(\tau)d\tau \tag{2.89}$$

where $G_0(x, t - \tau)$ is Green's function for an fPDE.

The detailed procedure for deriving Green's function will be discussed in chapters where solutions of fPDEs are presented. It is important to note that *both Green's function and the desired solution of the original problem (an fPDE or PDE and its IC and BCs) must satisfy a zero boundary condition at the origin* in order for the superposition or the convolution in Eq. (2.89) to make sense (Kevorkian 1990).

5.2 *Dirac Delta Function, S Function and the Heaviside Step Function*

As briefly stated above, the Dirac delta function has evolved from d'Alembert's time in 1747 to the modern era when information from experiments in different fields became so diverse that the basic mathematical concept had to be extended to handle practical issues. This forms the background of Dirac's approach to instantaneous signals in quantum mechanics where the use of an 'improper function' is required (Dirac 1947).

Dirac (1947) originally defined the S function as:

$$\int_{-\infty}^{\infty} S(x)dx = 1 \tag{2.90}$$

with $S(x) = 0$ for $x \neq 0$.

In literature, it is often written as:

$$\int_{-\infty}^{\infty} \delta(x)dx = 1 \tag{2.91}$$

The $S(x)$ function vanishes except inside a small domain of length ε and in the limit $S(x) \rightarrow \delta(x)$ as $\varepsilon \rightarrow 0$. Spiegel (1965) interpreted the function $\delta(x)$ in the following manner:
Consider a function $F_\varepsilon(t)$ as:

$$F_\varepsilon(t) = \begin{cases} \dfrac{1}{\varepsilon}, & 0 \leq t \leq \varepsilon \\ 0, & t > \varepsilon \end{cases}, \qquad \varepsilon > 0 \tag{2.92}$$

It is obvious that as $\varepsilon \rightarrow 0$, the height of the rectangle defined by Eq. (2.92) increases indefinitely and its width decreases in such a way that the area of the rectangle is equal to 1 as $\int_0^{\infty} F_\varepsilon(t)dt = 1$. With this property under consideration, physicists and engineers conceived a limiting function denoted by $\delta(x)$, approached by $F_\varepsilon(t)$ as $\varepsilon \to 0$. This limiting function is therefore called the unit impulse function or the Dirac delta function.

In line with Dirac's approach, the most common expression of this function in one dimension on a continuous domain is of the following form, which is also an important property of this function:

$$\delta(x) = \begin{cases} \infty, & x = 0 \\ 0, & x \neq 0 \end{cases} \tag{2.93}$$

Another important property of the Dirac delta function states that:

$$\int_{-\infty}^{\infty} f(x)\delta(x-a)dx = f(a) \tag{2.94}$$

which can be used in the derivation of Green's function or the fundamental solution (Evans et al. 1999).

Another essential application of the Dirac delta function is its Laplace transform used in solving PDEs or fPDEs, given as:

$$\mathcal{L}\{\delta(t)\} = \int_{0}^{\infty} e^{-st}\delta(t)dt = 1 \tag{2.95}$$

5.3 Heaviside Step Function and Its Relationship with Dirac Delta Function

Another special function associated with the Dirac delta function is the Heaviside step function. It is defined as (Knobel 2000):

$$H(x) = \begin{cases} 1, & x \geq 0 \\ 0, & x < 0 \end{cases} \tag{2.96}$$

The relationship between the Dirac delta function and the Heaviside step function is expressed as (Debnath and Bhatta 2007):

$$\int_{-\infty}^{x} \delta(y)dy = H(x) \tag{2.97}$$

and

$$\delta(x) = \frac{d}{dx}H(x) \tag{2.98}$$

which can be more generally written as (Evans et al. 1999):

$$\delta(x) \sim \{H'(x; n)\} \tag{2.99}$$

where $H'(x; n)$ is the sequence defining the Heaviside step function.

5.4 Kronecker's Delta Function

The Dirac delta function is used for a continuous domain, whereas, for a discrete numerical representation on a discrete domain, Kronecker's delta function δ_{jr} is used (Evans et al. 1999):

$$\delta_{jr} = \begin{cases} 1, & \text{if } j = r \\ 0, & \text{if } j \neq r \end{cases} \tag{2.100}$$

The discrete analogue of Kronecker's delta function to the integration of the Dirac delta function is:

$$f_j = \sum_{r=-\infty}^{\infty} f_r \delta_{jr} \tag{2.101}$$

5.5 Generalized Function and Its Applications in fPDEs

There is a close relationship between the delta function and its extension, Φ_β, named as the generalized function (Evans et al. 1999, Podlubny 1999). One form of the generalized function is defined as follows (Gel'fand and Shilov 1958, Saichev and Zaslavsky 1997, Podlubny 1999):

$$\Phi_{p+1}(t) = \begin{cases} \dfrac{t^p}{\Gamma(p+1)}, & t > 0 \\ 0, & t \leq 0 \end{cases} \tag{2.102}$$

where $\Gamma(p + 1)$ is the gamma function.

Saichev and Zaslavsky (1997) demonstrated the applications of the generalized function with an important identity in the form of the following limit, with $-\beta = p$ in Gel'fand and Shilov (1958):

$$\lim_{\beta \to 1} \frac{t^{-\beta}}{\Gamma(1-\beta)} = \delta(t) \tag{2.103}$$

where $\delta(t)$ is the Dirac delta function and β is the order of fractional derivatives in time. This relationship is very useful and important for the conversion of a fractional source term in the fPDE into a regular initial condition (IC) for the fractional diffusion equation in physics (Saichev and Zaslavsky 1997) or for water movement in hydrology (Su 2014).

Another key property (Podlubny 1999) of the generalized function for its relation with the derivatives of the Dirac delta function is:

$$\lim_{\beta \to -k} \Phi_\beta(t) = \Phi_{-k}(t)\delta^{(k)}(t), \qquad k = 0,1,2,\ldots \tag{2.104}$$

In addition to the definition and the identities above, many identities and properties of the Dirac function have been listed in Appendix 2.2.

6. Fractional Integration and Fractional Differentiation

Unlike the simple and clear definitions of integer integration and derivative, the definitions of fractional integration and differentiation differ among mathematicians such as Liouville, Riemann, Erdélyi-Kober, Hadamard, Grünwald-Letnikov, Caputo, Riesz and Weyl, etc. (Miller and Ross 1993, Podlubny 1999, Kilbas et al. 2006). In this book, we are mainly concerned with the definitions of fractional integration and fractional differentiation as given by Riemann-Liouville and Caputo. However, their connections with other definitions have also been briefly discussed.

The definitions of Riemann-Liouville (RL) fractional integration and fractional differentiation have three forms (Kilbas 2002a, 2006)—the RL fractional derivatives, the left-sided RL fractional derivatives and right-sided RL fractional derivatives.

6.1 Riemann-Liouville and Liouville Fractional Integrals

The symbols for fractional integration and fractional derivatives are also different in literature:

1. RL left-sided fractional integral of a function $f(x)$ of order λ: The symbols for the left-sided RL fractional integral include varying symbols like $(I_+^\lambda f)(x)$ (Samko et al. 1993, Kilbas et al. 2006), $_aD_t^{-\lambda} f(x)$ (Podlubny 1999), $_aJ_x^\lambda f(x)$ (Debnath and Bhatta 2007) and $^{RL}I_+^\lambda f(x)$ (Valério et al. 2013). In this book, we have chosen the simpler notation by Valério et al. (2013), however omitting the left superscript RL denoting RL fractional integrals, as:

$$I_{a+}^\lambda f(t) = \frac{1}{\Gamma(\lambda)} \int_a^t \frac{f(\tau)}{(t-\tau)^{1-\lambda}} d\tau, \qquad t \ge a \qquad (2.105)$$

where $a+$ signifies the left-sided fractional integration. The change of the lower integration limit from $a+$ to $a = -\infty$ yields the Liouville fractional integration, $^L I_+^\lambda f(t)$.

The RL left-sided fractional integral is the most widely used definition of fractional integral and is often referred to as the RL fractional integral. By omitting the word 'left-sided', the RL fractional integral can also be expressed as (Podlubny 1999, 65):

$$_aD_t^{-\lambda} f(t) = \frac{1}{\Gamma(\lambda)} \int_a^t \frac{f(\tau)}{(t-\tau)^{1-\lambda}} d\tau \qquad (2.106)$$

with $_aD_t^0 f(t) = f(t)$ as a special case.

2. RL right-sided fractional integral of a function $f(x)$ of order λ: The symbols for the right-sided RL fractional integral include $(I_-^\lambda f)(x)$ (Samko et al. 1993, Kilbas et al. 2006) and $I_{-a}^\lambda f(x)$ (Valério et al. 2013), which is defined as:

$$I_{b-}^\lambda f(t) = \frac{1}{\Gamma(\lambda)} \int_t^b \frac{f(\tau)}{(\tau-t)^{1-\lambda}} d\tau, \qquad b \ge t \qquad (2.107)$$

which, by setting $b = +\infty$ yields the Liouville right-sided fractional integral, $^L I_-^\lambda f(t)$, which is sometimes called the Wyle fractional integral (Valério et al. 2013).

6.2 Riemann-Liouville and Liouville Fractional Derivatives

There are several definitions of fractional derivatives, depending on how the order of derivatives is defined. The simplest RL left-sided and right-sided fractional derivatives are defined as (Samko et al. 1993, Kilbas et al. 2006, 69–73):

$$D_{a+}^{\lambda} f(t) = \frac{1}{\Gamma(1-\lambda)} \frac{d}{dt} \int_a^t \frac{f(\tau)}{(t-\tau)^{\lambda}} d\tau, \qquad t \geq a;\ 0 < \lambda < 1 \qquad (2.108)$$

and

$$D_{b-}^{\lambda} f(t) = -\frac{1}{\Gamma(1-\lambda)} \frac{d}{dt} \int_t^b \frac{f(\tau)}{(\tau-t)^{1-\lambda}} d\tau, \qquad b \geq t;\ 0 < \lambda < 1 \qquad (2.109)$$

For $\lambda \geq 1$, Samko et al. (1993) and Kilbas et al. (2006) iterated that:

$$D_{a+}^{\lambda} f(t) = D^n \left[I_{a+}^{n-\lambda} f(t) \right] = \frac{1}{\Gamma(n-\lambda)} \frac{d^n}{dt^n} \int_a^t \frac{f(\tau)}{(t-\tau)^{1-n+\lambda}} d\tau, \qquad t \geq a;\ \lambda \geq 1 \qquad (2.110)$$

and

$$D_{b-}^{\lambda} f(t) = (-D)^n \left[I_{b-}^{n-\lambda} f(t) \right] \frac{(-1)^n}{\Gamma(n-\lambda)} \frac{d^n}{dt^n} \int_t^b \frac{f(\tau) d\tau}{(\tau-t)^{1-n+\lambda}}, \qquad b \geq t;\ \lambda \geq 1 \qquad (2.111)$$

With the RL definitions, the initial condition for fDEs and fPDEs do not have a clear physical meaning. Hence, to overcome this problem, other definitions of fractional derivatives have been developed, the Caputo fractional derivatives for instance.

6.3 Caputo Fractional Derivatives

The RL fractional derivative has some unusual properties (Podlubny 1999) and two of these properties raise the issue of physical relevance:

1. RL fractional derivatives for an initial condition cannot be physically interpreted, and

2. RL fractional derivative of a constant is not zero (Podlubny 1999), i.e.,

$$_a D_t^{\beta} C = \frac{C t^{-\beta}}{\Gamma(1-\beta)} \qquad (2.112)$$

For the abovementioned and other reasons, Caputo (1967) introduced modifications in the definition of RL fractional derivatives to eliminate these problems. Caputo fractional derivatives are, therefore, based on the RL definitions. The left-sided Caputo fractional derivative, $^C D_{a+}^{\beta} f(t)$, and the right-sided Caputo fractional derivative, $^C_a D_t^{\beta} f(t)$, are defined in different forms. For a range $[a, b]$, these derivatives are written as (Valério et al. 2013):

$$^C D_{a+}^{\beta} f(t) = \frac{1}{\Gamma(n-\beta)} \int_a^t \frac{f^{(n)}(\tau)}{(t-\tau)^{\beta+1-n}} d\tau, \qquad t \geq a \qquad (2.113)$$

and

$$^{C}D_{b-}^{\beta}f(t) = \frac{(-1)^{n}}{\Gamma(n-\beta)}\int_{t}^{b}\frac{f^{(n)}(\tau)}{(\tau-t)^{\beta+1-n}}d\tau, \qquad t \le b \qquad (2.114)$$

where $f^{(n)}(\tau)$ denotes the *nth* derivative of the function $f(x)$.

The above definitions are the specialized forms of Caputo's original definition given as (Caputo 1967):

$$^{C}D_{a+}^{m+\beta}f(t) = \frac{1}{\Gamma(1-\beta)}\int_{a}^{t}\frac{f^{(m+1)}(\tau)}{(t-\tau)^{\beta}}d\tau \qquad (2.115)$$

where m is an integer, with i as the counter in the series expansion of the fractional integrand, such that $i = 1,2,\dots,m$.

Kilbas et al. (2006) discussed further forms of Caputo derivatives in different contexts, and their connections with other fractional derivatives. The relation between RL fractional derivatives and Caputo fractional derivatives is expressed as:

$$D_{a+}^{\beta}f(t) = {}^{C}D_{a+}^{\beta}f(t) + \sum_{k=0}^{n-1}\frac{f^{(k)}(a)}{\Gamma(k-\beta+1)}(t-a)^{k-\beta}, \qquad t \ge a; \beta \ge 1 \qquad (2.116)$$

and

$$D_{b-}^{\beta}f(t) = {}^{C}D_{b-}^{\beta}f(t) + \sum_{k=0}^{n-1}\frac{f^{(k)}(a)}{\Gamma(k-\beta+1)}(b-t)^{k-\beta}, \qquad t \le b; \beta \ge 1 \qquad (2.117)$$

and the above definitions can be simplified for $0 < \beta < 1$ as (Kilbas et al. 2006):

$$D_{a+}^{\beta}f(t) = {}^{C}D_{a+}^{\beta}f(t) + \frac{f(a)}{\Gamma(1-\beta)}(t-a)^{-\beta}, \qquad t \ge a; 0 < \beta < 1 \qquad (2.118)$$

$$D_{b-}^{\beta}f(t) = {}^{C}D_{b-}^{\beta}f(t) + \frac{f(b)}{\Gamma(1-\beta)}(b-t)^{-\beta}, \qquad t \le b; 0 < \beta < 1 \qquad (2.119)$$

Based on the concept of generalized functions, Podlubny (1999) showed that the connection between Caputo and RL fractional derivatives is as follows:

$$_{a}D_{t}^{\beta}f(t) = {}_{a}^{C}D_{t}^{\beta}f(t) + \sum_{k=0}^{n-1}\Phi_{k-\beta+1}(t-a)f^{(k)}(a) \qquad (2.120)$$

$$\Phi_{p+1}(t) = \begin{cases} \dfrac{t^{p}}{\Gamma(p+1)}, & t > 0 \\ 0, & t \le 0 \end{cases} \qquad (2.121)$$

when $\beta \to n$, the above relationship results in the following expression:

$$_{a}D_{t}^{\beta}f(t) = {}_{a}^{C}D_{t}^{\beta}f(t) + \sum_{k=0}^{n-1}\delta^{(n-k-1)}(t-a)f^{(k)}(a) \qquad (2.122)$$

or

$$\overline{f}^{(n)}(t) = f_C^{(n)}(t) + \sum_{k=0}^{n-1} \delta^{(n-k-1)}(t-a) f^{(k)}(a) \tag{2.123}$$

where

$\overline{f}^{(n)}(t)$ is the *nth* generalized derivative with $\overline{f}^{(n)}(t) = f(t)$ for $t \geq a$ and $\overline{f}^{(n)}(t) = 0$ for $t < a$;

$f_C^{(n)}(t)$ is the classic integer derivative;

$\delta^{(n-k-1)}(t-a)$ is the *(n–k–1)th* derivative of the Dirac delta function; and

$f^{(k)}(a)$ is the *kth* derivative of the function $f(a)$.

Mainardi (1996) showed the relationship between RL and Caputo fractional derivatives in terms of generalized functions, in the following different forms:

$$_a D_t^\beta f(t) = {}_a^C D_t^\beta f(t) + \sum_{k=0}^{n-1} f^{(k)}(0) \frac{t^{k-\beta}}{\Gamma(k-\beta+1)} \tag{2.124}$$

or

$$_a D_t^\beta f(t) = {}_a^C D_t^\beta \left(f(t) + \sum_{k=0}^{n-1} f^{(k)}(0) \frac{t^k}{k!} \right) \tag{2.125}$$

where $k! = \Gamma(k+1)$.

Recently, Caputo and Fabrizio (2015) proposed a new definition of fractional derivatives without a singular kernel (that is, a new definition with a kernel in an integral which does not have an infinite value for the variable in the kernel). In the new definition, an exponential function is used in place of a power function.

The Caputo fractional differentiation for time-fractional derivatives admits a bounded initial condition (Mainardi et al. 2001) that eliminates the problem with the RL fractional derivative.

6.4 *Symmetric Fractional Derivatives (SFDs)*

SFD of an independent variable or a function f, with respect to hydraulic head h or soil moisture ratio θ, is defined in the range $0 \leq x \leq L$ (Gorenflo and Mainardi 1998b, Jiang et al. 2012a) as:

$$\frac{\partial^\lambda f(x,t)}{\partial |x|^\lambda} = -\frac{1}{2\cos\left(\frac{\pi\lambda}{2}\right)} \left(\frac{\partial^\lambda f(x,t)}{\partial x^\lambda} + \frac{\partial^\lambda f(x,t)}{\partial |-x|^\lambda} \right) \tag{2.126}$$

or

$$\frac{\partial^\lambda f(x,t)}{\partial |x|^\lambda} = -\frac{1}{2\cos\left(\frac{\pi\lambda}{2}\right)} \left({}_0 D_x^\lambda f(x,t) + {}_x D_L^\lambda f(x,t) \right) \tag{2.127}$$

Further discussions on symmetric fractional calculus can be found in the works of Saichev and Zaslavsky (1997), Gorenflo and Mainardi (1998a, b), Mainardi et al. (2001) and Umarov and Gorenflo (2005b). Two of the most important properties of SFDs are as follows (Saichev and Zaslavsky 1997):

1. $\dfrac{\partial^\lambda f}{\partial |x|^\lambda} \neq 0$ for $x > 0$ and $\dfrac{\partial^\lambda f}{\partial |x|^\lambda} = 0$ for $x < 0$

2. $\dfrac{\partial^\lambda f}{\partial |-x|^\lambda} \neq 0$ for $x < 0$ and $\dfrac{\partial^\lambda f}{\partial |-x|^\lambda} = 0$ for $x > 0$

To be consistent with the notations used by Saichev and Zaslavsky (1997), $\dfrac{\partial f}{\partial |x|}$ is also retained when the derivative is an integer.

Umarov and Gorenflo (2005b) showed that the pseudo-differential operator for fractional derivatives is identical to the fractional power of the Laplace operator, or the fractional Laplacian.

$$D_0^\lambda = -(-\Delta)^{\lambda/2}, \qquad\qquad 0 < \lambda < 2 \qquad\qquad (2.128)$$

The fractional powers of Laplacian Δ:

$$-(-\Delta)^\lambda f \qquad\qquad\qquad (2.129)$$

is also called the fractional Laplacian of the function, f, was introduced by Bochner (1949) and analyzed by Feller (1971) and several other investigators later. Laplacian Δ is a linear operator, i.e., a second-order derivative of a function which can be in one, two or three dimensions.

For one-dimensional fractional derivatives of a known function f, Eq. (2.128) can also be expressed as:

$$D_0^\lambda f = -\left(-\frac{\partial^2}{\partial x^2}\right)^{\lambda/2} f \qquad\qquad (2.130)$$

For a finite domain $[0, L; 0, T]$ with the homogeneous boundary conditions $f(0, t) = 0$ and $f(L, t) = 0$, Jiang et al. (2012a) showed that the following equality holds in one dimension:

$$-\left(-\frac{\partial^2}{\partial x^2}\right)^{\lambda/2} f = -c_\lambda \left({}_0D_x^\lambda f + {}_xD_L^\lambda f\right) = \frac{\partial^\lambda f}{\partial |x|^\lambda} \qquad (2.131)$$

where ${}_0D_x^\lambda f$ and ${}_xD_L^\lambda f$ are the left-sided and the right-sided fractional derivatives respectively, and

$$c_\lambda = \frac{1}{2\cos\left(\dfrac{\lambda\pi}{2}\right)}, \qquad\qquad \lambda \neq 1 \qquad\qquad (2.132)$$

Comparing Eqs. (2.130) and (2.131), it is evident that the sign for SFDs $\dfrac{\partial^{\lambda} f}{\partial |x|^{\lambda}}$ and $D_0^{\lambda} f$ are identical. Hence, $D_0^{\lambda} f = \dfrac{\partial^{\lambda} f}{\partial x^{\lambda}}$ is used throughout the text as the synonym of symmetric fractional derivatives.

The fPDEs presented in the form $D_0^{\lambda} f = \dfrac{\partial^{\lambda} f}{\partial x^{\lambda}}$ in other chapters do not explicitly imply the backward motion. However, when we adopt the sign for Riesz symmetric fractional derivatives $\dfrac{\partial^{\lambda} \theta}{\partial |x|^{\lambda}}$ and $D_0^{\lambda} \theta$ being identical, $D_0^{\lambda} \theta = \dfrac{\partial^{\lambda} \theta}{\partial x^{\lambda}}$ is used as the synonym of SFDs.

Further information on the relation between Riesz SFDs and Riemann-Liouville fractional integrals and derivatives can be found in literature (Saichev and Zaslavsky 1997, Leith 2003, Jiang et al. 2012a, b, Ortigueira and Trujillo 2012).

7. Summary

This chapter serves as a preparation for essential mathematics used in this book. While many basic concepts and definitions are briefly discussed in this chapter to provide a common foundation, the specific concepts, functions and methods for solutions of fPDEs, etc., are detailed in subsequent chapters in association with water movement and solute transport in soils and aquifers and related topics. Some methods developed for solutions of fPDEs for one topic such as water movement in soils can also be extended to develop solutions for solute transport in soils, and water movement and solute transport in aquifers. To this end, the following chapters are complementary in terms of methodologies even though each chapter deals with one particular issue.

APPENDICES

Appendix 2.1: Laplace Transforms of Special Functions in Fractional Calculus and Related Topics

The following table lists the Laplace transforms of special functions frequently encountered in fractional calculus, fPDEs and related topics. This table has been prepared from a selection of published articles and books (Spiegel 1965, Stanković 1970, Gorenflo et al. 1999, Podlubny 1999, Debnath and Bhatta 2007, Srivastava and Tomovski 2009, Valério et al. 2013, Tomovski et al. 2014).

In this table, $f(t)$ is the function in the time domain and $\widetilde{f}(s)$ is its Laplace transform.

Table 2.1 Laplace transforms of selected functions.

Serial No.	$f(t)$	$\tilde{f}(s)$
1	1	$\dfrac{1}{s}$
2	t	$\dfrac{1}{s^2}$
3	$t^n, n = 0,1,2,\ldots$	$\dfrac{n!}{s^{n+1}}$
4	$t^a, a > -1$ is not an integer	$\dfrac{\Gamma(a+1)}{s^{a+1}}$
5	e^{at}	$\dfrac{1}{s-a}$
6	$t^n e^{at}$	$\dfrac{\Gamma(n+1)}{(s+a)^{n+1}}$
7	$\dfrac{t^{\beta-1}}{\Gamma(\beta)}$	$\dfrac{1}{s^\beta}$
8	$\dfrac{t^{n-1}e^{at}}{(n-1)!}$	$\dfrac{1}{(s-a)^n}, n = 0,1,2\ldots;0! = 1$
9	$\dfrac{t^{n-1}e^{at}}{\Gamma(n)}$	$\dfrac{1}{(s-a)^n}$ with $n > 0$
10	$\dfrac{1}{\sqrt{t}}$	$\sqrt{\dfrac{\pi}{s}}$
11	$2\sqrt{t}$	$\dfrac{1}{s}\sqrt{\dfrac{\pi}{s}}$
12	$\delta(t)$	1
13	$\delta(t-a)$	$e^{-as}, a > 0$
14	$\delta'(t-a)$	$se^{-as}, a \geq 0$
15	$\delta^{(n)}(t-a)$	$s^n e^{-as}$
16	$f(at)$	$\dfrac{1}{a}\tilde{f}\left(\dfrac{s}{a}\right)$
17	$f'(at)$	$s\tilde{f}(s) - f(0)$
18	$f^2(at)$	$s^2\tilde{f}(s) - sf(0) - f'(0)$
19	$f^{(n)}(t)$	$s^n\tilde{f}(s) - s^{n-1}f(0) - s^{n-2}f'(0)$ $\ldots\ldots -f^{(n-1)}(0)$

Table 2.1 contd. ...

... Table 2.1 contd.

Serial No.	$f(t)$	$\tilde{f}(s)$
20	$E_\beta(\pm at^\beta) = \sum\limits_{k=0}^{\infty} \dfrac{(\pm at)^k}{\Gamma(\beta k+1)}$	$\dfrac{s^{\beta-1}}{s^\beta \mp a}$
21	$t^{\beta-1}E_{\alpha,\beta}(\pm \lambda t^\alpha)$	$\dfrac{s^{\alpha-\beta}}{s^\alpha \mp \lambda}$
22	$t^{\beta-1}E_{\alpha,\beta}^\gamma(\pm \lambda t^\alpha)$	$\dfrac{s^{\alpha\gamma-\beta}}{\left(s^\alpha \mp \lambda\right)^\gamma}$
23	$\phi(\alpha,\beta;t)$	$s^{-1}E_{\alpha,\beta}\left(s^{-1}\right)$
24	$t^{\beta-1}\phi(\alpha,\beta;-\lambda t^\alpha)$	$s^{-\beta}\exp\left(-\lambda s^{-\alpha}\right)$
25	$t^{\rho-1}E_{\alpha,\beta}^{\gamma,k}\left(\lambda t^\sigma\right) = \sum\limits_{n=0}^{\infty}\dfrac{(\gamma)_{kn}t^n}{n!\Gamma(\alpha n+\beta)}$	$\dfrac{s^{-\rho}}{\Gamma(\gamma)}\sum\limits_{i=0}^{\infty}\dfrac{\Gamma(\rho+\sigma k)\Gamma(\gamma+ik)}{\Gamma(\beta+\alpha i)}\left(\dfrac{\lambda}{s^\sigma}\right)^i$
26	$t^{\alpha n}e_{\alpha,n}^{\lambda t}(\pm \lambda t^\alpha) = n!t^{\alpha(1+n)-1}E_{\alpha,(n+1)\alpha}^{n+1}(\pm \lambda t^\alpha)$	$\dfrac{n!}{\left(s^\alpha \mp \lambda\right)^{n+1}}$
27	$t^{\alpha k+\beta-1}\dfrac{d^k\left[E_{\alpha,\beta}(\pm \lambda t^\alpha)\right]}{d[\pm \lambda t^\alpha]^k}$	$\dfrac{k!s^{\alpha-\beta}}{\left(s^\alpha \mp \lambda\right)^{k+1}}$

Appendix 2.2: Additional Properties of the Dirac Delta Function

Dirac (1947) showed that the following property holds:

$$\int_{-\infty}^{\infty} f(x)\delta(x)dx = f(0) \qquad \text{(A2.1)}$$

which is applicable to the following convolutions (Evans et al. 1999):

$$\int_{-\infty}^{\infty} f(x)\delta(x-a)dx = f(a) \qquad \text{(A2.2)}$$

$$\int_{-\infty}^{\infty} f(a)\delta(x-a)dx = f(x) \qquad \text{(A2.3)}$$

and

$$\int_{-\infty}^{\infty} f(x)\delta(x-a)dx = \int_{-\infty}^{\infty} f(x)\delta(a-x)dx = f(a) \qquad (A2.4)$$

as the relationship $\delta(a - x) = \delta(x - a)$ holds true in these cases (Debnath and Bhatta 2007).

The convolution of the kth derivative of the Dirac delta function (DDF) $\delta^{(k)}$ is of the following form (Evans et al. 1999, Kilbas et al. 2006):

$$\int_{-\infty}^{\infty} f(x)\delta^{(k)}(x-a)dx = (-1)^k f^{(k)}(a) \qquad (A2.5)$$

The following properties of the DDF are also important (Dirac 1947, Evans et al. 1999):

$\delta(-x) = \delta(x)$	(Dirac 1947)	(A2.6)
$x\delta(x) = 0$	(Dirac 1947)	(A2.7)
$\delta'(-x) = -\delta'(x)$	(Dirac 1947)	(A2.8)
$x\delta'(x) = -\delta(x)$	(Evans et al. 1999)	(A2.9)
$\delta(ax) = \dfrac{1}{a}\delta(x), \quad (a > 0)$	(Dirac 1947)	(A2.10)
$\delta(ax + b) = \dfrac{1}{a}\delta(x + b/a), \quad (a > 0)$	(Evans et al. 1999)	(A2.11)
$\delta(x^2 - a^2) = \dfrac{1}{2a}[\delta(x + a) + \delta(x - a)], \quad (a > 0)$	(Dirac 1947)	(A2.12)
$a(x)\delta(x) = a(0)\delta(x)$	(Evans et al. 1999)	(A2.13)
$f(x)\delta(x - a) = f(a)\delta(x - a)$	(Dirac 1947)	(A2.14)
$\delta^{(n)}(t - \tau) = (-1)^n \delta^{(n)}(\tau - t)$	(Mainardi 1996)	(A2.15)
$\displaystyle\int_{-\infty}^{\infty} \delta(a-x)\delta(x-b)dx = \delta(a-b)$	(Dirac 1947)	(A2.16)

The Dirac delta function can also be extended to higher dimensions. Its application to two and three dimensions in a physical world is more relevant to hydrology and environmental processes. The two- and three-dimensional Dirac delta functions and their integration (Kevorkian 1990) are of the following forms respectively:

$$\delta_2(r) = \delta(x)\delta(y) \qquad (A2.17)$$

and

$$\iint_A \delta_2(r)dS = \begin{cases} 1, & \text{if } A \text{ contains the origin} \\ 0, & \text{otherwise} \end{cases} \qquad (A2.18)$$

for two dimensions, with ds as the element of area; and

$$\delta_3(r) = \delta(x)\delta(y)\delta(z) \tag{A2.19}$$

and

$$\iiint_V \delta_3(r)dV = \begin{cases} 1, & \text{if } V \text{ contains the origin} \\ 0, & \text{otherwise} \end{cases} \tag{A2.20}$$

for three dimensions, with dV as the element of volume.

In addition, the two-dimensional Dirac delta function is the product of two one-dimensional Dirac delta functions (Myint-U and Debnath 1987), i.e.,

$$\delta(x - \xi, y - \eta) = \delta(x - \xi)\delta(y - \eta) \tag{A2.21}$$

when the input is released on an element ξ from x and η from y, or

$$\delta(x, y) = \delta(x)\delta(y) \tag{A2.22}$$

when the input is released at the origin.

In addition to Dirac delta functions in different dimensions, when a source is introduced at the origin and at the initial time, i.e., at $x = 0$ and $t = 0$, combined Dirac delta functions can be used (Kevorkian 1990) in the form of the diffusion equation for water movement in a soil for instance:

$$\frac{\partial\theta}{\partial t} - D\frac{\partial^2\theta}{\partial z^2} = \delta(x)\delta(t) \tag{A2.23}$$

Appendix 2.3: Hypergeometric Series as the Generating Function for Other Special Functions

1. Introduction

The Mittag-Leffler function, the Wiman function and the generalized Prabhakar function outlined earlier are the most frequently used functions in fractional calculus and related topics. These functions and many other special functions appearing in literature can be derived from the generalized hypergeometric series (GHS) (Lavoie et al. 1976, Kiryakova 1997).

Although some of the special functions below were presented in modern times, the GHS, which forms the basis of these functions, appeared in 1303 in the works of Chu, was further proved in 1772 by Vandermonde, and is now known as the Chu-Vandermonde formula or the Vandermonde formula (Gasper and Rahman 2004).

The findings of ancient mathematicians summarized by Gasper and Rahman (2004) are less widely known to applied scientists and even modern mathematicians. The vast majority of the literature we find today deals with the series which were invigorated in the 18th century since Euler (1748), Pfaff (1797) and Gauss in particular (1813) (Gasper and Rahman 2004).

Here, we start our survey by citing the works of Erdélyi (1953a), which present the following hypergeometric equation (HE) (due to Poole 1936):

$$z(1-z)\frac{d^2u}{dz^2}+\left[c-(a+b+1)z\right]\frac{du}{dz}-abu=0 \tag{A2.24}$$

where a, b and c are independent of z, and its solutions are called the hypergeometric series (HS) (Erdélyi 1953a).

There are different forms of solutions of the HE in terms of the hypergeometric series, the most generic form being the generalized hypergeometric series (GHS) or the Gauss GHS (Erdélyi 1953a, Gasper and Rahman 2004, Kilbas et al. 2006) written as:

$$_pF_q(a_1,...a_p,b_1,...b_q;x)$$

$$=\sum_{k=0}^{\infty}\frac{(a_1)_k...(a_p)_k}{(b_1)_k...(b_q)_k}\frac{x^k}{k!}=\sum_{k=0}^{\infty}\frac{\prod_{i=1}^{p}(a_i)_k}{\prod_{j=1}^{q}(b_j)_k}\frac{x^k}{k!} \tag{A2.25}$$

where $\prod_{i=1}^{p}$ and $\prod_{j=1}^{q}$ are the products of terms from $i=1,2,...p$ and $j=1,2,...q$ respectively.

Two hypergeometric series are commonly used in fractional calculus and related analysis, which are special cases of the following GHSs (Kilbas et al. 2006):

1. The Gauss series:

$$_2F_1(a,b,c;x)=\sum_{k=0}^{\infty}\frac{(a)_k(b)_k}{(c)_k}\frac{x^k}{k!} \tag{A2.26}$$

2. The Kummer series:

$$_1F_1(a,c;x)=\sum_{k=0}^{\infty}\frac{(a)_k}{(c)_k}\frac{x^k}{k!} \tag{A2.27}$$

The GHS has been extensively investigated in literature (see Erdélyi 1953a, Fox 1928, 1961, Wright 1933, 1935, Mathai and Saxena 1978, Kilbas 2005) and given different names such as the generalized Wright function (GWF) (Kilbas 2005).

Lavoie et al. (1976) showed that the Gauss series in (A2.26) and the Kummer series in (A2.27) can be expressed as the fractional derivatives of the following elemental functions respectively:

$$_2F_1(a,b,c;x)=\frac{\Gamma(c)x^{1-c}}{\Gamma(b)}D_x^{b-c}\left[x^{b-1}(1-x)^{-a}\right] \tag{A2.28}$$

and

$$_1F_1(a;c;x)=\frac{\Gamma(c)x^{1-c}}{\Gamma(a)}D_x^{a-c}\left[e^xx^{a-1}\right] \tag{A2.29}$$

The GHSs with different orders are expressed as:

$$_{p+1}F_{q+1}\left[\begin{array}{c}a_1,...,a_p,c\\b_1,...,b_p,d\end{array}\middle|x\right]=\frac{\Gamma(d)x^{1-d}}{\Gamma(c)}D_x^{c-d}\left[x^{c-1}\,_pF_q\left[\begin{array}{c}a_1,...,a_p\\b_1,...,b_p\end{array}\middle|x\right]\right] \tag{A2.30}$$

2. The Fox H-function and the Wright Function

2.1 Introduction

The Fox H-function, also called the H-function, is a special form of the GHS. The Fox H-function forms a basis for a number of other functions, and, as we demonstrate below, is associated with the MLF, etc.

The GHS used by Fox (1928), often expressed as $_p\Psi_q$ and called the GWF (Kilbas et al. 2002a) or the Fox-Wright function (FWF) (Kilbas et al. 2006), is given as:

$$_p\Psi_q\left[\begin{matrix}(\alpha_1,\rho_1),\ldots,(\alpha_p,\rho_p)\\(\beta_1,\sigma_1),\ldots,(\beta_q,\sigma_q)\end{matrix}\bigg| x\right] = \sum_{k=0}^{\infty} \frac{\prod_{i=1}^{p}\Gamma(\alpha_i+k\beta_i)}{\prod_{j=1}^{q}\Gamma(\rho_i+k\sigma_i)}\frac{x^k}{k!} \tag{A2.31}$$

One special form of the Fox H-function is the GWF expressed as (Kilbas et al. 2006):

$$H_{p,q+1}^{1,p}\left[-x\bigg|\begin{matrix}(1-\alpha_1,\rho_1),\ldots,(1-\alpha_p,\rho_p)\\(0,1),(1-\beta_1,\sigma_1),\ldots,(1-\beta_q,\sigma_q)\end{matrix}\right] = \sum_{k=0}^{\infty}\frac{\prod_{i=1}^{p}\Gamma(\alpha_i+k\beta_i)}{\prod_{j=1}^{q}\Gamma(\rho_i+k\sigma_i)}\frac{x^k}{k!} \tag{A2.32}$$

implying that the Fox H-function $H_{p,q+1}^{1,p}$ in (A2.32) and the GWF $_p\Psi_q$ in (A2.31) are equivalent (Kilbas et al. 2002a, Kilbas 2005, Craven and Csordas 2006, Haubold et al. 2011).

$$_p\Psi_q\left[\begin{matrix}(\alpha_1,\rho_1),\ldots,(\alpha_p,\rho_p)\\(\beta_1,\sigma_1),\ldots,(\beta_q,\sigma_q)\end{matrix}\bigg| x\right] = H_{p,q+1}^{1,p}\left[-x\bigg|\begin{matrix}(1-\alpha_1,\rho_1),\ldots,(1-\alpha_p,\rho_p)\\(0,1),(1-\beta_1,\sigma_1),\ldots,(1-\beta_q,\sigma_q)\end{matrix}\right] \tag{A2.33}$$

Hence, the Fox H-function has also been given different names such as the Fox-Wright function (Kilbas et al. 2006), the GWF (Gorenflo et al. 1999, Kilbas et al. 2002a, Kilbas 2005), and the Wright generalized hypergeometric function (Shukla and Prajapati 2007), etc.

The properties of the Fox H-function have been amply discussed in literature (Kilbas and Saigo 1999b, Mainardi et al. 2005, Kilbas et al. 2006) and only the relevant major properties have been summarized here. The definition of the Fox H-function is given as (Kilbas and Saigo 1999b, Kilbas et al. 2006):

$$H_{p,q}^{m,n}(x) = H_{p,q}^{m,n}\left[x\bigg|\begin{matrix}\left(a_i,\alpha_i\right)_{1,p}\\\left(b_j,\beta_j\right)_{1,q}\end{matrix}\right]$$

$$= H_{p,q}^{m,n}\left[x\bigg|\begin{matrix}\left(a_i,\alpha_i\right),\ldots,\left(a_p,\alpha_p\right)\\\left(b_j,\beta_j\right),\ldots,\left(b_q,\beta_q\right)\end{matrix}\right] = \frac{1}{2\pi i}\int_c \mathcal{H}_{p,q}^{m,n}(s)x^{-s}ds \tag{A2.34}$$

where

$$\mathcal{H}_{p,q}^{m,n}(s) = \frac{\prod_{j=1}^{m}\Gamma(b_j+s\beta_j)\prod_{i=1}^{n}\Gamma(1-a_i+s\alpha_i)}{\prod_{i=n+1}^{p}\Gamma(a_i+s\alpha_i)\prod_{j=m+1}^{q}\Gamma(1-b_j+s\beta_j)} \tag{A2.35}$$

with $\prod_{j=1}^{m}$ as the product of the gamma functions $\Gamma(b_j + s\beta_j)$ with $j = 1,2,\ldots,m$; the same rules apply to the other product symbols.

Kilbas and Saigo (1999b) presented the power series expansion and the power logarithmic series expansion of the H-function, and the asymptotic properties of the H-function for $x \to 0$ and $x \to \infty$.

2.2 Connection between the GHS and the GWF

In special cases where $A_1 = B_1 = 1$, $i = 1,\ldots, p$ and $j = 1,\ldots, q$, the GHS, the FWF and $_pF_q$ are related as (Kilbas et al. 2006):

$$_p\Psi_q\left[\begin{matrix}(a_p,1)\\(b_q,1)\end{matrix}\bigg|x\right] = \frac{\prod_{i=1}^{p}\Gamma(a_i)}{\prod_{j=1}^{q}\Gamma(b_j)}\,_pF_q\left(a_1,\ldots,a_p;b_1,\ldots,a_q;x\right) \tag{A2.36}$$

2.3 Special Cases of the H-function: The Prabhakar Function and the Mittag-Leffler Function

A special form of the H-function $H_{1,2}^{1,1}$ (or $_1\Psi_1$, in term of the GHS) (Kilbas et al. 2004) is the Prabhakar function (PF) (Prabhakar 1971):

$$\frac{1}{\Gamma(\gamma)}H_{1,2}^{1,1}\left[-x\bigg|\begin{matrix}(1-\gamma,1)\\(0,1)(1-\beta,\alpha)\end{matrix}\right] = \frac{1}{\Gamma(\gamma)}\,_1\Psi_1\left[\begin{matrix}(\gamma,1)\\(\beta,\alpha)\end{matrix}\bigg|x\right] = E_{\alpha,\beta}^{\gamma}(x) \tag{A2.37}$$

when $\gamma = 1$, the PF in Eq. (A2.37) becomes the Wiman function in Eq. (2.41), or the two-parameter MLF. Moreover, when $\gamma = 1$ and $\beta = 1$ in (A2.37), it becomes the MLF in Eq. (2.30). This indicates that the H-function can generate the MLF, the WF, the PF, as well as the GWF and other functions.

3. The Hadid-Luchko Function or the Multinomial MLF

The following function presented by Hadid and Luchko (1996) is often called the multinomial or multivariate MLF:

$$E_{(\alpha_1,\ldots,\alpha_m),\beta}(x_1,\ldots,x_m) = \sum_{k=0}^{\infty}\sum_{l_1+l_2+\ldots+l_m=k}(k;l_1,\ldots,l_m)\frac{\prod_{i=1}^{m}x_i^{l_i}}{\Gamma(\beta+\sum_{i=1}^{m}\alpha_i l_i)} \tag{A2.38}$$

where the multinomial coefficient (Abramowitz and Stegun 1965) is given as:

$$(k;l_1,\ldots,l_m) = \frac{k!}{\prod_{i=1}^{m}l_i!} = \frac{k!}{l_1!\times l_i!\times\ldots\times l_m!} \tag{A2.39}$$

for

$$l_1 + l_2 + \ldots + l_m = k \tag{A2.40}$$

and

$$l_1 \geq 0, l_2 \geq 0 \ldots, l_m \geq 0 \tag{A2.41}$$

The multinomial MLF appears in solutions of fluid flow and solute transport in porous media of finite depths (Jiang et al. 2012a, b). For $m = 2$, Eq. (A2.38) is simplified as (Bhalekar and Daftardar-Gejji 2013):

$$E_{(\alpha_1,\alpha_2),\beta}(x_1,x_2) = \sum_{j=0}^{\infty}\sum_{k=0}^{j}\binom{j}{k}\frac{x_1^k x_2^{j-k}}{\Gamma[\beta+\alpha_1 k+\alpha_2(j-k)]} \tag{A2.42}$$

where the binomial coefficient is:

$$\binom{j}{k} = \begin{cases} \dfrac{j!}{k!(j-k)!} & \text{for } 0 \le k \le j \\ 0 & \text{for } 0 \le j < k \end{cases} \tag{A2.43}$$

Equation (A2.42) is an important function in solutions of water movement in unsaturated soils and saturated flow in aquifers which have dual porosities (Su 2017a, b).

In addition to the multinomial MLF in Eq. (A2.38), a generalized multinomial MLF was presented by Saxena et al. (2011), analogous to the extension of the Wiman function to the Prabhakar function through addition of the third set of parameters.

4. The Yu-Zhang Function with Two Variables

The one-parameter Mittag-Leffler function and many other related functions with multiple parameters have one variable only. Yu and Zhang (2006) defined the following function with two variables:

$$\varepsilon(t,y;\alpha,\beta,\gamma) = t^{\beta-1}E_{\alpha,\beta}\left(-D|y|^{\gamma} t^{\alpha}\right) \tag{A2.44}$$

where t is the time variable; y is the space variable which can be extended to an n-dimensional space with $y = (y_1, y_2,\ldots, y_n)$; $i = 1,2,\ldots,n$; α, β and γ are arbitrary real numbers; and D is a constant such as the diffusivity.

The Laplace transform of the Yu-Zhang function is given as:

$$\mathscr{L}\{\varepsilon(t,y;\alpha,\beta,\gamma)\} = \frac{s^{\alpha-\beta}}{s^{\alpha} + D|y|^{\gamma}} \tag{A2.45}$$

The function in Eq. (A2.44) is used for deriving the fundamental solution of the time-space fractional diffusion-wave equation in n dimensions (Yu and Zhang 2006). Equation (A2.44) with $\gamma = 0$ becomes the integral form of the Wiman function.

5. The Shukla-Prajapati Function and the Srivastava-Tomovski Function

The generalization of the Prabhakar function in Eq. (2.58) was made by Shukla and Prajapati (2007) as follows:

$$E_{\alpha,\beta}^{\gamma,k}(x) = \sum_{n=0}^{\infty}\frac{(\gamma)_{kn}x^n}{n!\Gamma(\alpha n+\beta)} \tag{A2.46}$$

and further investigation and extension of Eq. (A2.46) can be found in the works of Srivastava and Tomovski (2009).

Paneva-Konovska (2013) extended the parameters α, β and γ in the Srivastava-Tomovski function (STF) from real numbers to vectors, hence forming the multi-index MLF.

The Laplace transform of the STF is shown (Srivastava and Tomovski 2009) to be of the form:

$$\mathscr{L}\left[x^{\rho-1}E_{\alpha,\beta}^{\gamma,k}(\lambda x^{\sigma})\right] = \frac{s^{-\rho}}{\Gamma(\gamma)}\sum_{i=0}^{\infty}\frac{\Gamma(\rho+\sigma k)\Gamma(\gamma+ik)}{\Gamma(\beta+\alpha i)}\left(\frac{\lambda}{s^{\sigma}}\right)^{i} \tag{A2.47}$$

6. The Humber-Delerue Function

The two-variable MLF presented by Humbert and Delerue (1953) is expressed as:

$$E_{\alpha,\beta}(x,y) = \sum_{m=0}^{\infty}\sum_{n=0}^{\infty}\frac{x^{m+\frac{\beta(n+1)-1}{\alpha}}y^{n}}{\Gamma[m\alpha+(n+1)\beta]\Gamma[n\beta+1]} \tag{A2.48}$$

7. The Chak Function

Chak (1967) extended the Humbert-Delerue function with two variables the same way that Prabhakar extended the Wiman function, as:

$$E_{\alpha,\beta}^{k}(x,y) = \sum_{m=0}^{\infty}\sum_{n=0}^{\infty}\frac{x^{c}y^{n}}{\Gamma[m\alpha+(nk+1)\beta]\Gamma[n\beta+1]} \tag{A2.49}$$

with

$$c = m + \frac{\beta(nk+1)-1}{\alpha} \tag{A2.50}$$

8. The Srivastava Function

A further extension of the Chak function was given by Srivastava (1968) as:

$$E_{\alpha,\beta,\lambda,\mu}^{v,\delta}(x,y) = \sum_{m=0}^{\infty}\sum_{n=0}^{\infty}\frac{x^{m+\frac{\beta(vn+1)-1}{\alpha}}y^{n+\frac{\mu(\delta m+1)-1}{\lambda}}}{\Gamma[m\alpha+(vn+1)\beta]\Gamma[n\lambda+(\delta m+1)\mu]} \tag{A2.51}$$

Clearly, the Humbert-Delerue and the Chak functions are special cases of the Srivastava function (Srivastava 1968).

$$E_{\alpha,1,\lambda,1}^{0,0}(x,0) = E_{\alpha}(x) \tag{A2.52}$$

and

$$E_{\alpha,\beta,\lambda,1}^{0,0}(x,0) = E_{\alpha,\beta}(x) \tag{A2.53}$$

The above identities show that the Srivastava function is a generating function for the MLF, the WF, the PF and the GPF.

9. The Miller-Ross Function

A special case of the GPF is the Miller-Ross function (Miller and Ross 1993):

$$E_t(v, a) = \sum_{k=0}^{\infty} \frac{(at)^k t^v}{k!\Gamma(v+k+1)} = t^v E_{1,v+1}(at) \tag{A2.54}$$

which is equivalent to $\alpha = 1$, $\gamma = 1$ and $\beta = v + 1$ in the GPF.

Miller and Ross showed that Eq. (A2.54) is related to the incomplete gamma function as:

$$E_t(v, a) = t^v e^{az} \gamma^*(a, t) = t^v e^{at} \tag{A2.55}$$

10. The Kilbas-Saigo Function (KSF)

Kilbas and Saigo (1995) defined the following function for solutions of fractional integral equations as (Kilbas et al. 2004, 2006):

$$E_{\alpha,\beta,\lambda}(x) = \sum_{n=0}^{\infty} c_k x^k = 1 + \sum_{n=1}^{\infty} c_k x^k \tag{A2.56}$$

where

$$c_k = \prod_{i=0}^{n-1} \frac{\Gamma[\alpha(i\beta+\lambda)+1]}{\Gamma[\alpha(i\beta+\lambda+1)+1]} \tag{A2.57}$$

When $\beta = 1$, the following identity is valid (Kilbas et al. 2004):

$$E_{\alpha,1,\lambda}(x) = \Gamma(\alpha\lambda + 1) E_{\alpha,\alpha\lambda+1}(x) \tag{A2.58}$$

Miller and Samko (2001) showed that if $\alpha = 1$ and $\lambda > -1$, the following relationship holds:

$$E_{1,\beta,\lambda}(x) = \Gamma(\mu+1)E_{1,\mu+1}\left(\frac{x}{\beta}\right) \tag{A2.59}$$

with

$$\mu = \frac{x+1-\beta}{\beta} \tag{A2.60}$$

and it is completely monotonic if $\lambda \geq \beta - 1$. Other properties of the KSF have been discussed in Gorenflo et al. (1998).

11. The Yakubovich-Luchko Function

Yakubovich and Luchko (1994) introduced the following function:

$$E((\alpha, \beta)_m; x) = \sum_{k=0}^{\infty} \frac{x^k}{\prod_{j=1}^{m} \Gamma(\alpha_i k + \beta_i)} \tag{A2.61}$$

Kilbas et al. (2013) investigated this function in great detail.

12. The Al-Bassam-Luchko Function

The Al-Bassam and Luchko (1995) function (ABLF) is written as:

$$E_\rho((\alpha, \beta)_n; x) = E_\rho((\alpha_1 + \beta_1)...(\alpha_n + \beta_n); x)$$

$$= \sum_{k=0}^{\infty} \frac{(\rho)_k}{\prod_{j=1}^{m}(\alpha_i k + \beta_i)} \frac{x^k}{k!} \tag{A2.62}$$

where $(\rho)_k$ is the Pochhammer symbol. The ABLF, the YLF and the KSF have structural similarities.

13. The Kiryakova Function

The Kiryakova function (KF) is given as:

$$E_{(1/\rho_i),(\mu_i)}(x) = \sum_{k=0}^{\infty} \frac{x^k}{\prod_{i=1}^{m} \Gamma(\mu_i + k/\rho_i)} \tag{A2.63}$$

which can also be expressed as the Fox H-function in the following form:

$$E_{(1/\rho_i),(1/\mu_i)}(x) = H_{1,m+1}^{1,1}\left[-x \left| \begin{matrix} (0,1) \\ (0,1), (1 - \mu_i, 1/\rho_i)_1^m \end{matrix} \right. \right] \tag{A2.64}$$

Appendix 2.4: Some Special Functions Expressed as Fractional Derivatives

Following the earlier introduction to special functions and fractional derivatives in this chapter, we now briefly show that some commonly used special functions can be expressed as fractional derivatives (Osler 1971, Lavoie et al. 1976, Atangana and Secer 2013). These functions have been listed in the following table extracted from Table 1 in Osler (1971) and from some in Lavoie et al. (1976) where the Riemann-Liouville fractional derivative has been used.

Table 2.2 Special functions expressed as fractional derivatives.

	Fractional derivative representation	Notes
Hypergeometric function	$$_2F_1(a,b;c;z) = \frac{\Gamma(c)z^{1-c}}{\Gamma(b)} D_z^{b-c}\left[z^{b-1}(1-z)^{-a}\right]$$	Lavoie et al. 1976
Confluent hypergeometric function	$$_1F_1(a;c;z) = \frac{\Gamma(c)z^{1-c}}{\Gamma(a)} D_z^{a-c}\left(e^z z^{a-1}\right)$$	Lavoie et al. 1976
Generalized HF	$$_{p+1}F_{q+1}\left[\begin{matrix}a_1,...,a_p,c\\b_1,...,b_p,d\end{matrix}\middle| x\right] = \frac{\Gamma(d)x^{1-d}}{\Gamma(c)} D_x^{c-d}\left[x^{c-1}\,_pF_q\left[\begin{matrix}a_1,...,a_p\\b_1,...,b_p\end{matrix}\middle| x\right]\right]$$	Lavoie et al. 1976
Bessel function	$$J_v(z) = \frac{1}{\sqrt{\pi}\,(2z)^v} D_{z^2}^{-v-1/2}\left[\frac{\cos(z)}{z}\right]$$	Osler 1971
Modified Bessel function	$$I_v(z) = \frac{1}{\sqrt{\pi}\,(2z)^v} D_{z^2}^{-v-1/2}\left[\frac{\cosh(z)}{z}\right]$$	Osler 1971
Struve function	$$H_v(z) = \frac{1}{\sqrt{\pi}\,(2z)^v} D_{z^2}^{-v-1/2}\left[\frac{\sin(z)}{z}\right]$$	Osler 1971
Modified Struve function	$$L_v(z) = \frac{1}{\sqrt{\pi}\,(2z)^v} D_{z^2}^{-v-1/2}\left[\frac{\sinh(z)}{z}\right]$$	Osler 1971
Legendre function of the first kind	$$P_v(z) = \frac{1}{\Gamma(v+1)2^v} D_{1-z}^v\left(1-z^2\right)^v$$	Lavoie et al. 1976
Associated Legendre function of the first kind	$$P_v^u(z) = \frac{(1-z^2)^{u/2}}{\Gamma(v+1)2^v} D_{1-z}^v\left(1-z^2\right)^v$$	Osler 1971
Laguerre function	$$L_v^{(a)}(z) = \frac{\Gamma(a+v+1)}{\Gamma(v+1)\Gamma(-v)z^a} D_z^{-a-v-1}\left(e^z z^{-v-1}\right)$$	Osler 1971
Incomplete gamma function	$$\gamma(a,z) = \frac{\Gamma(a)}{e^z} D_z^{-a} e^z$$	Lavoie et al. 1976

Chapter 3

Essential Properties of Soils and Aquifers as Porous Media

1. Introduction: Soils and Aquifers as Porous Media

Porous media is the largest category of materials on the Earth, including numerous categories of soils and other geological strata, biological materials, and extensive types of artificial products. Soils and aquifers, two major porous media of geological origin, are the key domains discussed in this book.

Soil is the most important porous media on the Earth as it is biologically active and supports life on land. It is also the platform for material and energy exchanges with the atmosphere and the subsurface through a number of physical, chemical and biological processes. These processes include infiltration of water into the soil, evaporation of water from the soil, percolation of infiltrated water into aquifers, runoff on the land (which forms stream flows of different sizes), etc. Soils are either constantly unsaturated by water or variably saturated, depending on the local climatic conditions such as rainfall, and management options such as irrigation. In most of the cases reported in literature, there has been a tendency to regard soils as unsaturated porous media even though they can be saturated.

Aquifers are saturated porous media. Sometimes, aquifers are adjacent to soils in the deeper strata and are physically connected with varying levels of permeability for the exchange of water, solutes, gases, energy, etc., with the soils. In other cases, however, aquifers are isolated from soils by other geological strata.

The water in soils and aquifers differs significantly from the bulk fluid of water in classical hydraulics. The key aspects that distinguish flow in porous media from bulk flow in open channels, pipes and other forms of bulk flows include:

1. Dynamic inflow-outflow of water between pores of varying properties (size, shape, porosity, conductivity, etc.);

2. Dynamic absorption-desorption of solutes and gases in water onto porous particles; and

3. Presence of solid structure of porous media with varying properties, known as regions of excluded volume (Cushman and Ginn 1993).

The excluded volume includes both the bulk solid matrix and the inaccessible pores or immobile pores. Over spatial and temporal scales on which the excluded volume continuously evolves, the classical concept of diffusion fails and the concept for a more general process known as the non-local phenomenon applies, which is also interpreted as the scale effect (Cushman and Ginn 1993) and anomalous transport—this is where models based on fractional calculus suitably apply.

In this chapter, basic concepts widely used for the description of porous media have been briefly summarized. Many of these concepts and definitions have been defined and modified by researchers over the course of more than a century, hydraulic conductivity in Darcy's law since 1856 and some older concepts such as density and viscosity since Newton's time (1642–1727) for instance.

This chapter also reviews descriptive relationships, fractal concepts, scaling identities of some important hydrological variables and parameters, the fundamental governing equations for flow in porous media such as aquifers and unsaturated media such as soils.

2. Descriptive Concepts and Definitions of Soils and Aquifers

Soil has multiple components. It consists of air, water, solids (organic matter, colloids, clay, sand and other minerals), microbes and, in many cases, small creatures such as earthworms, etc. A subsurface porous media model is shown in Fig. 3.1A and an idealized soil profile is in Fig. 3.1B which can be used to illustrate how unsaturated soils and saturated aquifers relate to each other. The simplified, idealized 'soil' on the basis of which qualitative analyses have been developed in hydrology, soil physics and soil mechanics is mainly regarded as a three-component media, with air, water and solids coexisting, or a multiphase media if the energy in terms of heat is included.

Porous media of geological origin, such as soils and aquifers, are versatile in properties. These properties include various forms of porosities and elasticity or their swelling-shrinking property when wet or under external forces or other forms of energy gradients. Porous media are characterized by levels of heterogeneity in terms of spatial variability in physical (Kozeny 1927) and chemical properties (Michels et al. 2019), composition, and the existence of anisotropy in terms of changes in certain properties in different directions. Porous media are also subject to change

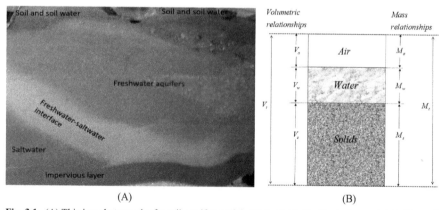

(A)   (B)

Fig. 3.1. (A) This is a photograph of a soil-aquifer model. (B) A schematic illustration and definitions of soil parameters in a three-component system (modified after Hillel 1998).

in properties as a result of thermal, acoustic and electromagnetic forces, etc. Owing to their importance as geological strata, and their widespread distribution, soils and aquifers are central topics in many disciplines such as hydrology, soil science, soil mechanics, engineering, environmental science, ecology, agricultural science, agronomy and civil engineering, etc.

Since we are mainly concerned with the concepts of porous media in relation to water flow, solute movement and thermal transfer, as well as their joint processes, we only discuss pertinent concepts and methods for their characterization here.

To quantify the properties of porous media, the first question to be answered is: *What scale or size of a sample is representative of the whole system?* One of the concepts which answers this is the representative elementary volume (REV). It is generally desirable for an REV to have two important properties (de Marsily 1986):

1. The REV is a volume large enough to contain a large number of pores so that the mean global properties can be defined while ensuring that fluctuation of a property from one pore to another is negligible. This criterion also defines the boundary beyond which the dominance of molecular effects may be considered insignificant (Bear 1972).

2. The REV is small enough to ensure that the parameter variations from one domain to the next may be approximated by continuous functions.

There are several simple quantities which are basic variables in hydrology, soil physics and soil mechanics. Some of these quantities have been frequently used throughout this book and briefly outlined here. Further details and information on more concepts can be found in the works of Marshall et al. (1996) and Hillel (1998).

2.1 Pores and Porosity, and Related Concepts

The quantity of pores in a given volume of porous medium is one of the most important parameters for qualifying it. This can be represented by different concepts such as porosity, pore-size distribution, pore surface area, particle size distribution, aggregate-size distribution, bulk density and other micro-morphological parameters (Perfect and Kay 1995). Porosity can be classified as total porosity ϕ, effective porosity ϕ_e, and, therefore, ineffective porosity ϕ_i, i.e., $\phi_i = \phi - \phi_e$. Total porosity is a measure of all the pores in a given volume of medium, whereas effective porosity is representative of connected pores directly conductive for flow or transport of particles in addition to the closed or inactive pores (measured by ϕ_i) which slowly respond to and interact with connected pores.

Total porosity ϕ is the ratio of the total volume of pores (or voids) V_p and the bulk volume of the medium V_t (Bear 1972, 43–44):

$$\phi = \frac{\text{pore volume}}{\text{bulk volume}} = \frac{V_P}{V_t} \tag{3.1}$$

where $V_p = V_a + V_w$, with V_a and V_w representing the volume of pores occupied by air and water respectively.

Effective porosity ϕ_e is defined as the ratio of the volume of interconnected pores (or effective pores) and the bulk volume:

$$\phi_e = \frac{\text{active pore volume}}{\text{bulk volume}} = \frac{V_e}{V_t} \tag{3.2}$$

The ratio of the volume of solids V_s and the bulk volume ϖ is (Henderson et al. 2010):

$$\varpi = 1 - \phi = \frac{\text{volume of solids}}{\text{bulk volume}} = \frac{V_s}{V_t} \tag{3.3}$$

The surface area of porous media is important, for it absorbs and adsorbs water, solutes and gases. One of the parameters for measuring the surface area of porous media is specific surface area S_t, given as early as 1927 by Kozeny (Henderson et al. 2010):

$$S_t = \frac{\text{interstitial surface area}}{\text{bulk volume}} = \frac{A_s}{V_t} \tag{3.4}$$

where A_s is the interstitial surface area per unit of bulk volume of an REV.

2.2 Void Ratio

The void ratio e for a medium is defined as the quotient of the volume of voids V_p and the volume of solids V_s (Bear 1972), which can also be expressed as a function of total porosity:

$$e = \frac{V_P}{V_s} = \frac{\phi}{1 - \phi} \tag{3.5}$$

The above equation can be rearranged to yield:

$$\phi = \frac{e}{1 + e} \tag{3.6}$$

2.3 Density of Solids (Mean Particle Density)

Mean particle density refers to the density of solids of a soil, which is measured as a ratio of the mass of solids M_s per unit volume of solids of the soil V_s (Hillel 1998):

$$\rho_s = \frac{M_s}{V_s} \tag{3.7}$$

2.4 Wet Bulk Density and Dry Bulk Density

The mass of a soil sample per unit of volume can be measured as either the wet bulk density (simply 'bulk density') γ_t or the dry bulk density γ_d (Hillel 1998).

Wet bulk density is the ratio of the total mass of a moist soil (under any condition) M_t per unit volume of the soil:

$$\gamma_t = \frac{M_t}{V_t} \tag{3.8}$$

where V_t is the total volume of the soil sample.

γ_t is generally measured using natural samples from the field. In some cases, the dry bulk density of a sample is needed for quantitative analysis and is defined as the ratio of the mass of solids M_s per unit volume of the soil:

$$\gamma_d = \frac{M_s}{V_t} \tag{3.9}$$

2.5 Gravitational Water Content and Volumetric Water Content

The gravitational water content θ_w for a soil sample is defined as the ratio of the mass of water M_w and the mass of solids in the sample M_s (Hillel 1998):

$$\theta_w = \frac{M_w}{M_s} \tag{3.10}$$

In many cases, the volumetric water content θ_v of a sample is more important and convenient, such as in two-phase flows and multiphase flows involving the flow of gases whose weights are negligibly small but the occupied volume is significant to the flow. In such cases, the volumetric water content of the soil sample is more important for its analysis.

θ_v is defined as the ratio of the volume of water and the total volume of the soil or the soil sample:

$$\theta_v = \frac{V_w}{V_t} = \frac{V_w}{V_s + V_p} \tag{3.11}$$

where V_p is the volume of pores in the soil sample, including the pores for water and air.

The relationship between θ_v and θ_w can be expressed as (Hillel 1998):

$$\theta_v = \frac{\gamma_d}{\rho_w} \theta_w \tag{3.12}$$

where ρ_w is the density of water and γ_d is the dry bulk density of the soil.

2.6 Degree of Saturation

The degree of saturation, or simply 'saturation', is the ratio of the volume of water and the volume of pores:

$$S = \frac{V_w}{V_a + V_w} = \frac{V_w}{V_p} \tag{3.13}$$

2.7 Other Identities and Their Relationships

2.7.1 Volumetric Water Content θ_v and Degree of Saturation S

The relationship between the volumetric water content θ_v for a soil sample and its degree of saturation S is given as:

$$\theta_v = s\phi \tag{3.14}$$

2.7.2 Porosity and Dry Bulk Density

For a soil sample, its total porosity ϕ, dry bulk density γ_d, and mean particle density ρ_s are related as:

$$\phi = \frac{\rho_s - \gamma_d}{\rho_s} = 1 - \frac{\gamma_d}{\rho_s} \tag{3.15}$$

2.7.3 Air-filled Porosity ϕ_a

This parameter is useful for measuring the magnitude of gases in unsaturated media and is defined as:

$$\phi_a = \frac{V_a}{V_a + V_w} \tag{3.16}$$

2.7.4 Double Porosity Concept

Pores in soils and aquifers are not uniform in size and shape. In addition to the above quantities, more methods are required for the quantitative analysis of material flow and energy transfer in porous media. The double porosity concept was originally introduced by Rubinshtein (1948) for investigating heat transfer in a heterogeneous medium by proposing a coupled model for heat transfer in a double porosity medium, followed by Pirson (1953) with the qualitative concept of double porosity for flow in oil reservoirs. Barenblatt et al. (1960) proposed a coupled model, the double porosity model, for describing the exchange of water between fractures for flow in two levels of porous media. Since then, different forms of double porosity models have been presented and in particular, the two-term advection-dispersion equation by Coats and Smith (1964) has been more familiar in the context of soil science and hydrology (Nielsen et al. 1986, Sardin et al. 1991).

2.8 Material Coordinate for Quantifying Swelling-shrinking Soils

In non-swelling soils, the conventional coordinate system is valid, wherein a fixed reference can be found. In swelling soils, however, the reference frame can change subject to swelling or shrinking of soil particles during wetting-drying cycles. With these types of soils, the use of a fixed coordinate such as the soil surface is not appropriate. In order to remedy this problem, a material coordinate was proposed in the 1960s (Raats 1965, Raats and Klute 1968, Smiles and Rosenthal 1968, Philip 1969b, Philip and Smiles 1969). In Chapter 5, there are extensive discussions on fPDEs developed for water flow in swelling-shrinking soils, using the material coordinate.

By neglecting the horizontal expansion of soil, resulting from an increase in the water content of swelling soils, the material coordinate m is defined as (Smiles and Rosenthal 1968, Philip 1969b, Philip and Smiles 1969):

$$m = \int_0^z (1+e)^{-1} dz_1 \tag{3.17}$$

where z is the usual spatial coordinate in the vertical direction and e is the void ratio given by Eq. (3.5).

Note that there are three cases with varying relationships between the void ratio e and the volumetric moisture content θ_v (Philip and Smiles 1969):

1. Normal change: This occurs in the two-component system of soil particles and water. Herein, the following relationship holds,

$$e = \frac{\theta_v}{1-\theta_v} \tag{3.18}$$

which implies that

$$\frac{de}{d\theta_v} = \frac{1}{(1-\theta_v)^2} \tag{3.19}$$

where θ_v is the volumetric water content or, equivalently, the moisture ratio defined as:

$$\theta = \frac{V_l}{V_s} \tag{3.20}$$

with V_l and V_s are the volume fractions of liquid and solid respectively.

2. Residual change: This occurs in the three-component system of soil particles, water and air, when the change in bulk volume produced by the removal of water from the three-component system is less than the volume of water removed. In this case, the following relationships hold:

$$e > \frac{\theta_v}{1-\theta_v} \tag{3.21}$$

$$0 < \frac{de}{d\theta_v} < \frac{1+e}{1-\theta_v} \tag{3.22}$$

or

$$0 < \frac{d(1+e)}{d[(1+e)\theta_v]} < 1 \tag{3.23}$$

3. Zero change: This case corresponds to a condition in the three-component system of soil particles, water and air, wherein changes in the water ratio produce no change in the volume of the soil particles, that is:

$$\frac{de}{d\theta_v} = 0 \tag{3.24}$$

and

$$e = \text{constant} = e_{min} > \frac{\theta_v}{1-\theta_v} \tag{3.25}$$

where e_{min} is the minimum value of e.

In earlier developments of fPDEs for water movement in soils (Su 2010, 2012, 2014), the first case of normal change was given for the definition of the material coordinate in a two-component system of water and soil particles with the effect of air being negligible.

3. Fundamental Equations of Flow in Soils and Aquifers

3.1 Introduction

For more than a century, fluid flow in different forms has been successfully described by a set of equations comprising those for the conservation of mass, momentum (or the equation of motion) and the equation of conservation of energy (de Marsily 1986). The development of this set of equations was the result of works of many scientists and mathematicians such as Euler (1707–1783). Two of these eminent researchers had their names associated with the final form of these equations of flow—Louis M.H. Navier (1785–1836) and George G. Stokes (1819–1905). This set of equations are hence known as the Navier-Stokes equations (NSEs) (Debler 1990), and are also central to the mathematical analysis of flow in porous media such as soils and aquifers.

The flow equations based on the NSEs (Eagleson 1970) accommodate viscoelasticity of the porous media and compressibility of the fluids, hence linking the three fields of groundwater hydrology, soil physics, and soil mechanics.

In groundwater hydrology, the NSE can be specifically formulated for both confined and unconfined aquifers, with compressibility as a crucial parameter for the consolidation of aquifers (de Marsily 1986).

In soil physics, the NSEs have been highly simplified and one of the most widely used PDE derived from the NSEs is called the Richards equation (1931) for water flow in unsaturated soils, which does not account for compressibility of the media and the fluid.

In soil mechanics, which is more concerned with elasticity, volume change and associated stresses of the soil, the NSEs can be specialized to relate stress and strain in addition to other approaches for the stress-strain relationship (Bemer et al. 2001, Liingaard et al. 2004, Mainardi 2012, Lai et al. 2016, Sun et al. 2016).

The two fields of soil physics and soil mechanics were arguably unnecessarily split in the 1930s, which was regarded as an impedance for the development of the two fields (Philip 1974). To address the integral physical processes without defining boundaries, this book endeavors to deal with the issues of water movement, solute transport and elasticity in soils and aquifers as well as flow overland.

3.2 Flow in Porous Media with Different Levels of Saturation by Water

3.2.1 Hydraulic Head for Flow in Saturated Media

Flow in porous media is driven by energy which, according to Bernoulli's theorem and Navier-Stokes equations, can be written as a sum of three components of energy. Hubbert (1940, de Marsily 1986, Hillel 1998) re-worked the generalization of a celebrated theorem stated by Daniel Bernoulli in 1738, relating elevation, pressure and velocity along a given flowline of a fluid in frictionless flow through the *potential* Φ:

$$\Phi = gz + \int_{p_0}^{p} \frac{dp}{\rho(p)} + \frac{V^2}{2} \tag{3.26}$$

where p is the pressure with p_0 as its reference, g is gravitational acceleration, z is the elevation, V is the velocity of water, and $\rho(p)$ is the density of water as a function of pressure that is often treated as a constant.

The Bernoulli equation is simplified when a suitable reference pressure p_0 is chosen such that, for $p_0 = 0$ and constant density ρ, it is expressed as:

$$\Phi = gz + \frac{p}{\rho} + \frac{V^2}{2} \tag{3.27}$$

while, in its generic form, Eq. (3.27) can be written as:

$$\Phi = gz + \frac{[\rho g(h-z) + p_0] - p_0}{\rho} + \frac{V^2}{2} = gh + \frac{V^2}{2} \tag{3.28}$$

where the flow potential Φ is dependent on the measured height of the fluid above a reference datum $(h - z)$, given the constant gravitational acceleration at a particular location. With Eq. (3.27), *hydraulic head h* is defined as (de Marsily 1986):

$$h = \frac{\Phi}{g} = z + \frac{p}{\rho g} + \frac{V^2}{2g} \tag{3.29}$$

On the right-hand side of Eq. (3.29), the first term is called the *elevation head*, the second term is the *pressure head*, and the third term is the *velocity head*. In hydraulics of flow in porous media, hydrology and soil science, the velocity head is normally very small compared to the other two terms. Hence, the *hydraulic head* is reduced to the *piezometric head* and, with the identity of specific weight of water with its density, it is given as:

$$\gamma = \rho g \tag{3.30}$$

The hydraulic head (or the piezometric head) without considering the velocity head is then written as (Bear 1972, de Marsily 1986):

$$h = \frac{p}{\gamma} + z \tag{3.31}$$

Now, it is clear that the hydraulic head in Eq. (3.29) is a summation of three components of the energy which drives fluid flow, while the piezometric head is a sum of two components, neglecting the kinetic head $\frac{V^2}{2g}$ as a consequence of the negligible flow of the fluid in some cases.

3.2.2 Hydraulic Head for Flow in Unsaturated Media

In unsaturated media, the pressure head is replaced by the *capillary pressure head h_c*, resulting in a hydraulic head of the following form (Bear 1972):

$$h = z + \frac{p_c}{\gamma} = z - h_c \tag{3.32}$$

with h_c having negative values. Depending on the choice of the origin of the vertical coordinate, the capillary pressure head h_c can also be written as a sum of the moisture potential $\psi(\theta)$ (which is expressed as a negative value and is a function of moisture ratio or moisture content θ), and the elevation (Philip 1969b, Bras 1990) in unsaturated soils as:

$$h = z + \psi(\theta) \tag{3.33}$$

3.3 General Governing Equations for Isothermal Flow of Water in Aquifers and Soils

Compared with the flow of bulk fluid such as in pipes or rivers, the fluid flow in porous media is subject to the effects of the internal structures of pores and excluded volume as discussed at the beginning of this chapter. Despite this difference, the 'complete' set of equations used for describing flow of a fluid in porous media consists of equations for conservation of mass, momentum (or the equation of motion) and the equation of conservation of energy (Eagleson 1970, Bear 1972, de Marsily 1986). In this section, these equations have been briefly discussed.

3.3.1 Equation of Motion and Darcy's Law

An empirical model given by Darcy (1856), relating the volumetric flow rate or discharge Q to the hydraulic gradient defines several important quantities. This empirical relationship states as follows:

$$Q = -KA\frac{h_2 - h_1}{L} \tag{3.34}$$

where the proportional coefficient K is either a constant or a function of other parameters and variables, A is the area across which the flow takes place, h_1 and h_2 are the hydraulic heads or piezometric heads at two locations at a distance L apart. In the differential form, Eq. (3.34) is written as:

$$Q = KJ = -KA\frac{dh}{dx} \tag{3.35}$$

where J is the hydraulic gradient and h is the hydraulic head or the piezometric head with different definitions for saturated and unsaturated flow.

Equation (3.35) can be written in terms of the velocity V or the specific discharge q as (Bear 1972):

$$V = \frac{Q}{\phi A} = \frac{q}{\phi} \tag{3.36}$$

that is,

$$V = -\frac{K}{\phi}\frac{dh}{dx} \tag{3.37}$$

with $q = \dfrac{Q}{A}$.

Here, porosity is used because the flow in reality only takes place in pores across the area. Hence, specific discharge is numerically equivalent to velocity. In literature, the term effective hydraulic conductivity also appears, thus giving rise to a relationship of the form $K_e = \dfrac{K}{\phi_e}$.

The expression establishing the relation between hydraulic conductivity and other parameters was a result of the works of Nutting (1930) (Bear 1972, de Marsily 1986):

$$K = \frac{kg\rho}{\mu} \tag{3.38}$$

where k is the permeability or intrinsic permeability (a property specific to a medium), g is the acceleration due to gravity, ρ is the fluid density, and μ is the dynamic viscosity. With the relationship between dynamic viscosity and kinetic viscosity v, given by $v = \mu/\rho$, Eq. (3.38) is rewritten as:

$$K = \frac{kg}{v} \tag{3.39}$$

A more general form of Darcy's law in multiple dimensions can be written in two equivalent forms as (de Marsily 1986):

$$\begin{aligned} \mathbf{V} &= -\mathbf{K}\nabla \cdot h \\ &= -\frac{k}{\mu}(\nabla \cdot p + \rho g \nabla \cdot z) \end{aligned} \tag{3.40}$$

where the gradient operator is grad$h = \nabla h$, with the del-operator ∇ for which the following identity holds (Bourne and Kendall 1977):

$$\nabla = \mathbf{i}\frac{\partial}{\partial x} + \mathbf{j}\frac{\partial}{\partial y} + \mathbf{k}\frac{\partial}{\partial z} = \left(\frac{\partial}{\partial x}, \frac{\partial}{\partial y}, \frac{\partial}{\partial z}\right) \tag{3.41}$$

As grad $z = 1$, with z as the vertical direction, Eq. (3.40) is in fact equivalent to the following:

$$\mathbf{V} = -\frac{k}{\mu}(\nabla \cdot p + \rho g) \tag{3.42}$$

In terms of energy, Hubbert (1940) showed that Darcy's hydraulic head is potential energy per unit weight of water (or other fluids). de Marsily (1986) demonstrated that the experimental form of Darcy's law in Eq. (3.35), and its generalization in Eq. (3.40) are results of the simplified NSEs, which can be derived from flow through a bundle of circular tubes conceptualized by Poiseuille.

Bear (1972) tabulated some 'typical' values of hydraulic conductivity and permeability for a range of aquifers and soils as well as rocks. A simplified table with his data is shown below in terms of logarithmic values of K and k.

Table 3.1. Approximate ranges of the logarithmic values of K and k for porous media of geological origin (after Bear 1972).

$-\log K(cm/s)$:	−2	−1	0	1	2	3	4	5	6	7	8	9	10	11
Permeability		Pervious			Semipervious			Impervious						
$-\log k(cm^2)$:		3	4	5	6	7	8	9	10	11	12	13	14	15	16

A porous medium is said to be of good permeability if its $-\log K(cm/s)$ lies in the range of -2 to 3, and of poor permeability if $-\log K(cm/s)$ ranges from 3 to 7. These values correspond to $-\log k(cm^2)$ in the range of 3 to 8 and 8 to 12 respectively.

3.3.2 Averaged Navier-Stokes Equations Governing Fluid Flow in Porous Media

To define the averaged NSEs for flow in porous media (Irmay 1958, Bear 1972), the total energy of flow per unit weight of fluid E, comparable with the hydraulic head h in Eq. (3.29), is expressed as:

$$E = h = z + \frac{p}{\rho g} + \frac{V^2}{2g} \tag{3.43}$$

Employing the averaging approach (Polubarinova-Kochina 1952, Irmay 1958) yields the relationship between the average hydraulic gradient $\bar{J}$ and flux q in one dimension (in the x direction) as:

$$\bar{J} = -\frac{\partial \bar{E}}{\partial x} = aq + bq^2 + c\frac{\partial q}{\partial t} \tag{3.44}$$

where

$$a = \beta \frac{(1-\phi)^2}{\phi^3} \frac{v}{gd^2} \tag{3.45}$$

$$b = \frac{1}{2} \frac{\alpha(1-\phi)}{\phi^3} \frac{1}{gd} \tag{3.46}$$

$$c = \frac{1}{\phi g} \tag{3.47}$$

A similar approach yields expressions for the y and z directions. Comparing Eq. (3.44) and Eq. (3.35), it is evident that Darcy's law is the linear part of the relationship resulting from the averaged NSEs which was derived by Polubarinova-Kochina (1952), that is:

$$q = \frac{1}{a}\bar{J} \tag{3.48}$$

where $K = \frac{1}{a}$ and the second-order and time-dependent terms in Eq. (3.44) have been neglected.

3.3.3 Conservation of Mass and General Equations of Flow

The fundamental equations governing flow in porous media can be written in a unified form as (Eagleson 1970):

$$\rho\left(\phi_e \frac{\partial s}{\partial t} + \phi_e s \beta \frac{\partial p}{\partial t} - s\alpha \frac{\partial \sigma_z}{\partial t}\right) = -\nabla \cdot \rho\mathbf{q} \tag{3.49}$$

where s is the degree of saturation or moisture ratio,

ϕ_e is the effective porosity,

σ_z is the normal intergranular stress on a horizontal plane, and

α is the vertical compressibility of the granular skeleton of porous media. This vertical compressibility is given as (Eagleson 1970, Bear 1972):

$$\alpha = -\frac{1}{U}\frac{dU}{d\sigma_z} \tag{3.50}$$

with U as the gross elemental volume (control volume), and β as the compressibility of the fluid, which is the reciprocal of bulk modulus of elasticity and is expressed as (Eagleson 1970, Bear 1972):

$$\beta = -\frac{1}{U}\frac{dU}{dp} = \frac{1}{\rho}\frac{d\rho}{dp} \tag{3.51}$$

with ρ being the density of water, and p the pressure head of pore water.

$\mathbf{q}$ is the specific discharge (Bear 1972) or apparent fluid velocity relative to porous grains (Eagleson 1970), in three dimensions. It is expressed as the ratio of volumetric discharge and flow area, that is, from Eq. (3.36):

$$\mathbf{q} = \frac{Q}{A} \tag{3.52}$$

and its three components are:

$$\left.\begin{array}{l} q_x = \phi_e V_x \\ q_y = \phi_e V_y \\ q_z = \phi_e V_z \end{array}\right\} \tag{3.53}$$

where q_x, q_y and q_z are specific discharges (with dimension $[LT^{-1}]$) in x, y and z directions respectively, and V_x, V_y and V_z are the velocities in the three dimensions. With the following identity:

$$-\nabla \cdot \rho\mathbf{q} = -\frac{\partial}{\partial x}(\rho q_x) - \frac{\partial}{\partial y}(\rho q_y) - \frac{\partial}{\partial z}(\rho q_z) \tag{3.54}$$

Equation (3.49) can be expanded as:

$$\rho\left(\phi_e \frac{\partial s}{\partial t} + \phi_e s \beta \frac{\partial p}{\partial t} - s\alpha \frac{\partial \sigma_z}{\partial t}\right) = -\frac{\partial}{\partial x}(\rho q_x) - \frac{\partial}{\partial y}(\rho q_y) - \frac{\partial}{\partial z}(\rho q_z) \tag{3.55}$$

Equation (3.49) or its expanded form in (3.55) is the equation for conservation of mass, one of the fundamental equations governing flow in porous media. The media in which flow takes place can be either unsaturated media (such as soils) when $s < 1$ or saturated media when $s = 1$. In both confined and unconfined aquifers, $s = 1$ applies.

3.3.4 Conservation of Mass for Flow in Unconfined Aquifers

The general flow equation in Eq. (3.55) can be simplified for flow in unconfined aquifers by applying the Dupuit assumption. Dupuit (1848, 1863) simplified the fundamental equation of flow by assuming that the flowlines are parallel to the direction of flow and that there is no vertical component of flow. This assumption allows the integration of Eq. (3.55) with the hydraulic head to reduce the three-dimensional equation to a two-dimensional equation of the following form (Bear 1972):

$$\phi_e \frac{\partial h}{\partial t} = \frac{\partial}{\partial x}\left(Kh\frac{\partial h}{\partial x}\right) + \frac{\partial}{\partial y}\left(Kh\frac{\partial h}{\partial y}\right) \tag{3.56}$$

where

h is the hydraulic head of the aquifer, with a dimension, $[L]$;

K is the saturated hydraulic conductivity of the aquifer, $[LT^{-1}]$;

ϕ_e is represents drainage porosity or specific yield or effective porosity (de Marsily 1986) and is dimensionless (see Table 3.1 for details);

x and y are the space variables in the two directions, $[L]$; and

t denotes time, $[T]$.

Equation (3.56) is known as Boussinesq's equation of groundwater flow in an unconfined aquifer (Boussinesq 1904).

3.3.5 Conservation of Mass for Flow in Confined Aquifers

In confined aquifers, $s = 1$ in Eq. (3.55) and, according to Terzaghi's consolidation theory, the change in hydraulic pressure in the pores of an aquifer produces an equal but opposite change in the intergranular stress throughout the medium. Hence, Eq. (3.55) becomes:

$$\rho(\alpha + \phi_e\beta)\frac{\partial p}{\partial t} = -\frac{\partial}{\partial x}(\rho q_x) - \frac{\partial}{\partial y}(\rho q_y) - \frac{\partial}{\partial z}(\rho q_z) \tag{3.57}$$

The temporal change in hydraulic pressure, $\dfrac{\partial p}{\partial t}$, follows the identity:

$$\frac{\partial p}{\partial t} = \gamma\frac{\partial h}{\partial t} \tag{3.58}$$

where γ is the specific weight of water. Equation (3.57) can therefore be written in terms of hydraulic head h as:

$$\rho S_e \frac{\partial h}{\partial t} = -\frac{\partial}{\partial x}(\rho q_x) - \frac{\partial}{\partial y}(\rho q_y) - \frac{\partial}{\partial z}(\rho q_z) \tag{3.59}$$

where

$$S_e = \gamma(\alpha + \phi_e\beta) \tag{3.60}$$

refers to as storativity (Bear 1972), specific storage (Eagleson 1972) or storage coefficient (de Marsily 1986).

Equation (3.59) applies to water flow in confined aquifers where changes in the density of water is appreciable due to factors such as seawater intrusion of aquifers. For a constant density or when the change in density is negligible compared to other larger quantities, Eq. (3.59) can be combined with Darcy's law to yield the following equation:

$$S_e \frac{\partial h}{\partial t} = -\nabla \cdot K\nabla \cdot h \tag{3.61}$$

for heterogeneous media, and

$$S_e \frac{\partial h}{\partial t} = -\boldsymbol{K}\nabla^2 \cdot h \qquad (3.62)$$

for homogeneous media.

3.3.6 Conservation of Mass for Flow in Unsaturated Media

For flow in unsaturated media such as soils with $s < 1$, Eq. (3.55) is written as (Eagleson 1970):

$$\frac{\partial \theta}{\partial t} = -\nabla \cdot \mathbf{q} \qquad (3.63)$$

where the specific discharge of flow in unsaturated media is given by Darcy's law, depending upon moisture content or moisture ratio. Combining Eq. (3.59) and Darcy's law yields:

$$\frac{\partial \theta}{\partial t} = -\nabla \cdot \left(\frac{k}{\mu} (\nabla p + \rho g) \right) \qquad (3.64)$$

for isothermal, compressible, laminar flow in non-elastic unsaturated soils. When the hydraulic head is expressed in terms of moisture content or moisture ratio θ, Eq. (3.64) can be written in a compact form as (Klute 1952, Philip 1969a):

$$\frac{\partial \theta}{\partial t} = \nabla \cdot (D \nabla \theta) \pm \frac{dK}{d\theta} \frac{\partial \theta}{\partial z} \qquad (3.65)$$

which, through a functional relationship of $\psi(\theta)$ with θ, can be rewritten as:

$$\frac{d\theta}{d\psi} \frac{\partial \psi}{\partial t} = \nabla \cdot (K \nabla \psi) \pm \frac{dK}{d\theta} \frac{\partial \psi}{\partial z} \qquad (3.66)$$

where diffusivity D and hydraulic conductivity K are related through the following relationship:

$$D = K \frac{d\psi}{d\theta} \qquad (3.67)$$

Both D and K are functions of θ, and may also be functions of solute concentrations and other variables under an isothermal condition. Notice that the sign $\pm$ is used: when the vertical coordinate z is chosen to be positive downwards, a negative sign is used (Philip 1969b).

It can be observed that Eqs. (3.65) and (3.66) are not completely equivalent in some cases due to the effect of air, such as when ψ exceeds the air-entry value in which case Eq. (3.66) would apply, whereas Eq. (3.65) would not (Philip 1969b, 222).

The one-dimensional forms of Eqs. (3.65) and (3.66) are respectively written as:

$$\frac{\partial \theta}{\partial t} = \frac{\partial}{\partial x} \left(D \frac{\partial \theta}{\partial x} \right) - \frac{dK}{d\theta} \frac{\partial \theta}{\partial z} \qquad (3.68)$$

and

$$\frac{d\theta}{d\psi}\frac{\partial\psi}{\partial t} = \frac{\partial}{\partial x}\left(K\frac{\partial\psi}{\partial x}\right) - \frac{dK}{d\theta}\frac{\partial\psi}{\partial z} \tag{3.69}$$

Equation (3.65) or (3.66) is known as Richards equation (Richards 1931) even though, to a large extent, a similar formulation had been presented, a decade before Richards, by Gardner and Widstoe (1921) who, in the meantime, also developed an equation for infiltration, which is essentially the Horton equation (Horton 1939) as pointed out by Philip (1954b) and Eagleson (1970).

4. Applicability of Darcy's Law

Over the past several decades, the integer calculus-based PDEs summarized above have formed the conventional mathematical models for analyzing water flow and solute transport in soils and aquifers. Darcy's law is valid for laminar flow (judged by a criterion called the Reynolds number Re) in porous media Re (Bear 1972, de Marsily 1986). For bulk flow in pipes and other regular geometries, the Reynolds number can be easily calculated. For flow in porous media, Re takes a similar form (Bear 1972):

$$Re = \frac{qd}{v} \tag{3.70}$$

where d is some characteristic length of porous media, v is kinetic viscosity, and q is specific discharge (numerically equivalent to the flow velocity).

Despite the definition in Eq. (3.70), the difficulty in determining Re in porous media is that choosing a characteristic length d is less straightforward because of the irregularities of pores in terms of size and shape. For this reason, different empirical formulae have appeared in literature to relate d to other parameters. One of the methods is the use of the mean grain diameter of the medium to determine Re and, when Re is determined from the mean grain diameter, it is generally accepted that Darcy's law is valid as long as Re belongs to the range of 1 to 10 (Bear 1972, de Marsily 1986). When Re lies between 10 and 100, there is a beginning of transient flow when the inertial forces cannot be neglected and Darcy's law ceases to be valid (de Marsily 1986). It should be noted that these values of Re apply to integer calculus-based PDEs.

As it will be evident from Chapter 4 onwards, fractional PDEs for water flow and solute transport can be derived from the continuous-time random walk theory, and the fundamental equations of flow, including Darcy's law, are not essential for such derivations. For the application of fPDEs, we are less concerned about the values of Re. We are rather interested in how the flow patterns evolve subject to the dynamic balance in orders of fractional derivatives in time and space, which determines whether the flow pattern is sub-diffusion, super-diffusion or normal diffusion.

5. Traditional and New Parameters for Hydraulic Properties

In addition to the models based on fPDEs, various important parameters in the models have been discussed in different chapters of this book. Some of these parameters appear frequently in literature and are called traditional parameters, whereas some are comparatively less frequently used, such as fractal parameters. Moreover, many new parameters presented here are related to fractional fPDEs.

Hydraulic conductivity K and porosity ϕ are two important parameters which influence many other hydraulic properties during water flow and solute transport. It is observed in Eq. (3.38) that hydraulic conductivity depends on the properties of the medium represented by the permeability k, and of the fluid characterized by its density ρ and dynamic viscosity μ. All these parameters could be functions of other variables such as temperature, solute concentration, pressure, etc. While pure fluid properties can be measured, soil water is not pure water; it is instead a solution with varying concentrations of solutes and gases. It is, therefore, difficult to precisely quantify the real values of these parameters; even permeability is difficult to quantify precisely.

Darcy's empirical formula initiated investigations into 'permeability' and its relationships with other parameters. Carman (1937, 1939) had briefly summarized studies of up to that time. Works by Dullien (1992) and more recent efforts for characterizing hydraulic properties of porous media are continuously reported using various methods.

In this section, some well-known models for hydraulic parameters have been discussed in association with the concept of fractal porous media. The relationships between water content (water ratio or saturation) and hydraulic head (or matrix potential) have been successfully described by widely-used models such as those by Brooks-Corey (1964, 1966), its fractal counterpart by Tyler and Wheatcraft (1990), and the one by van Genuchten (1980), with fractal parameters incorporated later. Ghanbarian-Alavijeh et al. (2011) reviewed mathematical models on water content and potentials developed by other researchers (Rieu and Sposito 1991, Perrier et al. 1996, Perfect 1999, Bird et al. 2000, Millán and González-Posada 2005, Cihan et al. 2007), whereas Perfect and Kay (1995) and Giménez et al. (1997) reviewed a number of fractal models for various parameters of flow and porous media.

Applications of the fractal concept outlined in this section are generally grouped into three broad categories (Perfect and Kay 1995)—(a) describing physical properties of soil, (b) modeling physical processes in soils, and (c) quantification of spatial variability.

5.1 Fractals and Their Application to Porous Media and Hydraulic Parameters

The concept of fractals is a result of departure from the classical definition of integer-based Euclidean geometry. Fractal objects, whether physical objects such as rigid and poroelastic materials or information such as time series data of different kinds, exhibit three major attributes (Perrier et al. 1996):

1. Similar structure over a range of length scales;
2. Intrinsic structure that is scale-independent; and
3. Irregular structure that cannot be entirely captured by classical geometric concepts.

The structures and properties of porous media resemble those of fractals. Some fractal models for hydraulic parameters and fractal parameters of certain traditional models for porous media have achieved satisfactory success in capturing hydraulic and geometric properties.

As will be explained later, Zaslavsky's (2002) scaling parameters for fPDEs in Chapter 4 bridge important concepts in the fractal geometry of porous media and the orders of fractional derivatives in time and space when fractional calculus-based models are used. This is one reason why the concept of fractals has been briefly discussed here.

The concept of fractal is a measure of self-similar or scale-invariant property of an object or a data series. Self-similarity or scale-invariance has been found in properties of soils and other geological materials with scales ranging from microscopic to laboratory and field scales (Tyler and Wheatcraft 1992b). The fractal approach can be applicable to one dimension such as a line or time series data, two dimensions such as a surface of an object, or higher dimensions such as irregular particles.

Mandelbrot (1983) showed that the relationship between length $L(\varepsilon)$ and unit measurement scale ε can be expressed as:

$$L(\varepsilon) = k\varepsilon^{1-D} \tag{3.71}$$

where k is the proportionality constant and D is the fractal dimension.

5.2 Kozeny, Carman and Other Early Models for Hydraulic Conductivity

There are many empirical formulas for hydraulic conductivity K, the most widely used mathematical formula being the one attributed to Kozeny (1927). In theory, for a set of formulas for hydraulic conductivity, Kozeny hypothesized that the porous media is treated as a bundle of capillary tubes of equal length, and solved simultaneously the Navier-Stokes equations for all channels passing through a cross-section normal to the flow in the porous medium (Bear 1972). Kozeny (1927) thus arrived at an equation with total porosity and surface area as the key variables which essentially modified Darcy's law as:

$$V = \frac{c\gamma\phi^3}{\mu S^2} J \tag{3.72}$$

where γ is the specific weight of water, μ is the dynamic viscosity of water, ϕ is the total porosity, S is the specific surface area of the porous medium, J is the hydraulic gradient, and c is a constant. Comparing Darcy's law with Kozeny's formulation in Eq. (3.72), it can be seen that hydraulic conductivity K can be expressed as:

$$K = \frac{c\gamma\phi^3}{\mu S^2} \tag{3.73}$$

By considering two types of flow paths which define tortuosity ξ, Kozeny (1927) also presented the following formula incorporating the effective particle diameter d:

$$K = \frac{c\gamma L_s d^2 \phi^3}{36\mu L_w (1-\phi)^2} \tag{3.74}$$

where L_s is the length of the apparent easier flow path, and L_w is the length of the more tortuous real flow path, and $(L_s/L_w) < 1$ (Kozeny 1927) which is a measure of tortuosity.

Kozeny (1932) later modified his original formula in Eq. (3.73) by including effective porosity ϕ_e for clayey soils. Hydraulic conductivity can, therefore, be given by incorporating the dynamic viscosity μ as (Carman 1937, 1939):

$$K = \frac{g\phi_e^3}{\tau\mu[S(1-\phi)]^2} \tag{3.75}$$

where g is the acceleration due to gravity, and

$$\tau = 2.5\left(\frac{L_e}{L}\right)^2 \tag{3.76}$$

with L_e being the actual length of the flow path, and L the direct length between the two points of interest, thus making τ a measure of tortuosity. Note that in Eq. (3.75) by Carman, the total porosity ϕ is used in the denominator while the effective porosity ϕ_e is in the numerator.

5.3 Recent Modified Fractal Kozeny-Carman Models for Hydraulic Conductivity

Since Kozeny's work was reported in 1927, many similar investigations have followed. Among various modifications of Kozeny's model are the modified Kozeny-Carman (KC) models (Xu and Yu 2008). Here, we briefly discuss two fractal models related to Kozeny's model, which capture much of the information on porous media.

5.3.1 Xu-Yu Model for Hydraulic Conductivity

One of the recent modifications of the KC model incorporates the fractal dimension for tortuosity D_T and fractal dimension for pore areas D_f (Xu and Yu 2008). Here, we call this model the KCXY model, which is of the following form (Xu and Yu 2008):

$$K = C_f \lambda_{max}^2 \left(\frac{\phi}{1-\phi}\right)^{(1+D_T)/2} \tag{3.77}$$

where

$$C_f = \frac{(\pi D_f)^{(1-D_T)/2}[4(2-D_f)]^{(1+D_T)/2}}{128(3+D_T-D_f)} \tag{3.78}$$

and

$$\lambda_{max} = d\left(\frac{\phi}{1-\phi}\right)^{1/2} \tag{3.79}$$

with d as the diameter of particles, and λ_{max} being the diameter of the largest particles.

For 2-dimensional pores, $1 < D_f < 2$, whereas $2 < D_f < 3$ for 3-dimensional pores. The KCXY model has an interesting property about the tortuosity that depends on the fractal dimension (Xu and Yu 2008):

$$\tau = \frac{L_t(\lambda_0)}{L_0} = \left(\frac{L_0}{\lambda_0}\right)^{D_T-1} \tag{3.80}$$

where $L_t(\lambda_0)$ is the tortuosity length of the pores with a diameter λ_0 and L_0 is the length of the straight line along which $L_t(\lambda_0)$ is measured. The fractal dimension of tortuosity changes depending upon the porosity; D_T is a constant for a given porosity and $D_T = 1$ for the straight capillary model.

5.3.2 Henderson et al.'s Generalized Kozeny Model for Hydraulic Conductivity

Henderson et al. (2010) also modified the Kozeny model to yield the following form:

$$\sqrt{\frac{K}{\phi}} = \frac{\xi\phi^{(\zeta+2)/2}}{(1-\phi)^\eta} \tag{3.81}$$

or

$$K = \frac{\xi^2\phi^{\zeta+3}}{(1-\phi)^{2\eta}} \tag{3.82}$$

where $\zeta = D_\tau \geq 0$, $\eta = D_{1/Sa} > 0$ and $\xi = \dfrac{C_\tau^{1/2}C_{1/S_a}}{f} \geq 0$ with S_a as the specific surface area.

5.4 Fractal Relationships between Water Content and Matrix Potential

To relate the moisture content to the matrix potential in a soil, two forms of empirical formulae are frequently used in soil science and hydrology. One of these empirical models is accredited to Brooks and Corey (1964, 1966) and the other one to van Genuchten (1980).

With the dimensionless water content S, which is also called relative saturation or reduced saturation,

$$S = \frac{\theta - \theta_r}{\theta_s - \theta_r} \tag{3.83}$$

where θ_s and θ_r are saturated and residue (or minimum) values of the water content θ. Hence, the Brooks-Corey model is of the form (Brooks and Corey 1964):

$$S = \left(\frac{h}{h_a}\right)^{-\lambda} \tag{3.84}$$

where h is the matrix potential, h_a is the air entry pressure and λ is the pore-size distribution index.

The van Genuchten (1980) model is of the following form:

$$S = \left(\frac{1}{1+(\alpha h)^n}\right)^m \tag{3.85}$$

where h is the matrix potential, $S = \dfrac{\theta - \theta_r}{\theta_s - \theta_r}$ (sometimes replaced with $S = \dfrac{\theta}{\theta_s}$, without considering the residue moisture) is the relative saturation (Deinert et al. 2008), and α, m and n are shape parameters.

The relationship between the parameters in the Brooks-Corey model and the van Genuchten model (van Genuchten 1980) is expressed as:

$$\left.\begin{array}{l} n = \lambda + 1 \\ m = 1 - \dfrac{1}{n} \end{array}\right\} \tag{3.86}$$

based on Mualem's theory (1976), and as:

$$\left.\begin{array}{l} n = \lambda + 2 \\ m = 1 - \dfrac{2}{n} \end{array}\right\} \tag{3.87}$$

based on Burdine's theory (1953).

In both cases for the van Genuchten model, λ is given as:

$$\lambda = 3 - D \tag{3.88}$$

for three-dimensional fractal media (Perrier et al. 1996), and

$$\lambda = 2 - D \tag{3.89}$$

for two-dimensional fractal media (Tyler and Wheatcraft 1990).

There is a connection between the parameters in the van Genuchten model and the fractal dimensions—the relationship between λ and D holds for three-dimensional fractal media as:

$$\left.\begin{array}{l} n = 4 - D \\ m = 1 - \dfrac{1}{4 - D} \end{array}\right\} \tag{3.90}$$

based on Mualem's theory, and

$$\left.\begin{array}{l} n = 5 - D \\ m = 1 - \dfrac{2}{5 - D} \end{array}\right\} \tag{3.91}$$

based on Burdine's theory.

5.5 *Hydraulic Conductivity and Relative Hydraulic Conductivity*

With the above fractal dimensions, the original van Genuchten model (1980) can be specified in terms of fractal parameters. Two forms of relative hydraulic conductivity are generally used—(a) based on water content

$$K_r(\theta) = \frac{K(\theta)}{K_s(\theta)} \tag{3.92}$$

and (b) based on hydraulic head

$$K_r(h) = \frac{K(h)}{K_s(h)} \tag{3.93}$$

where $K_s(\theta)$ and $K_s(h)$ are saturated hydraulic conductivities expressed as water content and potential (head) respectively.

5.5.1 The Brooks-Corey Model Based on Burdine's Theory

The relative hydraulic conductivity in this case is given as (van Genuchten 1980):

$$K_r(\theta) = S^{3+2/\lambda} \tag{3.94}$$

and

$$K_r(h) = (\alpha h)^{-2-3/\lambda} \tag{3.95}$$

where λ is as described in section 5.4 and subsequent relationships, and the diffusivity is expressed as:

$$D(\theta) = \frac{K_s}{\alpha\lambda(\theta_s - \theta_r)} S^{2+1/\lambda} \tag{3.96}$$

5.5.2 The Brooks-Corey Model Based on Mualem's Theory

In this case, the relative hydraulic conductivity is given as (van Genuchten 1980):

$$K_r(\theta) = S^{5/2+2/\lambda} \tag{3.97}$$

in terms of water content, and

$$K_r(h) = (\alpha h)^{-2-5\lambda/2} \tag{3.98}$$

in terms of water potential.

Herein, diffusivity is expressed as:

$$D(\theta) = \frac{K_s}{\alpha\lambda(\theta_s - \theta_r)} S^{3/2+1/\lambda} \tag{3.99}$$

5.5.3 The van Genuchten Model Based on Mualem's Theory and Their Fractal Relations

According to Mualem's theory, the relative hydraulic conductivity in this case is given as (van Genuchten 1980):

$$K_r(\theta) = S^2[1 - (1 - S^{1/m})^m]^2 \tag{3.100}$$

and

$$K_r(h) = \frac{\left\{ \left[1-(\alpha h)^n\right]^{n-1} \left[1+(\alpha h)^n\right]^{-m} \right\}^2}{\left[1+(\alpha h)^n\right]^{m/2}} \tag{3.101}$$

Diffusivity here is expressed as:

$$D(\theta) = \frac{(1-m)K_s}{2\alpha m(\theta_s - \theta_r)} S^{1/2-1/m} \left[(1-S^{1/m})^{-m} + (1-S^{1/m})^m - 2\right] \tag{3.102}$$

with

$$m = 1 - \frac{1}{n}, \; 0 < m < 1 \tag{3.103}$$

For a large hydraulic head h, the van Genuchten model approaches the following relation:

$$S = (\alpha h)^{-nm} \tag{3.104}$$

which implies that the Brooks-Corey and the van Genuchten model parameters are related as: $\lambda = nm$.

5.5.4 The van Genuchten Model Based on Burdine's Theory and Their Fractal Relations

This set of models can be expressed as a function of either moisture content $K_r(\theta)$ or matrix potential $K_r(h)$ (van Genuchten 1980):

$$K_r(\theta) = S^2[1 - (1 - S^{1/m})^m] \tag{3.105}$$

and

$$K_r(h) = \frac{1 - (\alpha h)^{n-2}\left[1 + (\alpha h)^n\right]^{-m}}{\left[1 + (\alpha h)^n\right]^{2m}} \tag{3.106}$$

Diffusivity is thus expressed as:

$$D(\theta) = \frac{(1-m)K_s}{2\alpha m(\theta_s - \theta_r)} S^{(3-1/m)/2}\left[(1 - S^{1/m})^{-(m+1)/2} - (1 - S^{1/m})^{(m-1)/2}\right] \tag{3.107}$$

Hence, from the above expressions, it can be observed that:

$$m = 1 - \frac{2}{n}, \quad 0 < m < 1, \, n > 2 \tag{3.108}$$

In the above equations, the values of the parameters $\lambda = 2 - D$ in Tyler and Wheatcraft (1992b) model (TW) in Eq. (3.89) have been determined by Rawls and Brakensiek (1995) for different soils. Once D is calculated, $\lambda = 3 - D$ in Eq. (3.88) is determined from the data of Rawls and Brakensiek (1995) for three-dimensional media. These data are listed in Table 3.2.

5.6 Other Fractal Models for Porous Media and Their Parameters

5.6.1 Deinert et al.'s Model for Water Content and Hydraulic Potential

In addition to the above models, Deinert et al. (2008) proposed a more generic model of the following form:

$$\theta_s - \theta = \frac{V_0}{V}\left(1 - \frac{h_s}{h}\right)^\lambda \tag{3.109}$$

with

$$V_0 = \frac{1}{(3 - D_v)}(FD_v)^{(3-D_v)/3} \tag{3.110}$$

Table 3.2. Fractal hydraulic parameters (the first 6 columns of the following data are from Table 1 in Rawls and Brakensiek (1995), and Tyler and Wheatcraft (1992a)).

Soil texture	Porosity ϕ	Bubbling pressure h_b	Pore-size parameter λ	Largest matrix pore radius R_m	Macroscopic capillary length λ_m	2-D fractal dimension D	Perrier et al.'s 3D fractal dimension D
		cm		$cm \cdot 10^{-2}$	cm		
Sand	0.437	7.26	0.694	1.93	2.83×10^{-2}	1.41	2.41
Loamy sand	0.437	8.69	0.553	1.71	2.06×10^{-2}	1.53	2.53
Sandy loam	0.453	14.66	0.378	1.00	9.92×10^{-3}	1.68	2.68
Loam	0.463	11.15	0.252	1.33	4.63×10^{-3}	1.75	2.78
Silt loam	0.501	20.76	0.234	0.72	1.11×10^{-3}	1.78	2.75
Sandy clay loam	0.398	28.08	0.319	0.54	5.83×10^{-3}	1.79	2.75
Clay loam	0.464	25.89	0.242	0.58	4.50×10^{-3}	1.81	2.81
Silty clay loam	0.471	32.56	0.177	0.45	3.84×10^{-3}	1.83	2.85
Sandy clay	0.430	29.17	0.233	0.51	3.31×10^{-3}	1.85	2.83
Silty clay	0.479	34.19	0.160	0.43	3.02×10^{-3}	1.87	2.87
Clay	0.475	37.30	0.165	0.40	2.77×10^{-3}	1.87	2.87

and

$$\lambda = \frac{3 - D_v}{3 - D_s} \qquad (3.111)$$

where θ is the water content, θ_s is the saturated value of θ, h and h_s are capillary pressures corresponding to θ and θ_s respectively, D_v is the pore volume fractal dimension, D_s is the pore surface fractal dimension, and F is a constant.

   This model generalizes several other models proposed by Brooks and Corey (1964, 1966), Tyler and Wheatcraft (1990), and Rieu and Sposito (1991). The advantage of this model is that it considers both fractal pore space and pore surface, characterized by D_v and D_s respectively, which, to some extent, is similar to the KCXY model with two fractal parameters. When $D_s = 2$ and $\frac{V_0}{V} = \theta_s$, this model becomes the fractal Brooks-Corey model by Perrier et al. (1996), which considers fractal volumes and non-fractal surfaces. This model (Perrier et al. 1996) is also applicable to two-phase flows of wetting and non-wetting fluids such as the water-air or water-oil systems.

5.6.2 Mass-Volume Relationship

Tarasov (2005a, 2005b) demonstrated that with the definition of heterogeneity in terms of fractals, the mass of fractal medium, M, in the range of $R_0 < R < R_m$ where R_0 is the characteristic scale and R_m is the scale of the object, scales as:

$$M = k \left(\frac{R}{R_0} \right)^{D_m} \tag{3.112}$$

where D_m is the fractal dimension of the medium, and the proportionality constant k is given as:

$$k = \frac{2^{5-D} \pi \Gamma(3/2)}{D\Gamma(D/2)} \rho_0 \tag{3.113}$$

with ρ_0 being the density of the medium, and $\pi = 3.1416$.

The mass of the medium, M, and its volume V also scale as:

$$M \sim V^{D/n} \tag{3.114}$$

Tyler and Wheatcraft (1992b) showed that fractal mass-volume relationship is of a similar form:

$$\frac{M}{M_T} = \left(\frac{R}{R_L} \right)^{3-D} \qquad r < R \tag{3.115}$$

where R_L is the radius of the upper size limit and its mass is M_T. At $R = R_L$, the critical radius is equal to the mean pore scale λ_m, that is, $R_L = \lambda_m$.

5.6.3 Fractal Bulk Density

Pachepsky and Rawls (2003) showed that fractal scaling of bulk density takes the following form:

$$\gamma = aL^{D_m - 3} \tag{3.116}$$

with $2 < D_m < 3$; laboratory measurements showed the range to be $2.85 < D_m < 2.95$. Based on Eq. (3.116), Pachepsky and Rawls (2003) also showed that the ratio of bulk densities measured in the laboratory, γ_L, and in the field, γ_F, depends on the ratio of the corresponding sample radii:

$$\frac{\gamma_L}{\gamma_F} = \left(\frac{R_F}{R_L} \right)^{3-D_m} \tag{3.117}$$

Pachepsky and Rawls (2003) also discovered that the following relationship holds for data from laboratory measurements:

$$\left(\frac{15}{2.9} \right)^{0.05} < \left(\frac{R_F}{R_L} \right)^{3-D_m} < \left(\frac{15}{2.2} \right)^{0.15} \tag{3.118}$$

which implies that $1.09 < \left(\dfrac{R_F}{R_L} \right)^{3-D_m} < 1.33$.

Furthermore, they deduced that when the gravimetric water contents are the same for samples from the laboratory and from the field, for a given soil water matric potential, the ratio of the volumetric water contents, θ_L/θ_F, is the same as the ratio of bulk densities.

5.6.4 Fractal Porosity

Porosity is one of the most intensively investigated parameters. In association with the KCXY model for hydraulic conductivity in Eq. (3.77), one of the fractal models for porosity is owed to Yu and Li (2001). This model is written as:

$$\phi = \left(\frac{\lambda_{min}}{\lambda_{max}}\right)^{D_E - D_f} \tag{3.119}$$

where λ_{max} and λ_{min} are the largest and the smallest diameters of pores, and D_E and D_f are the Euclidean dimension ($D_E = 1,2,3$) and the fractal dimension respectively. The total number of pores with the above parameters, N, is given as (Xu and Yu 2008):

$$N = \left(\frac{\lambda_{max}}{\lambda_{min}}\right)^{D_f}, \qquad\qquad \varepsilon \geq \lambda_{min} \tag{3.120}$$

where ε is the length scale used as the unit to determine these quantities. The probability density function (pdf) for the pore-size distribution in fractal media, $f(\lambda)$, is given as (Xu and Yu 2008):

$$f(\lambda) = D_f \lambda_{max}^{D_f} \lambda^{-(D_f + 1)} \tag{3.121}$$

where λ is the diameter of pores.

6. Similarity, Scales, Models and Measurements

6.1 Mechanical Similarity and Scaling in Hydraulics

In hydraulics and fluid mechanics of bulk flow, such as flow in pipes and other regular geometries, the scale effects for a particular physical phenomenon increase with the scale ratio or the scale factor λ_r (Heller 2011) expressed as:

$$\lambda_r = \frac{L_p}{L_m} \tag{3.122}$$

where L_p and L_m are the length of the prototype and that of the model respectively, and the inverse of scale ratio is the scale $1 : \lambda_r$.

A physical scale model is said to be completely similar to its real-world counterpart and involves no scale effects if it satisfies the *mechanical similarity*, which has three components (Heller 2011): *geometrical similarity, kinetic similarity* and *dynamic similarity* (Yalin 1971, Heller 2011).

For water flow, solute transport and energy transfer in porous media, more variability results from the heterogeneities originating from structural differences at different levels of pores of the medium. The heterogeneities of the medium are also responsible for variation in measurable quantities in different directions of flow, termed as *isotropy* if the measured quantities are the same irrespective of directions, or as *anisotropy* when the measured quantities vary in different directions.

6.1.1 Geometrical Similarity or Similitude

Geometrical similarity as one of the three types of similitude scales and scale ratios (Daugherty et al. 1989, Heller 2011) is also relevant to water flow and solute transport in porous media. Geometric similarity or similitude requires the model and its prototype to be identical in shape. Model lengths, areas and volumes, therefore, scale with λ_r, λ_r^2 and λ_r^3 respectively, in relation to the prototype (Heller 2011).

6.1.2 Kinetic Similarity or Similitude

Kinetic similarity, in addition to geometric similitude, requires that the ratio of velocities at all corresponding points along the flow should be the same. Velocity ratio is expressed as (Daugherty et al. 1989):

$$V_r = \frac{V_{pr}}{V_{mr}} \tag{3.123}$$

where V_{pr} and V_{mr} are the velocities of the prototype and the model respectively. Accordingly, time ratio follows as $T_r = \frac{L_r}{V_r}$.

6.1.3 Dynamic Similarity or Similitude

Dynamic similitude states that if two systems are dynamically similar, then the corresponding forces in the two systems must be the same (Daugherty et al. 1989). These forces include those due to *gravity, pressure, viscosity, elasticity, surface tension* and *inertia*.

6.2 Types of Scale Effects

With physical models in hydraulics, there are three types of problems associated with a model's representation of the real-world prototype (Heller 2011), namely, *model effect, scale effect* and *measurement effect*. These effects are relevant to the analysis and measurement of variables and parameters for porous media as the governing equations of flow and transport in porous media are special forms of the fundamental equations in hydraulics, namely the NSEs. These three problems associated with the differences between a model and its real-world prototype mean that both mathematical models and physical models can have these types of problems.

6.2.1 Model Effect

The model effect originates from incorrect representations of prototype features or fluid properties and is relevant to the study of flow in porous media as laboratory soil columns, flumes or artificial aquifers are often unable to represent the real-world complexities accurately due to the incorrect packing of porous media for experimental purposes.

6.2.2 Scale Effect

The scale effect arises when the relevant force ratios are not held constant between the model and the prototype. These ratios include those of the model and prototype's *Froude number, Reynolds number, Weber number, Cauchy number* and *Euler number* (Heller 2011). When a model is not properly set up, such as with an insufficient size, the scale effect generally increases with the scale ratio or scale factor L_r. This means that the smaller the model, the larger the scale effect, as evident from Eq. (3.122). The scale effect clearly manifests in mathematical models for water flow and solute transport due to differences in the ratios of these forces between a model and the real world, and the incomplete model variables and parameters.

6.2.3 Measurement Effect

The measurement effect results from the non-identical measurement techniques used for data collection in the model to represent the prototype. Comparative studies with measurements of solute dispersion in a laboratory and in the field form a typical example of this 'effect'. In the former case, the data of interest for solute transport and/ or water flow in a plume or small physical model is easily and 'precisely' measured, whereas, in case of the latter, it is practically impossible to estimate the quantity of a solute or a tracer that has reached the expected point where measurements are made.

6.3 Scales of Interest: Space-Time Scales in Hydrology, Soil Science and Porous Media Flow

As discussed in section 6.2, scale effects in classical hydraulics and fluid mechanics in general refer to the differences created by a physical model which incompletely represents a real-world prototype. In subsurface hydrology, which includes soil science and groundwater hydrology, the term *scale effect* has only been used to widely refer to the differences in certain flow and transport parameters such as dispersivity, diffusion coefficient or hydraulic conductivity measured at different locations (Lallemand-Barrés and Peaudecerf 1978, de Marsily 1986, Gelhar et al. 1992, Gelhar 1993) or from different sizes of experimental settings such as on laboratory and field scales.

As the experimental or theoretical analysis of flow and transport processes can be carried out at different spatial and temporal scales, the desired details of the problem determine the level of these scales. Bonnet (1982) (also see van der Heijde 1988) defined the spatial scales in hydraulics of subsurface flow at which different processes operate.

The focus of a problem in research on porous media, as illustrated in Fig. 3.2, determines whether the issue is to be investigated at the level of individual particles,

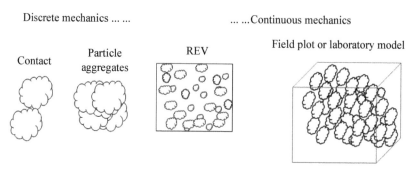

Fig. 3.2. The physical scales (after Keller et al. 2013).

aggregates, a sample of medium representing the bulk media (known as the REV) (Bear 1972), or macro-scale physical models.

In the meantime, the temporal scale is also important and depends on the interest of the time span for the problem and also the discipline. Dooge (1986) outlined the spatial and temporal scales in several fields, including turbulent flow in hydraulics (length scale $L = 10^{-2}$ metres, and time scale $T = 10^{-2}$ seconds) and hydrology (length scales ranging from $L = 10$ metres for plots to $L = 10^4 \sim 10^5$ metres for catchment experiments, and time scales ranging from $T = 10$ seconds for plots to $T = 10^4$ seconds for catchment experiments).

Dooge's data is plotted in Fig. 3.3 on a logarithmic scale.

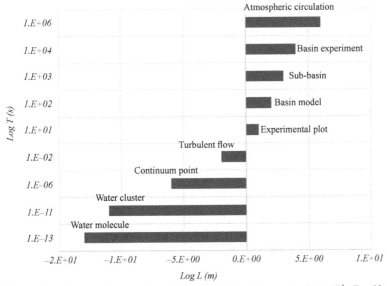

Fig. 3.3. Spatial and temporal scales for related studies ($E + 00 = 1$, $E + 01 = 10^1$, $E + 02 = 10^2$, $E - 06 = 10^{-6}$ and so on).

Dooge's time scales, however, do not include long-term processes such as decadal and centenary time scales. Similar to Dooge's data, Blöschl and Sivapalan (1995) and Skøien et al. (2003) showed that time and space scales are related as:

$$T \sim L^\alpha \tag{3.124}$$

where the scaling exponent α varies with the type of flow: $\alpha = 0.5$ for precipitation, $\alpha = 0.8$ for run-off from small catchments, and $\alpha = 1.2$ for run-off from large catchments. For flow in porous media, $\alpha = 0.8$ for groundwater and $\alpha = 0.5$ for soil moisture.

6.4 Additional Similarity for Flow in Porous Media

The three components of mechanical similarity or similitude in sections 6.1 and 6.2 are standard terminologies in fluid mechanics of bulk flow. For flow in porous media, being different as a consequence of heterogeneities and excluded volume, additional criteria of similarity are required.

6.4.1 Haines Similarity, Hubbert Geometric Similarity and Miller-Miller Similitude

Prior to the widely-regarded Miller-Miller similarity (1955a, 1956) and the concept of geometric similarity used by Hubbert (1940), Miller and Miller (1955b) stated that Haines (1927, 1930) had first used the concept of reduced variables as a measure of similarity to investigate the surface tension of water. In literature, however, the similarity of porous media has been regarded as a contribution of Miller and Miller (1955a, b, 1956) for investigating geometric media. The Miller-Miller scaling relationships (Sposito 1998) are listed in Table 3.3.

Table 3.3. The Miller-Miller similitude.

Soil water property or parameters	Original variable	Dimension	Scaling relationship
Volumetric content	θ	[1]	$\theta_* = \theta$
Matrix potential	ψ	[L]	$\psi_* = \lambda\psi$
Hydraulic conductivity	K	$[LT^{-1}]$	$K_* = \dfrac{K}{\lambda^2}$

Notes: The asterisk (*) denotes scaled quantities.

In addition to the quantitative scaling relationship, the Miller-Miller similitude also assumes that soil is rigid and there is no movement of soil particles (Miller and Miller 1955a, b). Nielsen et al. (1997) showed that the Miller-Miller similitude can be applied to different levels of soil particles, with examples of hydraulic conductivity and water saturation (water content).

6.4.2 Hydraulic Similitude for Water Flow in Porous Media

Philip (1954b) postulated the following hydraulic similitude:

$$\frac{\Phi_1}{\Phi_2} = \frac{L_1}{L_2} = \frac{\gamma_1 \cos(\theta_1\rho_1)}{\gamma_2 \cos(\theta_2\rho_2)} \tag{3.125}$$

where, at two length scales L_1 and L_2, Φ_1 and Φ_2 are the total moisture potentials measured, ρ_1 and ρ_2 are the densities of the fluids, γ_1 and γ_2 denote the surface tension, and θ_1 and θ_2 represent the contact angles of the liquid-air mixture under the condition of liquid advance.

6.4.3 Scale-heterogeneity

Philip (1967) extended Miller and Miller's idea of scaling in porous media with the term *scale-heterogeneity* to relate two porous media a and b whose internal geometry is uniform while the characteristic internal length scale is free to vary spatially. With the Miller-Miller scaling relationship, the hydraulic conductivities of the two media, K_a and K_b scale as

$$\frac{K_a(\theta)}{\lambda_a^2} = \frac{K_b(\theta)}{\lambda_b^2} \tag{3.126}$$

and the matrix potentials of the two media scale as

$$\lambda_a\Psi_a(\theta) = \lambda_b\Psi_b(\theta) \tag{3.127}$$

where λ_a and λ_b are their characteristic length scales. Philip extended the above relationships for one-dimensional scale-heterogeneous media as:

$$K(\theta, x) = [\lambda(x)]^2 K_*(\theta, x) \tag{3.128}$$

and

$$\Psi(\theta, x) = \frac{\Psi_*(\theta, x)}{\lambda(x)} \tag{3.129}$$

where $K_*(\theta, x)$ and $\Psi_*(\theta, x)$ are the reduced hydraulic conductivity and matrix potential respectively, and the scale-dependent variability of $K(\theta, x)$ and $\Psi(\theta, x)$ are wholly embodied in the spatial variability of $\lambda(x)$.

With the scaling factor as a function of the space variable, Broadbridge (1987, 1988) investigated the scale-heterogeneous model by incorporating the scale parameter as a space-dependent function in a non-linear diffusion equation for water flow in soils.

6.4.4 Warrick-Nielsen Similitude

For positive scaling parameters α_p and ω_p, the relationships established by the Warrick-Nielsen similitude are listed in Table 3.4 (Sposito 1998).

Table 3.4. The Warrick-Nielsen similitude.

Soil water property or parameter	Original variable	Dimension	Scaling relationship
Volumetric content	S	$[1]$	$S_f = S_p^b$
Matrix potential	ψ_p	$[L]$	$\psi_f = \alpha_p \psi_p$
Hydraulic conductivity	K_p	$[LT^{-1}]$	$K_f = \dfrac{K_p}{\omega_p^2}$
Diffusivity	D_p	$[L^2 T^{-1}]$	$D_f(S) = \dfrac{\alpha_p D_p(S)}{\omega_p^2}$

Notes: The subscripts f and p denote the quantity at a field scale and the plot scale respectively.

6.5 Scales at the Pore Level with Respect to Water Flow

At the pore-scale, Philip (1985) introduced two parameters for characterizing the structure of pores—*macroscopic capillary length* and *characteristic mean pore size* (White and Sully 1987).

Macroscopic capillary length is defined as:

$$\lambda_c = \left[K(\psi_0) - K(\psi_n) \right]^{-1} \int_{\psi_n}^{\psi_0} K(\psi) d\psi$$

$$= \left[K(\psi_0) - K(\psi_n) \right]^{-1} \int_{\theta_n}^{\theta_0} D(\theta) d\theta \tag{3.130}$$

where ψ_n is the initial water potential, ψ_0 is the water potential at the supply surface of infiltration, $K(\psi_n)$ and $K(\psi_0)$ are hydraulic conductivities at water potentials ψ_n and ψ_0 respectively, θ is water content or ratio, θ_n is the initial water content, θ_0 is the water content corresponding to ψ_0, and $D(\theta)$ is the diffusion coefficient.

Characteristic mean pore size is expressed as:

$$\lambda_m = \frac{\sigma}{\rho g \lambda_c} \tag{3.131}$$

where σ is the surface tension of water at standard conditions. White and Sully (1987) determined that for pure water at 20°C, Eq. (3.131) yields $\lambda_m \approx 7.4 \lambda_c$ with both λ_m and λ_c in mm.

For both transient and steady-state flows, the temporal scale t_c is given as (White and Sully 1987):

$$t_c = \frac{(\theta_0 - \theta_n)\lambda_c}{K(\psi_0) - K(\psi_n)} \tag{3.132}$$

With the concept of fractal scaling, the fractal relationships between the *macroscopic capillary length and characteristic mean pore size* were, respectively, given by Tyler and Wheatcraft (1992a) as:

$$\lambda_c = \frac{\psi_b}{(7 - 3D_f)} \tag{3.133}$$

$$\lambda_m = \frac{\sigma(7 - 3D_f)}{\rho g \psi_b} \tag{3.134}$$

where h_b is the bubbling pressure (see Table 3.2) and D_f is the fractal dimension of the soil, based on Sierpinski's carpet model (Bear 1972).

7. Other Forces Coupled with the Flow of Fluids in Porous Media

Water flow in soils and aquifers as porous media is not an isolated process. There are, instead, multiple processes coupled with further other processes such as:

1. Heat transfer obeying Fourier's law;
2. Solute movement (diffusion, dispersion, convection and reactions);
3. Gas movement that can be described by Leibenzon's law (1929) (see Su 2004 for a review);
4. Electric effects (de Marsily 1986) and magnetic effects (Geindreau and Auriault 2002);
5. Energy-related reciprocal mechanics: The *thermal diffusion process* (known as the *Soret effect*) that is the fluid motion initiated or enhanced by the thermal gradient, and the *diffusion thermal process* that is the heat transfer initiated or enhanced by the gradient of diffusive mass concentrations (known as the *Dufour effect*);

6. Chemical and biochemical reactions with redox potentials (Eh), that influence mass and energy transfer;

7. Seismic waves, which can also change the permeability of geological strata (Elkhoury et al. 2006);

8. Elasticity, strain and load: Two concepts are key to the elasticity of porous media—stress and strain. Stress (σ) is the force per unit of area applied on the medium, whereas strain (ε) is the deformation due to stress. Mainardi and Spada (2011) reviewed various models of the relationship between stress and strain. A simple, yet generic, relationship is:

$$\sigma = f(\varepsilon) \tag{3.135}$$

Soils and geological strata which form aquifers are poroelastic materials and flow in these forms of media is characterized by the dominant viscous flow. In addition to their constituents, various other factors can change the properties of viscoelastic materials, such as pressure, temperature, liquid content, solute concentration, compaction, magnetic force, etc. These issues will be briefly discussed in Chapter 11.

8. Heterogeneities and Isotropy

The term 'heterogeneity' is often discussed in literature exclusively as the variation of physical properties in space or from one location to another such as those discussed in sections on scaling and scale effects. In fact, the nature of heterogeneity can be physical, chemical or biological, such as the variation of solute concentration over an area or along the profile of a soil or aquifer, and changes in microbial communities across a soil profile due to their preference to light, oxygen, temperature or other factors (Noyes and Field 1974). However, there is more to heterogeneity than simply the spatial variability of a particular variable or parameter. Experimental and theoretical analyses indicate that the movement of chemicals in natural porous media may be dominated by larger-scale variations in both physical and chemical properties (Dagan 1989, Gelhar 1993, Cushman et al. 1995). The previous sections in this chapter only overview physical heterogeneity which affects the movement of water and solutes in soils and aquifers, and chemical heterogeneity in terms of non-equilibrium solute transport, with no mention of biological or other forms of heterogeneities.

In most of field and laboratory investigations on flow and transport in porous media, the isotropy is clearly assumed. In reality the difference in physical and chemical variables and parameters can change in different directions, resulting in the anisotropy.

9. Summary

Together with Chapter 2 for essential mathematics used in this book, this chapter outlines a background to the concepts and descriptions of porous media, fundamental equations governing water flow and solute transport in porous media, poroelasticity and the parameters of these fundamental equations and related topics. The fundamental equations of flow in porous media consist of three types of equations: [1] the equation of motion based on the Navier-Stokes equations and their simplified form as Darcy's law; [2] the equation of conservation of mass, and [3] the conservation of energy. With isothermal flow, the energy (or heat as a convenient variable) is not considered, and a more detailed discussion of the coupled flow of water and heat transfer is considered in Chapter 5.

In addition to the outlines of the selected topics in this chapter, some of the topics addresses in this chapter are also active research topics such as the issues of similarity, fractals, scale issues and the fundamental equations too.

Chapter 4

Transition from Classical Diffusion to Anomalous Diffusion
The Evolution of Concepts and Ideas

1. Introduction

Quantitative description of many physical processes, biological phenomena, and social and economic behaviors requires several concepts, including rate of change or other forms of definitions for changes in time and space, and their relationships. Ordinary differential equations (ODEs) have been a fundamentally important tool for these purposes, and ODEs as a field of mathematics have been important since the time of Isaac Newton (1642–1727). Partial differential equations (PDEs) soon emerged in the 18th century when ODEs failed to describe physical processes such as vibration of strings, propagation of waves, sounds in liquids and in gravitational fields, although PDEs were not consciously developed as a subject of study originally (Evans et al. 1999).

The use of partial derivatives appeared in the times of Newton, Leibniz and the Bernoulli family (Cajori 1923). Partial derivatives were significantly used by Leonhard Euler (1707–1783) in the field of hydrodynamics in 1734, and later extended by d'Alembert (1717–1783) in 1744 and 1745 with respect to the study of dynamics (Cajori 1928, Evans et al. 1999). The modern notation of partial derivative is accredited to Legendre in 1786 (Cajori 1923).

With PDEs as a tool, Euler and Lagrange (1736–1813) investigated waves in strings, and Daniel Bernoulli (1700–1782) and Euler investigated potential theory. Further developments in PDEs by Legendre (1752–1833) and Laplace (1749–1827) also followed. A significant development in PDEs, with lasting impact on applied mathematics and physics associated with heat and mass transfer in porous media, is the analytical theory of heat transfer by Joseph Fourier (1768–1830).

The one-dimensional equation for heat conduction was originally given by Fourier (1807, 1822). In modern representation of partial derivatives, it is of the following form:

$$\frac{\partial v}{\partial t} = \frac{k}{CD} \frac{\partial^2 v}{\partial x^2} \qquad (4.1)$$

where v represents temperature, x is the distance, t denotes time, and k, C and D are thermal conductivity (or conducibilité), specific heat capacity, and density (of the solid) respectively (Fourier 1822, Narasimhan 1999, 2009). The three-dimensional equation of heat conduction was given as (Fourier 1822):

$$\frac{\partial v}{\partial t} = \frac{k}{CD} \left(\frac{\partial^2 v}{\partial x^2} + \frac{\partial^2 v}{\partial y^2} + \frac{\partial^2 v}{\partial z^2} \right) \qquad (4.2)$$

with corresponding heat equations in other coordinates as well. It was Fourier's monumental landmarks in 1822 which were received by the scientific community as a rational framework for generalization in other fields. Fourier's work considerably boosted further investigations in seemingly unconnected fields such as energy and mass transfer, electron movement, flow in porous media, genetics, economics and other branches of physics (Narasimhan 1999). The diffusion equation, also referred to as Fick's second law, is a consequence of applying Fourier's concept to the movement of mass (rather than heat). In addition to this set of PDEs with widespread applications, Fourier's work also initiated several major mathematical concepts of historical significance, which include separation of variables for solving an ODE, eigenvalue problems, Fourier series, and Fourier transform which is the origin of other integral transforms such as Laplace transform, etc. (Debnath and Bhatta 2007).

Table 4.1 is an extract of some major developments in diffusion-related processes relevant to hydrology, soil physics and fluid mechanics, compiled by Narasimhan (1999).

The diffusion equation credited to Fick (1855a, b), also called Fick's second law, is a result of applying one of Fourier's PDEs as an analogue. The diffusion equation is mathematically identical to the heat equation, and is the foundation for the mainstream mathematical models widely in use today for the analysis of mass diffusion, heat transfer and many other related processes based on integer PDEs.

The heat equation and diffusion equation are based on integer calculus. The integer PDEs, despite their crucial roles and great success in many fields for over two centuries, cannot always satisfactorily describe some processes which have a memory- or history-dependence during material or energy transfer in a medium (Caputo 1999, 2000, Di Giuseppe et al. 2010). In mathematics and some branches of science and engineering, fractional ODEs (fDEs) and fractional PDEs (fPDEs) appeared in response to these memory-dependent phenomena. Physically, the memory-dependent movement of particles such as solutes in soils and aquifers is now known as the non-local effect (Zhang et al. 2009, Benson et al. 2013).

Table 4.1. Major milestones in theoretical and experimental development in diffusion-related processes in porous media since Fourier's landmark contributions (majority of the following data is based on Table 1 in Narasimhan (1999))

Name of investigator(s)	Year of occurrence	Description of events
Fourier	1807	PDE describing heat conduction in solids
Fourier	1822	Analytical theory of heat transfer, which provides a set of mathematical methods
Green	1828	Formal definition of 'potential'
Graham	1833	Law governing diffusion of gases
Poiseuille	1846	Experimental studies on water flow through capillaries
Graham	1850	Experimental studies on diffusion in liquids
Fick	1855	Fourier's model applied to diffusion in solids
Darcy	1856	Law governing water flow in porous media
Dupuit	1863	Potential theory applied to flow in groundwater
Forchheimer	1886	Potential theory applied for solving seepage problems by defining the concept of flow nets
Roberts-Austen	1896	Experimental measurement of solid diffusion
Einstein	1905	Brownian motion and diffusion equation
Pearson	1905	Random walk concept
Buckingham	1907	Soil water potential due to capillarity
Green and Ampt	1911	First model of infiltration into soils
Gardner	1922	Measurement of potential in multiphase flow in porous media
Terzaghi	1924	Seepage in deformable clays analogous to heat transfer
Richards	1931	PDE for water flow in unsaturated porous media
Taylor	1953	Hydrodynamic dispersion of solutes in porous media
Rubinshtein	1948	A coupled model for heat transfer in double porosity media
Barenblatt et al.	1960	Dual-pore model for flow in porous media

2. The Inception of Models Based on Fractional Calculus in Geoscience and Related Fields

2.1 From Pure Mathematical Curiosity to a Physical Model of Anomalous Diffusion

2.1.1 The Inception of the Concept Starting as a Question in 1695

fDEs and fPDEs are the consequences of the concept of fractional calculus which started in a correspondence dated 30th September, 1695 when Leibniz (1646–1716) replied to a question raised by I'Hopital (1661–1704), regarding the meaning of a derivative of the order $\frac{1}{2}$, i.e., $\frac{\partial^{1/2} y}{\partial x^{1/2}}$ (Podlubny 1999, Kilbas and Trujillo 2001, Valério et al. 2014).

2.1.2 The Answer in 1800 and Onwards

The correspondence between Leibniz and l'Hopital in 1695 marked the start of the theory of derivatives and integrals of arbitrary orders. The answer to l'Hopital's original query for Leibniz was not given until 1819 when Lacroix showed that for the function $y(x) = x$, the following relationship is the answer:

$$\frac{\partial^{1/2} y(x)}{\partial x^{1/2}} = D^{1/2} y(x) = 2\sqrt{\frac{x}{\pi}} \tag{4.3}$$

which is the same as given by the Riemann-Liouville definition of a fractional derivative (Miller and Ross 1993).

For over three centuries since the inception of fractional calculus, theories of fractional derivatives were mainly developed as a field of pure mathematics of limited scope (Podlubny 1999). Among mathematicians, Abel (1823) was the first to use fractional operations (Miller and Ross 1993), who applied fractional calculus to the problem of a tautochrone curve (Valério et al. 2014). Abel's solution to the tautochrone problem led to the fractional integral known as the Abel's integral equation (AIE).

In applied mathematics, according to Kilbas and Trujillo (2001), O'Shaughnessy (1918) was probably the first to propose and consider the differential equation of the form $\frac{\partial^{1/2} y}{\partial x^{1/2}} = \frac{y}{x}$, with solutions provided in a discussion by Post (1918, 1919), and Pitcher and Sewell (1938) were the first to consider a non-linear fDE of the form $D^{\alpha}_{a+} y(x) = f[x, y(x)]$ (Kilbas et al. 2006).

Davis (1924) extensively investigated fractional integral equations (fIE) by considering both the Volterra equation and the AIE. Tamarkin (1930) proved that the solution of AIE $I^{\alpha}_x u(x) = f(x)$, where $p - 1 < \alpha < p$ (with p as an integer), is $u(x) = D^{\alpha}_x f(x)$, which clearly demonstrates the inverse operations connecting the fractional AIE and fDE. In fact, as seen in Chapter 2, all the concepts of various definitions of fractional derivatives appear as a fractional integral of some kind, which are alternative forms of the AIE (Gorenflo 1987).

2.1.3 In Applied Fields Since 1823

In addition to the pioneering work by Abel (1823), the mechanics of viscoelastic media is a fertile ground where the applications of fractional calculus and some important ideas originated. Blair and Coppen (1943) proposed a non-linear fDE of the form:

$$\frac{d^n \sigma}{dt^n} = \frac{\Gamma(k)}{\Gamma(k-n)} t^{k-n} \tag{4.4}$$

for describing the variation of strain σ over time t. Strain is the change in a material in terms of shape or other parameters under a stress (force per unit area). In Eq. (4.4), $\Gamma(k)$ and $\Gamma(k-n)$ are gamma functions with k as the dissipation coefficient and n as the order of fractional derivatives (see also Rogosin and Mainardi 2014). As an implementation of Volterra's idea, Rabotnov (1948) developed a fractional integral approach to viscoelastic problems (Valéria et al. 2014). This approach is important in terms of fractional integral equations and their applications.

The application of models to processes in the subsurface evolved from the use of integer-order PDEs, physically known as the advection-dispersion equation (ADE) or the Fokker-Planck equation (Fokker 1914, Planck 1917) in the 1910s to an era when fPDEs were pioneered for elastic media or deformation, by Gerasimov (1948). Gerasimov was in fact the first who applied fPDEs in a field close to groundwater hydrology and soil science (Kilbas et al. 2006), and his fPDEs for viscous flow was the most relevant to flow in porous media, such as water flow and solute transport, at that time.

On the same topic of viscoelastic media, Caputo (1966, 1967) made very important contributions in his time by not only proposing and solving an fPDE for the stress-strain relationship, but also defining a new fractional derivative to eliminate the problems in the physical interpretation of the initial condition associated with the Riemann-Liouville definition. Caputo's definition of the fractional derivative (Caputo 1967) is now named after him and is parallel to other definitions of fractional derivatives, such as the Riemann-Liouville definition of fractional derivatives. Caputo (1966) also proposed a multi-term time-fractional model for stress-strain relations (Eq. (2)) and investigated elastic waves in spherical coordinates. Caputo and Mainardi (1971a, b) proposed a time-fractional generic model and investigated dissipation properties of elastic materials in comparison with experimental data.

2.1.4 *Efforts Since the 1900s*

As Sandev et al. (2015) surveyed, it was Richardson (1926) who provided a rigorous mathematical and physical exploration of the mechanics of diffusion or flow that resulted in the discovery of anomalous diffusion. While investigating turbulent diffusion in the atmosphere, Richardson (1926) found that the classical Brownian motion did not apply. Instead, the following non-linear diffusion equation was pertinent and he named it as non-Fickian diffusion:

$$\frac{\partial q}{\partial t} = \frac{\partial}{\partial l}\left(F(l)\frac{\partial q}{\partial l} \right) \tag{4.5}$$

where q is the weighted mean concentration of particles in continuous diffusion, l is the spatial range of diffusion in the atmosphere ($1\ m < l < 10\ km$), and $F(l)$ is the non-Fickian diffusivity [cm^2/s] which is expressed as:

$$F(l) = 0.6l^{4/3} \tag{4.6}$$

Following Richardson's non-Fickian diffusion, the terminology 'anomalous diffusion' has been widely used and different forms of mechanisms and mathematical formulations have been put forward to explain these phenomena (Metzler and Klafter 2004, Sandev et al. 2015), including fractional Brownian motion (fBM), generalized Langevin equation (GLE), sub-diffusion in the framework of continuous-time random walks (CTRWs), Lévy flights and Lévy walks, and time-fractional diffusion (Metzler and Klafter 2004). Although these different concepts and models have different properties to quantify anomalous diffusion, the most commonly used characteristic in physical sciences and stochastics is mean square displacement (MSD) $\langle x^2 \rangle(t)$, and, in limited cases, *variance* $\langle (\Delta x)^2 \rangle(t)$ (Bouchaud and Georgers 1990, Metzler et al. 1998, Metzler and Klafter 2004).

The MSD $\langle x^2 \rangle(t)$ and variance $\langle (\Delta x)^2 \rangle(t)$ are related by the following relationship (Metzler et al. 1998, Metzler and Klafter 2004):

$$\langle (\Delta x)^2 \rangle(t) = \langle (x - \langle x \rangle)^2 \rangle$$
$$= \langle x^2 \rangle(t) - \langle x \rangle^2(t) = 2dDt^\beta \tag{4.7}$$

where d is the spatial dimension ($d = 1,2,3$), D is the diffusion constant, and β is the anomalous diffusion exponent. Different expressions for variance can be found in literature, such as $\langle (\Delta x)^2 \rangle = \text{var}\{x\} = \{\sigma[x]\}^2$, or simply σ^2, with $\sigma[x]$ being the standard deviation (Gardiner 1985).

For radial diffusion in disordered media, it has been found that the MSD is of the form (Havlin and Ben-Avraham 1987) $\langle R^2 \rangle \sim t^\beta$, which corresponds to diffusion with a time-fractional PDE (to be discussed later). For $\beta = 1$, Eq. (4.7) represents the mean square displacement for classical diffusion, which, in one dimension due to Brownian motion, has been known since Einstein (1905) (Risken 1996). For classical diffusion, the MSD and variance are equal (Metzler et al. 1998), that is,

$$\langle x^2 \rangle = 2Dt \tag{4.8}$$

and the variance or the deviation from the MSD is given as:

$$\langle (\Delta x)^2 \rangle = \langle (x - \langle x \rangle)^2 \rangle = 2Dt \tag{4.9}$$

When the particle movement due to anomalous diffusion is in a moving fluid or a moving frame (velocity V), and subject to different dispersion patterns, there are different forms of the assemble mean displacement $\langle x \rangle(t)$, the MSD, and its variance (Compte 1997, Compte et al. 1997, Metzler et al. 1998). For instance, one of the patterns considered by Metzler et al. (1998, Case D) is a more generalized situation with a finite "advection time scale" τ_a introduced by Compte (1997) which is for dispersive particle transport with a constant velocity V in sticking media. For this case, the three measures are given as follows:

$$\left.
\begin{aligned}
\langle x \rangle(t) &= \frac{AV}{\Gamma(1+\beta)} t^\beta \\
\langle x^2 \rangle(t) &= \frac{2D}{\Gamma(1+\beta)} t^\beta + \frac{2(AVt^\beta)^2}{\Gamma(1+2\beta)} \\
\langle \Delta x^2 \rangle(t) &= \frac{2D}{\Gamma(1+\beta)} t^\beta + (AVt^\beta)^2 \left[\frac{2}{\Gamma(1+2\beta)} - \frac{1}{\Gamma(1+\beta)^2} \right]
\end{aligned}
\right\} \quad 0 < \beta < 1 \tag{4.10}$$

where $A = \dfrac{\tau_a}{\tau^\beta}$ in the normal advection transport with the time scale τ as opposed to the anomalous transport with the advection time scale τ_a (Compte 1997).

The three components in Eq. (4.10) are identical to those in Eq. (4.7). Clearly, $\langle (\Delta x)^2 \rangle(t) = \langle x^2 \rangle(t)$ when $V = 0$ implying that the MSD and the variance is the same when the transport is due to diffusion only. In fact it can be seen that $\langle x \rangle(t)$ and $\langle (\Delta x)^2 \rangle(t)$ in Eq. (4.10) are, respectively, the expectation (or the mean) and the variance for the time fractional diffusion-advection equation (Zhang et al. 2008).

Bochner (1949) investigated fPDEs in stochastics and suggested that the operator for an integer-order PDE can be replaced by a fractional counterpart or a functional operator such as:

$$\frac{\partial c}{\partial t} = A^p c, \qquad\qquad 0 < p < 1 \tag{4.11}$$

and

$$\frac{\partial c}{\partial t} = \left[(A+k)^{1/2} - k^{1/2} \right] c \tag{4.12}$$

where

$$A = -\left(v\frac{\partial c}{\partial x} - D\frac{\partial^2 c}{\partial x^2} \right) \tag{4.13}$$

is the ADE operator. Bochner's terminology is clearly the fractional Laplacian widely used in literature.

Barrett (1954) investigated an fDE with a constant boundary condition defined as a fractional derivative or the fractional Cauchy problem, and presented a unique solution. According to Kilbas et al. (2006), Al-Bassam (1965) was arguably the first to investigate a non-linear fDE and proved the solution based on the method of contractive mapping. Erdélyi (1965) investigated a system of axially symmetrical PDEs and fractional integration. Dzhrbashyn and Nersesyan (1968) analyzed a multi-term fDE which is also called the sequential differential equation (Miller and Ross 1993, Podlubny 1999).

Electrochemistry is another field where fractional calculus found applications. Valéria et al. (2014) found that Nigmatullin and Belavin (1964) first used semi-derivatives and performed fractional operations on integration and differentiation of order $\frac{1}{2}$ in electrochemistry. This was later used and called semi-differentiation by Oldham and Spanier (1970) in electronic chemistry and diffusion, which is also fundamentally important for solute movement in porous media.

From Gerasimov (1948) to Oldham and Spanier (1970), the use of fPDEs on interesting topics in applied fields was not physically and mathematically proven. It is known that widespread research and applications of fPDEs in physics and applied mathematics occurred since the 1970s when Monin and Yaglom (1971), Nigmatullin (1986), Wyss (1986), Klafter et al. (1987), Schneider and Wyss (1989) and other researchers proposed different forms of fDEs and fractional diffusion-wave equations (fDWEs). At this time, the explorative transition from integer PDEs to fPDEs seems to be complete in both mathematics and applied fields.

2.1.5 Re-launch of Fractional Calculus in 1974

Fractional calculus slowly evolved from 1695 till around 1974 when it was re-launched, marked by three elements: publication of the first monograph on fractional calculus by Oldham and Spanier (1974), the first PhD conferred to Bertram Ross on a topic in fractional calculus, and the first *International Conference on Fractional Calculus and Its Applications* (organized by the inaugural PhD recipient in fractional calculus) held at University of New Haven, Connecticut (Ross 1975, Miller and Ross 1993, Machado et al. 2011).

The slow evolution of fractional calculus over 279 years from 1695 to 1974 saw many mathematicians and scientists in different fields applying fractional calculus to investigate various processes.

2.1.6 Models for Particle Transport Processes since the 1990s

One important development relevant to water flow and solute transport in porous media, and possibly in other branches of geosciences, is accredited to Lenormand (1992) for introducing fractional derivatives to describe flow in heterogeneous porous media and deriving the following fPDE as a 'general transport equation':

$$\frac{\partial c(x,t)}{\partial t} = \frac{A}{2}\Gamma(\alpha+1)\frac{\partial^{1-\alpha}}{\partial t^{1-\alpha}}\left[\frac{\partial^2 c(x,t)}{\partial x^2}\right] + \frac{B}{6}\Gamma(\beta+1)\frac{\partial^{1-\beta}}{\partial t^{1-\beta}}\left[\frac{\partial^3 c(x,t)}{\partial x^3}\right] \qquad (4.14)$$

Lenormand's derivation and analysis showed that $\beta < 1$ and the "heterogeneity parameter" α varies from $\alpha = 1$ for dispersive flow to $\alpha = 2$ for convective flow; $\sigma^2 = \langle x^2\rangle(t) = At^\alpha$ and $\langle x^3\rangle(t) = Bt^\beta$.

Compte (1997) further demonstrated the connection between the fADE and continuous-time random walk (CTRW) theory by presenting time-fractional and space-fractional ADEs. Before Compte's work, CTRW was associated mainly with the rigorous demonstration of anomalous diffusion and how it behaves in different scenarios. Compte (1997) derived two forms of generalized ADEs (Compte 1997) which are essentially the time and space fPDEs, subject to different conditions for CTRW models. The time fPDE Compte (1997) derived was:

$$\frac{\partial^\beta \rho(x,t)}{\partial t^\beta} = D\nabla^2\rho(x,t) - A\nabla\cdot[v(x)\rho(x,t)] + \frac{t^{-\beta}}{\Gamma(1-\beta)}\delta(x) \qquad 0 < \beta < 1 \qquad (4.15)$$

where $\rho(x, t)$ is the probability density that can be used as the hydraulic head for water flow or the solute concentration in porous media, and D is diffusivity expressed as:

$$D = \frac{\sigma^2}{\tau^\beta} \qquad (4.16)$$

with σ^2 as the variance and τ as the waiting time in the CTRW model; $v(x)$ as velocity and A as the scaling parameter in velocity, with significant implications in the form of:

$$A = \frac{\tau_a}{\tau^\beta} \qquad (4.17)$$

τ_a being the advection time scale which, in classical Brownian motion, is coincident with mean waiting time; for anomalous motion, however, its meaning is yet to be explained (Compte 1997, Metzler and Klafter 2004); $\delta(x)$ represents the vector form of the Dirac delta function.

However, as the backward motion of a particle is a feature of the CTRW model, τ_a can be explained as the time scale for backwater effect, permitting the fractional derivatives in the fPDE to be symmetrical. The advection velocity incorporating the scaling parameter is $V(x) = Av(x)$.

The space fPDE derived by Compte (1997, Eq. (34)) is written as:

$$\frac{\partial \rho(\mathbf{x},t)}{\partial t} = D\nabla^{2\eta} \rho(\mathbf{x},t) - A\nabla \cdot [v(\mathbf{x})\rho(\mathbf{x},t)] \tag{4.18}$$

with

$$D = \frac{\sigma^{2\eta}}{\tau} \tag{4.19}$$

and

$$A = \frac{\tau_a}{\tau} \tag{4.20}$$

For $\eta = 1$, Compte's fPDE in Eq. (4.18) becomes the standard ADE and the diffusivity becomes $D = \frac{\sigma^2}{\tau}$. Similar multi-dimensional space fPDEs have later been investigated by Meerschaert et al. (1999) and Umarov and Gorenflo (2005b).

After Lenormand (1992) and Compte's (1997) developments, other important works relevant to hydrology and soil science also appeared. The first such development is attributed to He (1998) for developing three fPDEs in space for water flow in aquifers, which He named as seepage flow in porous media, by generalizing Darcy's law with fractional derivatives in three dimensions. Of the three types of fPDEs for flow in porous media, the most general form is given as (He 1998):

$$S_0 \frac{\partial P}{\partial t} = \frac{\partial^{\beta_1}}{\partial x^{\beta_1}} \left(K_x \frac{\partial^{\alpha_1} P}{\partial x^{\alpha_1}} \right) + \frac{\partial^{\beta_2}}{\partial y^{\beta_2}} \left(K_y \frac{\partial^{\alpha_2} P}{\partial y^{\alpha_2}} \right) + \frac{\partial^{\beta_3}}{\partial z^{\beta_3}} \left(K_z \frac{\partial^{\alpha_3} P}{\partial z^{\alpha_3}} \right) \tag{4.21}$$

where P is hydraulic potential or head; K_x, K_y, and K_z are the hydraulic conductivities in x, y, and z directions respectively; and S_0 or $\frac{1}{v}$ is the effective porosity for an unconfined aquifer or the effective storativity for a confined aquifer. In his work, He specified the following orders of fractional derivatives:

$0 < \alpha_1, \alpha_2, \alpha_3 < 1$;
$0 < \beta_1, \beta_2, \beta_3 \leq 1$, and
$0 < (\alpha_1 + \beta_1), (\alpha_2 + \beta_2), (\alpha_3 + \beta_3) \leq 2$.

Equation (4.21) as an fPDE in three dimensions is a generalization of the equation for flow in aquifers from integer calculus (Bear 1979). For confined and unconfined aquifers, these equations are specified differently. Here the 'seepage flow' used by He (1998) is in fact diffusion in porous media indeed.

The two other developments in the same year of 1998 were made by Benson (1998), who modified Zaslavsky's symmetrical fADE and applied it for investigating solute transport, and by Chaves (1998), for he presented a symmetrical space-fractional diffusion equation.

To this extent, the decade of the 1990s was a turning point in hydrology, soil science and other related areas in geosciences, for extensive research and applications of fPDEs led to milestone achievements in these areas (Debnath 2003a, b, 2004, Zhang et al. 2009, Benson et al. 2013).

To date, many literature surveys on fractional calculus and related models have been published. A chorological bibliography of fractional calculus till 1974 was provided by Oldham and Spanier (1974). More recent surveys include those by Machado et al. (2011), which summarized some of the major events and documents in fractional calculus since 1974 up to 2011; by Mainardi (2012), on linear viscoelasticity; and by Machado and Kiryakova (2017), who provided more complete highlights of major documents and events in the field of fractional calculus from 1974 to 2017. Valério et al. (2014) provided a survey of 10 eminent pioneers who applied fractional calculus to different scientific and engineering disciplines, including Niels Abel (1802–1829), Andrew Gemant (1895–1983), Andrey Gerasimov, Oliver Heaviside (1850–1925), Paul Lévy (1886–1971), Scott Blair (1902–1987), Cole (1900–1984) and his brother Cole (1914–1990), Yuri Rabotnov (1914–1985) and Nigmatullin (1923–1991).

2.2 *Experimental and Physical Investigation of Anomalous Processes*

As seen from the previous section, development of calculus and related concepts was initiated more than 300 years ago, but application of the concepts in fractional calculus was only a matter of interest until 1823 due to Abel, and in 1943 due to Blair and Coppen, with regards to deformation.

Darcy (1856) developed the first mathematical model based on an integer ODE and later expanded it to a PDE for relating the flow rate of water to the hydraulic gradient in porous media. Green and Ampt proposed the first infiltration model based on an ODE 1911. This was followed predominantly by extensive investigations of water flow in rigid soils with a PDE known as the Richards equation (1931). The applications of fDEs and fPDEs to liquid flow and particle transport in porous media were seen much later.

With extensive laboratory and field experiments using tracers since the 1950s and 1960s (Lallemand-Barres and Peaudecerf 1978, Gelhar et al. 1992, Berkowitz et al. 2006), the quantification of solute transport processes followed, which prompted widespread application of the ADE in hydrology and soil science (Nielsen et al. 1986). The ADE in use over the past decades is a PDE with integer orders in time and space. In geosciences, including hydrology and soil science, the traditional integer-based models, which contributed to a significant understanding of quantitative relationships between many processes and related factors, are however unable to explain the mechanics of anomalous diffusion in the field.

In addition to Richardson's (1926) work on non-Fickian diffusion, Scheidegger (1954) clearly identified non-Brownian flow in porous media, which confirmed the fractional generalization of the ADE as reported by Bochner (1949). Other reports introduced an empirical exponent in the diffusion equation formulated for rigid media, for analyzing anomalous flow (Küntz and Lavallee 2001, Lockington and Parlange 2003, Gerolymatou et al. 2006, Valdes-Parada et al. 2007).

Numerous reports on anomalous flow and solute transport in hydrology and soil science have appeared over the past two decades since 1990 (Benson 1998, He 1998, Benson et al. 2000a, b, 2001, Pachepsky et al. 2000, Meerschaert et al. 2002, Schumer et al. 2003a, b, Baeumer et al. 2005, Baeumer and Meerschaert 2007, Krepysheva et al. 2007, Logvinova and Neel 2007, Martinez et al. 2007, Zhang et al. 2009, Benson and Meerschaert 2009, Su 2009a, b, 2010, 2012, 2014, 2017a, b, Schumer et al. 2011, Benson et al. 2013, Su et al. 2015, Atangana 2018). One particular form of fPDEs is the distributed-order fractional partial differential equation (dofPDE) (Schumer et al. 2003b). dofPDEs are important for analyzing flow and transport in porous media as they represent mobile and immobile zones first investigated by Rubinshtein (1948) and then popularized by Barenblatt et al. (1960) and other investigators. The double porosity model has been investigated for over 7 decades in soil science and flow in porous media, and the further introduction of dofPDEs is particularly interesting in porous media, including water movement in unsaturated soils (Su 2012, 2014, 2017b) and in groundwater as saturated media (Su et al. 2015, Su 2017a).

2.3 Explanation and Justification for Anomalous Water Flow and Solute Transport in Porous Media

The term 'anomalous diffusion' refers to the mechanisms described by an fDE or an fPDE, in contrast to the processes described by the classical diffusion equation ('normal' diffusion) (Zaslavsky 2002, Zhang et al. 2009). Water flow in soils, which includes infiltration into soils, and solute transport by water are further complicated by the fact that many natural soils have certain degrees of swelling tendencies and joint energy or heat transfer as well as the reciprocal effects.

The physical proof and justification for the application of anomalous diffusion in the interpretation of flow and transport processes in various media have been provided in many published reports such as those by Kündtz and Lavallee (2001), Gerolymatou et al. (2006) and Valdes-Parada et al. (2007). Timashev et al. (2010) explained that anomalous diffusion is a stochastic component in the dynamics of complex processes. It is the heterogeneity of porous media with irregular pores which causes the flow in the media to generate random fluctuations in the velocity field that is responsible for the stochastic component of the flow dynamics. In addition to different quantitative methods in geosciences (Christakos 2012), Compte (1997) demonstrated how diffusion and advection with different forms of random walks in a convective field can generate time- and space-fractional PDEs (also see Metzler et al. 1998).

As explained by Morales-Casique et al. (2006), Zhang et al. (2009) and Benson et al. (2013), the two fractional derivatives in fPDEs are the statements of the non-locality property of transport processes in heterogeneous media. *Non-locality* is used to explain how the concentration of a tracer, fluid parcels, or any other quantity at previous times and/or upstream locations contributes to the variation of the concentration at the point of observation due to the uncertain velocity field. *Spatial non-locality* refers to the change in the solute concentration or moisture ratio at the point of observation depending on its upstream value, whereas *temporal non-locality* implies that the change in the concentration or ratio at the point of observation depends on the prior concentration loading (Benson et al. 2013).

These non-localities indicate scale-dependent and history-dependent phenomena which can also be interpreted using other models such as non-linear PDEs, etc. The non-locality or scale-effect for infiltration was mathematically illustrated by Voller (2011), who showed that the integer PDE $\dfrac{\partial}{\partial z}\left(K(z)\dfrac{\partial h}{\partial z}\right)=0$

with $K(z)=\dfrac{1}{\lambda}\Gamma(1+\lambda)z^{1-\lambda}$ yields an identical result for the wetting front derived

using a fractional model under the same boundary conditions. This fact justifies that non-local effects can either be represented by an fPDE or a non-linear PDE of integer order, and mathematically explains how the fractional model interprets the physical process of infiltration.

Wheatcraft and Meerschaert (2008) presented a space-fractional PDE for the conservation of mass for flow in porous media as:

$$\Gamma(\lambda+1)\Delta x^{1-\lambda}\rho(\beta_s+\phi\beta_w)\frac{\partial p}{\partial t}=-\rho\nabla^\lambda\cdot\mathbf{q} \tag{4.22}$$

where λ is the order of space-fractional derivatives, Δx is the elemental size (hence the term $\Delta x^{1-\lambda}$ represents the scale effect due to non-local flow processes), and β_s and β_w are the coefficients of compressibility of porous media and water respectively, which are clearly the fractional form of Eq. (3.49).

Equation (4.22) implies that the spatial fractional flux and the appearance of a scaling parameter $\Delta x^{1-\lambda}$ are related to the order of spatial fractional derivatives. The scaling factor renders the fractional equation scale-invariant and the scaling captures the fractal nature of the porous medium, which means that $\Delta x^{1-\lambda}\to 1$ as $\lambda\to 1$.

Furthermore, Eq. (4.22) by Wheatcraft and Meerschaert (2008) is a one-sided fPDE which has been shown by Olsen et al. (2016) as a special form of a two-sided fPDE.

Chapters 5 to 11 of this book present a collection of fPDEs and related models for water flow and solute transport in soils, aquifers and overland flow, as well as for elasticity of porous media.

3. Theory, Models and Parameters for Water Flow and Solute Transport in Porous Media

Compte (1997) made crucial progress in understanding transport processes by considering anomalous diffusion in a velocity field which resulted in time and space fPDEs in terms of probability as in Eqs. (4.15) and (4.18). His generic approach, as elaborated in his opening statement, is relevant to many industrial and environmental processes which include sediment transport and pollutant transport in addition to anomalous hydraulics of flow. In his approach, Compte separately considered time and space fPDEs. As we will see in this section, the separate time and space fPDEs are special cases of the time-space fPDE for anomalous diffusion and advection, which can be derived from the CTRW theory.

This section briefly demonstrates the connection between fPDEs and the CTRW theory for water flow and solute transport in porous media, considering both anomalous diffusion and advection in a velocity field. It shows the connections between multi-dimensional fPDEs and CTRW, and relevant parameters in fPDEs. Furthermore, the interrelationships between the parameters in fPDEs which include spatial and temporal scaling parameters, and some key hydraulic variables such as flow velocity are also examined here. Both one-dimensional and multi-dimensional CTRW concepts have been briefly discussed in this section.

3.1 Random Walks and Continuous-Time Random Walks in Brief

Although fractional calculus was conceived over 320 years ago, its development and applications did not parallel those of integer calculus. Despite this disparity, some concepts and methods developed for fractional calculus, such as fPDEs and their association with stochastics and fluid mechanics, etc., have found significant applications. The examples of these developments include the application of the CTRW theory in the derivation of the governing equations for flow and transport in porous media, without resorting to the traditional method of mass conservation (Lenormand 1992, Compte 1997, Gorenflo and Mainardi 1998a, b, 2001, 2005, 2009, Zaslavsky 2002, Uchaikin and Saenko 2003, Gorenflo et al. 2007). A very important feature of the fPDEs so derived is associated with the transport exponent (Zaslavsky 2002), which defines the pattern of flow and transport by the two parameters in the CTRW model, and is also related to the scaling parameters which are in turn related to fractal space-time structures.

The concept of random walk was first mathematically illustrated by Crofton (1865) in terms of random flight, and later used by Pearson (1905). Problems in solid-state physics associated with random walks on periodic space lattices were first discussed by Pólya (1921), according to Montroll and Weiss (1965). The concept of CTRW is a further development from random walks, and according to Kilbas et al. (2006), the CTRW concept was first proposed by Gnedenko and Kolmogorov in 1949 in the Russian language (English published in 1954) (Gnedenko and Kolmogorov 1954). Gorenflo and Mainardi's (2005) brief review showed that the concept of generalized functions by Gel'fand and Shilov (1964) enables the Montroll-Weiss equation (Montroll and Weiss 1965) to form the basis for the concept of CTRW, which was further popularized by Montroll and Scher (1973) to relate the concepts of anomalous transport and fPDEs (Gorenflo and Mainardi 2005, Berkowitz et al. 2006, Cvetkovic 2012).

CTRW as a concept can be interpreted in the framework of the classical renewal theory (Cox 1967, Gorenflo et al. 2007). The phenomenon described by CTRW is anomalous, which is a generalization of transport processes, with classical diffusion as a special case (Gorenflo et al. 2007). It has been applied to investigate particle motion in many physical, geological and biological systems and processes (Metzler and Klafter 2000a, 2004, Kutner and Masoliver 2017, Masoliver and Lindenberg, 2017), and to model solute movement in porous media 60 years ago (Saffman 1959).

In recent years, the CTRW theory has been more widely applied to model flow and transport in different geological formations such as aquifers and soils (Lenormand 1992, Berkowitz and Scher 1995, Benson 1998, Benson et al. 2000b, Berkowitz et al. 2002, Dentz and Berkowitz 2003, Schumer et al. 2003a, Berkowitz et al. 2006, Benson et al. 2013, Su 2014, 2017a, Dentz et al. 2015).

The CTRW theory models particle motion with two probabilities for the two stages of random particle movements: one probability relates to the motion length and the second to the waiting time of the particles before the next movement. When the motion of solute particles in porous media is explained successfully using this theory, it is natural and logical to apply the same concept to water flow as water is the original driving force for the motion of solute particles in the subsurface. Detailed discussions of the connection between CTRW and fPDEs can be found in literature (Compte 1997, Zaslavsky 2002, Uchaikin and Saenko 2003, Gorenflo and Mainardi 2005, 2009, 2012, Gorenflo et al. 2007, Abdel-Rehim and Gorenflo 2008, Tejedor and Metzler 2010, Meerschaert 2012).

There have been extensive reports about the CTRW theory, its numerous variants and applications. Section 3.5 briefly overviews these concepts, of which major basic types of CTRWs (Abdel-Rehim and Gorenflo 2008) are particularly important here. The *first type*, considering independent time and space steps, is known as 'decoupled or separable CTRW' and treats time and space steps as *independent identically distributed* (iid) random variables. The *second type* of CTRWs is called 'coupled CTRW' or 'non-separable CTRW', which treats time and space steps as dependent on each other. With these two major distinctions, other forms of conditions can be considered to yield different categories of CTRW-based models as briefly listed later. For a long-time and large-distance limit, the decoupled CTRW model of iid variables leads to an fDE or fDWE (Gorenflo et al. 2007, Abdel-Rehim and Gorenflo 2008). In the following section, the decoupled CTRW concept has been discussed.

3.2 CTRW and its Connection with the fPDE for Anomalous Diffusion

3.2.1 One-dimensional CTRW Theory and the fPDE

CTRW interprets the motion of a particle in a sequence of two states: the jump and the wait preceding the subsequent jump. For decoupled CTRWs, the jump length and the waiting time are independent random variables and each probability is an iid distribution function (Gorenflo et al. 2007, Tejedor and Metzler 2010). The iid positive waiting times are denoted by $T_1, T_2, T_3,\ldots$, each having the same probability density function (pdf), $\phi(t)$, $t > 0$; the iid random jumps are denoted by $X_1, X_2, X_3,\ldots$ in the real domain, $\mathbf{R}$, and have the same pdf, $w(x)$, $x \in \mathbf{R}$. With these definitions, the probability density of the particle (or water parcel) movement in the porous media is $p(x, t)$, which is represented by the following series (Gorenflo et al. 2007):

$$p(x,t) = \Psi(t)\delta(x) + \sum_{n=1}^{\infty} v_n(t)w_n(x) \tag{4.23}$$

where $\delta(x)$ is the Dirac delta function and $\Psi(t)$ is the survival function given as:

$$\Psi(t) = \int_t^{\infty} \phi(t')dt' \tag{4.24}$$

$v_n(t)$ and $w_n(x)$ being repeated convolutions in time and space respectively, that is, $v_n(t) = (\Psi * \phi^{*n})(t)$ and $w_n(x) = (w^{*n})(x)$.

Based on Eq. (4.24), Gorenflo and Mainardi (2005), Gorenflo et al. (2007) and Gorenflo and Mainardi (2012) showed that:

$$\hat{\tilde{u}}(\kappa, s) = \int_0^{\infty} \left[\exp(-t_* |\kappa|^{\alpha} i^{\omega \, sign\kappa})\right] \left[s^{\beta-1} \exp(-t_* s^{\beta})\right] dt_* \tag{4.25}$$

where κ and s are the Fourier and Laplace transform variables respectively, β is the exponent for the probability of the waiting time intervals between two consecutive steps, λ is the exponent for the probability of the length of steps for the random walks (Zaslavsky 2002), and ω is the skewness acting on the space variable, $|\omega| \le \min\{\lambda, 2 - \lambda\}$.

For the limiting case wherein the scaling factors approach zero, Gorenflo and Mainardi (2005) showed that Eqs. (4.23) to (4.25) lead to the following result:

$$\hat{p}(\kappa, s) \sim \frac{s^{\beta-1}}{s^\beta + |\kappa|^\lambda \, i^{\omega \, sign \kappa}} = \tilde{\hat{u}}(\kappa, s) \tag{4.26}$$

For $\omega = 0$, Scalas et al. (2004) and Gorenflo and Mainardi (2005, 2009) showed that Eq. (4.26) is the joint Laplace-Fourier transform of the following fDWE:

$$\frac{\partial^\beta u(x,t)}{\partial t^\beta} = D \frac{\partial^\lambda u(x,t)}{\partial |x|^\lambda}, \qquad u(x, 0) = \delta(x) \tag{4.27}$$

where β is called the order of the time (or temporal) fractional derivatives and λ is the order of the space (or spatial) fractional derivatives.

The case of $\omega = 0$ is called symmetrical (Mainardi et al. 2005), which corresponds to the definition of fractional Laplacian given by Bochner (1949) (who introduced the concept of symmetrical fPDEs) and used by others since (Sachev and Zaslavsky 1997, Jiang et al. 2012a, b). Symmetrical fractional derivatives are also written as:

$$_x D_0^\alpha = -\left(-\frac{\partial^2}{\partial x^2}\right)^{\lambda/2} \tag{4.28}$$

The delta function in the initial condition of Eq. (4.27) can be incorporated in the fDWE (Saichev and Zaslavsky 1997, Zaslavsky 2002), resulting in the following form:

$$\frac{\partial^\beta u(x,t)}{\partial t^\beta} = \frac{\partial^\lambda u(x,t)}{\partial |x|^\lambda} + \frac{t^{-\beta}}{\Gamma(1-\beta)} \delta(x) \tag{4.29}$$

as $\lim_{\beta \to 1} \dfrac{t^{-\beta}}{\Gamma(1-\beta)} = \delta(t)$.

With $\omega = 0$ and diffusivity D included, Eq. (4.27) may be simply written as a symmetrical fPDE of the form:

$$\frac{\partial^\beta u(x,t)}{\partial t^\beta} = \frac{\partial^\lambda u(x,t)}{\partial x^\lambda}, \qquad u(x, 0) = \delta(x) \tag{4.30}$$

or in a compact form as:

$$_t D_*^\beta u(x, t) = _x D_0^\lambda u(x, t), \qquad u(x, 0) = \delta(x) \tag{4.31}$$

For a homogeneous BC and non-zero IC, Eq. (4.27) is equivalent to Eqs. (4.29), (4.30) and (4.31) (Kevorkian 1990).

3.2.2 Multi-dimensional CTRW Theory and the fPDE

The above derivations are for one-dimensional cases, which are most commonly encountered in literature. In fact, multi-dimensional CTRWs are more realistic in physical terms. Pólya (1921) investigated two-dimensional random walks about a century ago. Chandrasekhar (1943) investigated multi-dimensional random walks and Levy flights; Mogul'skii (1976) also investigated multi-dimensional random walks before others (Samorodnitsky and Taqqu 1994, Gorenflo and Mainardi 1998b).

In addition to the separate multi-dimensional fPDEs in Eqs. (4.15) and (4.18) by Compte (1997), in the symmetrical case of $\omega = 0$ for n-dimensional CTRWs, Uchaikin and Saenko (2003) showed that CTRW in the case of anomalous diffusion leads to:

$$\hat{\tilde{u}}(\kappa, s) = \frac{s^{\beta-1}}{s^{\beta} + D|\kappa|^{\lambda}} \tag{4.32}$$

with

$$D = \frac{\Gamma(n/2)\Gamma(1 - \lambda/2)A}{2^{\lambda}\Gamma[(\lambda+n)/2]\Gamma(1-\beta)B} \tag{4.33}$$

where A and B are coefficients in the transitional probability density functions, that is,

$$P(X > x) = Ax^{-\lambda} \tag{4.34}$$

and

$$P(J > t) = Bt^{-\beta} \tag{4.35}$$

with n as the number of dimensions. The Fourier-Laplace inversion of Eq. (4.32) results in an fPDE in the multi-dimensional form as (Uchaikin and Saenko 2003):

$$\frac{\partial^{\beta} u(\mathbf{x}, t)}{\partial t^{\beta}} = -D(-\Delta)^{\lambda/2} + \frac{t^{-\beta}}{\Gamma(1-\beta)}\delta(\mathbf{x}) \tag{4.36}$$

or

$$\frac{\partial^{\beta} u(\mathbf{x}, t)}{\partial t^{\beta}} = -D(-\Delta)^{\lambda/2}, \qquad u(\mathbf{x}, 0) = \delta(\mathbf{x}) \tag{4.37}$$

Equation (4.36) and Eq. (4.31) result from the asymptotic or long-time approximation of the CTRW model with two transitional probability distribution functions for the length of jumps, $P(X > x)$, and the waiting time intervals, $P(J > t)$, obeying power laws, which are given in Eqs. (4.34) and (4.35) respectively.

It should be noted that D in Eqs. (33) and (36) for water movement in unsaturated soils is dependent on the moisture ratio or content, which requires that A and B in the transitional cumulative distribution functions be functions of the moisture ratio or content.

As briefed in Chapter 2, $D_0^{\lambda} = -(-\Delta)^{\lambda/2}$, and the one-dimensional form of Eq. (4.37) is equivalent to:

$$\frac{\partial^\beta u(\mathbf{x},t)}{\partial t^\beta} = D\frac{\partial^\lambda u(\mathbf{x},t)}{\partial|\mathbf{x}|^\lambda}, \qquad\qquad u(\mathbf{x},0) = \delta(\mathbf{x}) \qquad\qquad (4.38)$$

Compared to the classical diffusion with $D = \left\langle\frac{1}{2}x^2\right\rangle/\langle T\rangle$ (Uchaikin and Saenko 2003), the anomalous diffusivity in Eq. (4.37) is now expressed in terms of the jump length, X, and the waiting time, T:

$$D \sim \frac{\langle X^\lambda\rangle}{\langle T^\beta\rangle} \qquad\qquad (4.39)$$

where the sign $\langle\rangle$ represents the ensemble averaging of the distance or displacement, x, over repeated observations called realizations in stochastics and probability. For classical diffusion with $\beta = 1$ and $\lambda = 2$, Eq. (4.39) becomes $D \sim \frac{\langle X^2\rangle}{\langle T\rangle} = \frac{\sigma^2}{\tau}$, where σ^2 is the variance and τ is the diffusion time scale (Compte 1997) with the dimensions of D being $[L^2T^{-1}]$. Generally, the dimension of D is $[L^\lambda T^{-\beta}]$.

The derivation of the fPDE by Uchaikin and Saenko (2003) is of great significance wherein the multi-dimensional problem is considered and diffusivity is exactly given as a function of the orders of fractional derivatives in time and space in Eq. (4.33) or the ensemble quantities in time and space in Eq. (4.39). These results offer a method for the determination of diffusivity.

In different dimensions, Eq. (4.33) has different forms as follows:

$$D = \frac{\pi^{1/2}\Gamma(1-\lambda/2)A}{2^\lambda\,\Gamma[(\lambda+1)/2]\Gamma(1-\beta)B}, \qquad n = 1 \qquad\qquad (4.40)$$

$$D = \frac{\Gamma(1-\lambda/2)A}{2^\lambda\,\Gamma[(\lambda+2)/2]\Gamma(1-\beta)B}, \qquad n = 2 \qquad\qquad (4.41)$$

and

$$D = \frac{\pi^{1/2}\Gamma(1-\lambda/2)A}{2^{\lambda+1}\,\Gamma[(\lambda+3)/2]\Gamma(1-\beta)B}, \qquad n = 3 \qquad\qquad (4.42)$$

For classical diffusion in the case of n dimensions, which corresponds to $\lambda = 2$ and $\beta = 1$, Eqs. (4.32) and (4.33) become Eqs. (4.43) and (4.44) respectively as:

$$\hat{\tilde{u}}(\kappa,s) = \frac{1}{s + D|\kappa|^2} \qquad\qquad (4.43)$$

$$D = \frac{\Gamma(n/2)\pi^{1/2}A}{4\Gamma[(2+n)/2]B} \qquad\qquad (4.44)$$

The inverse Fourier-Laplace transforms of Eq. (4.43) yields (Uchaikin and Saenko 2003):

$$u(\mathbf{x}_n,t) = \frac{1}{(4\pi Dt)^{n/2}}\exp\left(-\frac{\mathbf{x}_n^2}{4Dt}\right) \qquad\qquad (4.45)$$

which is a well-known solution of the classical diffusion equation:

$$\frac{\partial u(\mathbf{x}_n,t)}{\partial t} = D\frac{\partial^2 u(\mathbf{x}_n,t)}{\partial \mathbf{x}_n^2} \tag{4.46}$$

subject to the delta function condition $u(\mathbf{x}_n, 0) = \delta(\mathbf{x}_n)\delta(t)$, where $\delta(\mathbf{x}_n)$ and $\delta(t)$ are the n-dimensional and the one-dimensional Dirac delta functions respectively. It is obvious that Eqs. (4.33) and (4.44) provide alternatives for determining the diffusivity and the two constants, A and B, in the CTRW model.

To derive Eq. (4.37) or Eq. (4.38), many reports present the case of $0 < \beta \le 1$ and/or $0 < \lambda \le 2$. However, Zaslavsky (2002) showed that the restriction of $0 < \beta \le 1$ is unnecessary. For example, the case of $1 < \beta \le 2$ and $0 < \lambda \le 2$ corresponds to anomalous transport with two time scales (Becker-Kern et al. 2004, Baeumer and Meerschaert 2007) and the scaling limit of a decoupled CTRW (Caceres 1986, Meerschaert et al. 2010), which is an important property relevant to the two-term distributed-order fPDE for flow in soils (to be discussed later).

The left-hand side of the above fPDE is the Caputo fractional derivative with respect to time t, whereas the right-hand side is the Riesz-Feller fractional derivative with respect to space x. Note that the Riesz-Feller fractional derivative becomes the Liouville fractional derivative for $\omega = \pm\alpha$, with the positive sign for the forward fractional derivative and the negative sign for the backward fractional derivative when more generic fractional derivatives are considered (Ortigueira and Trujillo 2012).

The above presentation means that the CTRW model in Eq. (4.43) is identical to the fDWE in Eq. (4.46) in the Laplace-Fourier domain. In other words, as shown by Zaslavsky (2002), Scalas et al. (2004), and Gorenflo et al. (2007), under the asymptotic or long-time limit condition, the process modelled by a CTRW converges to a simpler probability density function which solves the fDWE. This approach of deriving a fractional diffusion-wave equation from the CTRW perspective is called the diffusion limit or the hydrodynamic limit (Scalas et al. 2004) as the jumping steps and the waiting time intervals are scaled by scaling parameters such as h and r respectively, both of which are made to vanish in a way so that the 'random walks' approach 'random diffusion'. The transition from random walks to random diffusion is analogous to the transition from discrete to continuous motion of particles. This connection provides an alternative approach for deriving the fDWE without resorting to the classical mass balance method for its derivation, in addition to the method of renormalization group of kinetics (Zaslavsky 2002) which also yields the fDWE.

3.3 *fPDEs Based on CTRW for Anomalous Diffusion and Convection*

Referring to the above developments, it is logical to extend the multi-dimensional time-space fPDEs to water movement and solute transport in porous media by incorporating convection in a moving field or a flow. The extension for solute transport for instance can also be done either by including a convection term in Eq. (4.36) or by introducing time-fractional derivatives in the PDEs derived by Compte (1997) which is given by Eq. (4.15):

$$\frac{\partial^{\beta} c(\mathbf{x},t)}{\partial t^{\beta}} = \mathbf{D}\nabla^{\lambda} c(\mathbf{x},t) - A\nabla \cdot [\nu(\mathbf{x})c(\mathbf{x},t)] + \frac{t^{-\beta}}{\Gamma(1-\beta)}\delta(\mathbf{x}) \tag{4.47}$$

which, with the initial condition $c(\mathbf{x}, t) = \delta(\mathbf{x})$, $t = 0$, can be written as:

$$\frac{\partial^{\beta} c(\mathbf{x},t)}{\partial t^{\beta}} = \mathbf{D}\nabla^{\lambda} c(\mathbf{x},t) - A\nabla \cdot [\nu(\mathbf{x})c(\mathbf{x},t)] \tag{4.48}$$

where A and B are the coefficients in the transitional probability density functions, respectively, and A is the advection scaling parameter (Compte, 1997), given now as $A = \frac{\tau_a^{\lambda}}{\tau^{\beta}}$ and the diffusivity in one dimension given as $D = \frac{\sigma^{\lambda}}{\tau^{\beta}}$ which are comparable to Eqs. (4.17) and (4.16) respectively.

The above fPDEs are presented as asymmetrical so that any degree of skewness in any dimension can be accounted for.

3.4 Connections between the Scaling Parameters, Fractals and the Orders of Fractional Derivatives

3.4.1 Anomalous Flow in Non-swelling Media

With the concept of renormalization group of kinetics (RGK), and space and time scaling parameters L_s and L_t (with x and $|x|$ identical in our cases), Zaslavsky (2002) showed that the space-time fPDE in Eq. (4.38) can be written, by including a diffusion coefficient, as:

$$\frac{\partial^{\beta} u(x,t)}{\partial t^{\beta}} = \left(\frac{L_s^{\lambda}}{L_t^{\beta}}\right)^{n} D\frac{\partial^{\lambda}}{\partial x^{\lambda}} u(x,t) \tag{4.49}$$

where L_s and L_t are space and time scaling parameters, n is the number of renormalization transformation in the RGK approach, and D is diffusivity.

When the space-time fPDE in Eq. (4.49) 'survives' the RGK transformation and if the condition $\lim_{n\to\infty} (L_s/L_t) = 1$ is met, the following relationship results:

$$\frac{\beta}{\lambda} = \frac{\ln L_s}{\ln L_t} = \frac{\mu}{2} \tag{4.50}$$

where μ is called the transport exponent which determines the flow patterns as (Zaslavsky 2002):

sub-diffusion when $\mu < 1$;

classical when $\mu = 1$, and

super-diffusion when $\mu > 1$.

The transport exponent μ connects the orders of fractional derivatives and the scaling parameters, L_s and L_t, as:

$$\mu = \frac{2\ln L_s}{\ln L_t} \tag{4.51}$$

and

$$\mu = \frac{2\beta}{\lambda} \tag{4.52}$$

The moment equation for Eq. (4.38) is given as (Zaslavsky 2002):

$$\left\langle x^{\lambda} \right\rangle = \frac{\Gamma(1+\lambda)}{\Gamma(1+\beta)} Dt^{\beta} \tag{4.53}$$

which is a very important result and can be derived from an instantaneous input or Dirac delta function input (Saichev and Zaslavsky 1997).

The mean displacement in diffusion is a measure of the average distance repeated, measured from a reference. In hydrological terms, it is the ensemble average from the diffusion front to the source, or solute leaching from a reference datum such as the surface to the front. For horizontal water movement, it is the ensemble mean depth from the wetting front to the source where water is applied.

The fractional derivative models for water movement account for non-Fickian flow processes. The *time-fractional* derivatives in the model account for partitioning of water parcels on sticky porous surfaces resulting in slowing processes (sub-diffusive), whereas the *space-fractional* (or mass-fractional) derivatives describe the flow processes in the media with higher velocity flow paths of long spatial correlation leading to super-diffusion. The resulting pattern of these two competing processes is measured by the transport exponent given by Zaslavsky (2002) as $\mu = \frac{2\beta}{\lambda}$ in Eq. (4.52) which is also related to the two scaling parameters.

For classical diffusion ($\lambda = 2$ and $\beta = 1$), the classical moments are recovered from Eq. (4.53). Examples of effects on flow in soils by the orders of fractional derivatives can be seen in Pachepsky et al. (2003) who demonstrated the variability of $\beta \neq 1$ for $\lambda = 2$ in a time-fractional diffusion equation for moisture movement.

Zaslavsky showed the connections between temporal and spatial scaling parameters with the orders of fractional derivatives in time and space. As fractals are defined as self-similar structures with similarity at all levels, Zaslavsky's scaling parameters, in fact, can be measuring units for fractals.

It can now be seen that the two parameters, β and λ, have three meanings:

1. orders of temporal fractional derivatives and spatial fractional derivatives in the fPDE;

2. exponents for the two probability density functions in the CTRW model; and

3. the critical exponents characterizing the fractal structures of space-time fractals (Zaslavsky 2002).

As such, these two parameters span and connect the three distinctive fields of fractional calculus, stochastics/probability, and fractal geometry. This connection is very valuable for evaluating these parameters using different methods in these fields.

3.4.2 Anomalous Flow in Swelling Media

Following the above developments for non-swelling soils, similar relationships can be developed for swelling soils (Su 2014):

$$\frac{\partial^\beta \theta(m,t)}{\partial t^\beta} = \left(\frac{L_m^\lambda}{L_t^\beta}\right)^n D \frac{\partial^\lambda}{\partial m^\lambda} \theta(m,t) \tag{4.54}$$

and

$$\frac{\beta}{\lambda} = \frac{\ln L_m}{\ln L_t} = \frac{\mu}{2} \tag{4.55}$$

where L_m is the mass scaling parameter. Similarly, the following entities also hold:

$$\mu = \frac{2\ln L_m}{\ln L_t} \tag{4.56}$$

and

$$\mu = \frac{2\beta}{\lambda} \tag{4.57}$$

which establishes the connection among β, λ and the scaling parameters for swelling soils. The moment for flow in swelling media is thus given as:

$$\langle m^\lambda \rangle = \frac{\Gamma(1+\lambda)D_m}{\Gamma(1+\beta)} t^\beta \tag{4.58}$$

The relationships of these parameters for swelling porous media resemble those for non-swelling media and the explanations are similar except for the fact that swelling porous media accommodates random walks of particles in an expanding or shrinking domain subject to the moisture ratio (or content).

3.4.3 Order of Fractional Derivatives and its Relationship with Other Hydraulic and Physical Parameters

Butera and Di Paola (2014) showed that the time-dependent effective velocity $V_{eff}(t)$ of a fluid through porous media in the form of a Lichtenberg tree is related to the order of Caputo fractional derivatives, β, and the fractal dimension d, with the scaling parameter ε. They showed that this relationship is expressed as:

$$V_{eff}(t) = \frac{\bar{V}}{\log \varepsilon} \left[1 - \left(\frac{\bar{T}}{t}\right)^\beta \right] \tag{4.59}$$

where the scaled velocity $\bar{V}$ and the scaled travel time $\bar{T}$, across the distance or thickness h of a section of porous media, are given as:

$$\bar{V} = \left(\frac{\rho g l_0^2}{32 v h}\right) p \tag{4.60}$$

$$\bar{T} = \left(\frac{32 v h^2}{\rho g l_0^2}\right) \frac{1}{p} \tag{4.61}$$

and

$$\beta = 1 + \frac{1}{\beta}(2-d) \tag{4.62}$$

with p as the fluid pressure or hydraulic potential, β being the order of fractional derivatives used for effective velocity, d as the fractal dimension, ρ being the fluid density, v being viscosity, and l_0 as the characteristic length. Equation (4.62) can be rearranged to yield:

$$d = 2 - \beta(\beta - 1) \tag{4.63}$$

which can be used to determine the fractal dimension when the order of fractional derivatives is known and vice versa.

Tarasov (2013) also established the relationship between fractal dimensions and velocity and provided a methodology for measuring fractal dimensions. Tatom (1995) examined the relationships between fractional calculus and fractals, and Podlubny (2002) explained the geometric and physical interpretation of fractional differentiation and integration.

3.5 *CTRW and Related Concepts*

The CTRW theory has several versions under different assumptions regarding the dependence or correlation between two successive jumps and waits as well as the relationships between jumps and waits. There are modifications with correlated waiting times or correlated jump lengths (Chechkin et al. 2009, Tejedor and Metzler 2010), and coupled CTRW with the sum of random jump lengths dependent on the random waiting times immediately preceding each jump (Schumer et al. 2011, Comlli and Dentz 2017), or the delay between particle jumps affecting the subsequent jump magnitude (Meerschaert et al. 2002), etc.

For more background on CTRW models and their applications, the readers may refer to literature (Klafter et al. 1987, Balescu 1995, Compte 1997, Zaslavsky 2002, Gorenflo and Mainardi 2005, 2009, 2012, Berkowitz et al. 2006, Gorenflo et al. 2007, Meerschaert 2012, Sandev et al. 2015, Kutner and Masoliver 2017 and Masoliver and Lindenberg 2017). Among these developments, the following CTRW concepts and related models are more relevant to water flow and solute transport in porous media:

1. Noisy CTRW model or nCTRW (Jeon et al. 2013): It considers that the random walk $x(t)$ is a combination of the ordinary CTRW $x_\lambda(t)$ and a noise $\eta z(t)$ such that $x(t) = x_\lambda(t) + \eta z(t)$;

2. Ageing CTRW or aCTRW (Barkai and Cheng 2003, Schulz et al. 2014, Chechkin et al. 2017): It describes a CTRW process which began at time $t = -t_\lambda$, and is relevant to trapping of particles in porous media;

3. Heterogeneous CTRW or hCTRW (Grebenkov and Tupikina 2018): This takes into consideration the shapes of the paths which determine the travel time probability;

4. CTRW with an expanding probability and in expanding media: an expanding probability (Vot et al. 2017) is introduced to account for diffusion in a swelling media and an expanding coordinate (Le Vot and Yuste 2018) is introduced to account for swelling media;

5. Cluster CTRW or cCTRW (Jurlewicz et al. 2008, 2011): It is useful for analyzing preferential flow in large pores of porous media;

6. Random walks with shrinking steps (Rador and Taneri 2006): This version can explain the transport processes in a heterogeneous medium;

7. Chechkin-Gonchar random walks: This considers random walk discrete in time and continuous in space among other forms of random walk models (Gorenflo and Mainardi 2001);

8. Fully discrete CTRW model: Herein, both time and space are discrete (Gorenflo and Vivoli 2003);

9. Persistent CTRW (Masoliver and Lindenberg 2017) and fractal time random walks (Hilfer 1995); and

10. Decoupled and coupled CTRWs (Metzler et al. 1998).

4. Relationships and Differences between Anomalous Diffusion and Scale-dependent and Time-dependent Transport Processes

4.1 fPDEs and Non-linear ODEs

As shown in section 3, the two orders of fractional derivatives in fPDEs are a consequence of the non-locality property of transport processes in heterogeneous media (Morales-Casique et al. 2006, Zhang et al. 2009, Benson et al. 2013). Physically, the term non-locality is used to explain that the concentrations of a tracer (or moisture ratio or any other quantity) at previous times and/or larger upstream locations contribute to the variation of its concentration at the point of observation due to the uncertain velocity field and/or physico-chemical factors such as absorption/desorption, etc. *Spatial* non-locality means that the concentration change at the point of observation depends on upstream concentrations, whereas *temporal* non-locality implies that the concentration change at the point of observation depends on the earlier concentration loading at that point (Benson et al. 2013).

4.2 fPDEs and Scale-dependent Nonlinear PDEs

While the fractional calculus-based approach is explained here, non-fractional models with functional parameters can also account for heterogeneity, such as scale-dependent processes. Voller (2011) showed that the integer PDE:

$$\frac{\partial}{\partial z}\left(K(z)\frac{\partial h}{\partial z}\right) = 0 \tag{4.64}$$

with

$$K(z) = \frac{1}{\eta}\Gamma(1+\eta)z^{1-\eta} \tag{4.65}$$

yields an identical result for hydraulic head with a moving wetting front, derived using the following fractional model under the same boundary conditions:

$$\frac{\partial}{\partial z}\left(\frac{\partial^{\eta}h}{\partial z^{\eta}}\right) = 0 \tag{4.66}$$

Voller's work indirectly justifies the approach to modelling solute transport processes in the subsurface using classical PDEs with a space- and time-dependent dispersion coefficient (Su et al. 2005) and also implies that the connection between fPDEs and classical PDEs for water movement in soils deserves further investigation.

4.3 fPDEs and Time-dependent Nonlinear PDEs

For a given scale of heterogeneity of porous media, Dieulin (1980) observed the time effect (de Marsily 1986), which can be described through a non-linear diffusion equation of the following form (Lenormand 1992):

$$\frac{\partial c(x,t)}{\partial t} = \frac{2D}{\beta}t^{\beta-1}\frac{\partial^{\lambda}c(x,t)}{\partial x^{\lambda}} \tag{4.67}$$

which yields variance (Lenormand 1992) as $\langle(\Delta x)^2\rangle(t) = \langle x^2\rangle(t) - \langle x\rangle^2(t) = Dt^{\beta}$ (or an MSD as given by Eq. (4.7)), which implies that $2d = 1$ in Eq. (4.7) in Lenormand's terminology.

The MSD of the form $\langle(\Delta x)^2\rangle(t) = Dt^{\beta}$ can be also derived from a time-fractional diffusion-wave equation of the form:

$$\frac{\partial^{\beta}c(x,t)}{\partial x^{\beta}} = D\frac{\partial^2 c(x,t)}{\partial x^2} \tag{4.68}$$

The physical equivalence of the non-linear PDE in Eq. (4.67) and the fDWE in Eq. (4.68) for the same variance or MSD provides another interpretation of transport in heterogeneous media.

4.4 Anomalous Diffusion and 'Generalized Self-diffusion'

With the convolution form of the anomalous diffusion equation, Cushman and Ginn (2000) showed that the anomalous diffusion equation is a generalized fractional diffusion which includes the space-fractional PDE. However, the convolution approach by Cushman and Ginn (2000) and the earlier work by Cushman (1991) pose technical challenges for practitioners in hydrology and soil science for their complex formulation in the Fourier-Laplace space. Cushman (1991) showed that the generalized self-diffusion coefficient defined by Zwanzig (1961, 1964) is expressed in the Fourier-Laplace space as:

$$\hat{\tilde{D}}(k,s) = \frac{A}{s^{d-1}(1+ck^2s^{-d})} \tag{4.69}$$

where $\hat{\tilde{D}}(k, s)$ is the joint Fourier-Laplace transform of the generalized self-diffusion coefficient $D(x, t)$, and k and s are the Fourier and the Laplace transform variables respectively;

$$A = \frac{D\Gamma(d+1)}{d} \tag{4.70}$$

and

$$c = \sqrt{3A} \tag{4.71}$$

with D being the classical Brownian diffusion coefficient, and d as the fractal dimension.

The inverse Laplace transforms of Eq. (4.69) (Kilbas et al. 2004) gives:

$$D(k, t) = At^{d-2}E_{d,d-1}(-ck^2t^d) \tag{4.72}$$

Furthermore, the inverse Fourier transform of Eq. (4.72) (Debnath 2003a) yields:

$$D(x,t) = \frac{D\Gamma(d+1)}{d} \int_{-\infty}^{\infty} G(x-\xi,t)d\xi \tag{4.73}$$

which is dependent on the kernel in system analysis, known as memory in stochastic analysis (Tarasov 2013).

$$G(x,t) = \frac{1}{\pi} \int_{-\infty}^{\infty} t^{d-2} E_{d,d-1}(-ck^2t^d)\cos(kx)dk \tag{4.74}$$

which determines the shape of $D(x, t)$.

$G(x, t)$ in Eq. (4.74) can be evaluated using the Laplace transform as (Debnath 2003a):

$$\begin{aligned}\tilde{G}(x,s) &= \frac{1}{\pi} \int_{-\infty}^{\infty} \frac{As\cos(kx)}{(s^d + ck^2)}dk = \frac{1}{\pi} \int_{-\infty}^{\infty} \frac{As^{1-d}\cos(kx)}{1+cs^{-d}k^2}dk \\ &= \frac{D\Gamma(d+1)s^{1-d}}{d\sqrt{4c}}\exp\left[-\frac{|x|}{\sqrt{c}}s^{d/2}\right]\end{aligned} \tag{4.75}$$

which can be inverted (Gorenflo et al. 1999), with $c = \sqrt{3A}$ given by Eq. (4.71), to yield:

$$G(x,t) = \frac{D\Gamma(d+1)}{2d}\left(\frac{d}{3D\Gamma(d+1)}\right)^{1/4}\frac{|x|}{t^{d-2}}\phi\left(\frac{d}{2}, d-1;-\xi\right) \tag{4.76}$$

where $\phi\left(\frac{\beta}{2}, \beta-1;-\xi\right)$ is the Wright function with

$$\xi = \left(\frac{d}{3D\Gamma(d+1)}\right)^{1/4}\frac{|x|}{t^{d/2}} \tag{4.77}$$

The Wright function is written as:

$$\phi\left(\frac{d}{2}, d-1; -\xi\right) = \sum_{n=0}^{\infty} \frac{1}{n![\Gamma[(nd/2)+d-1]]}\left[-\left(\frac{d}{3D\Gamma(d+1)}\right)^{1/4} \frac{|x|}{t^{d/2}}\right]^n \tag{4.78}$$

It can be shown (Stanković 1970) that for $(d-1)<1$ (or $d<2$), Eq. (4.76) has a large-time asymptote of the following form:

$$G(x,t) \sim D_0 \frac{|x|}{t^{d-2}}, \qquad\qquad t \to \infty \tag{4.79}$$

with

$$D_0 = \frac{D\Gamma(d+1)}{2d}\left(\frac{d}{3D\Gamma(d+1)}\right)^{1/4}\frac{\sin[(3-d)\pi]\Gamma(d-2)}{\pi} \tag{4.80}$$

In the half space which is physically relevant to a flow and transport domain $[0, \infty)$ or finite domain $[0, x]$, where x is the length of the flow path (the sign '[' denotes the inclusion of the boundary value 0, while ')' denotes the exclusion of the boundary value at the infinity), Eq. (4.73) can be solved with Eq. (4.76) or (4.79), that is, Eq. (4.73) can be considered in the range $[0, x]$:

$$D(x,t) = \frac{D\Gamma(d+1)}{d}\int_0^x G(x-\xi,t)d\xi \tag{4.81}$$

With the kernel given by Eq. (4.76) and by following the procedures given by Kilbas (2005), Eq. (4.81) can be integrated to yield:

$$D(x,t) = D_0\left(\frac{|x|}{t^{d-2}}\right)^2 \phi\left(\frac{d}{2}, 1+d; -\xi\right) \tag{4.82}$$

where

$$D_0 = \frac{D\Gamma(d+1)}{2d}\left(\frac{d}{3D\Gamma(d+1)}\right)^{1/4} \tag{4.83}$$

and ξ is given by Eq. (4.77).

The generalized diffusivity in Eq. (4.82) appears as the generalized Wright function. One approximation of this function can be achieved by retaining the first term in the Wright function in Eq. (4.78) to give:

$$D(x,t) = \frac{D_0}{\Gamma(d-1)}\left(\frac{|x|}{t^{d-2}}\right)^2 \tag{4.84}$$

This analysis indicates that the space- and time-dependent generalized diffusivity in Eq. (4.82) is a result of a memory function in the form of the Wright function. The relationship in Eq. (4.81) can also be written as a spatial fractional integral equation (Tarasov 2013) for a fixed time.

A PDE with the generalized self-diffusion coefficient in a form similar to Eq. (4.84) can determine the space- and time-dependent trajectory of solutes in an integer PDE which has been demonstrated using experimental data (Su et al. 2005).

Following the procedures by Stanković (1970), the asymptotic result of Eq. (4.82) can be derived as:

$$D(x, t) \sim D_0 t^{2-d}, \qquad\qquad t \to \infty \qquad\qquad (4.85)$$

where D_0 is given by Eq. (4.80).

With $\alpha = 2 - d$, Eq. (4.85) is exactly the asymptotic result given earlier by Philip (1986), based on the Lagrangian autocorrelation analysis where $0 < \alpha < 1$ is the exponent in the autocorrelation function, which means that:

$$d = 2 - \alpha \qquad\qquad (4.86)$$

or $1 \leq d \leq 2$ as a result of $0 < \alpha < 1$. Equation (4.86) implies that Zwanzig's (1961, 1964) generalized dispersion belongs to diffusion in two-dimensional fractal media.

The form of the generalized diffusivity in Eq. (4.85) has been widely applied to investigate diffusion processes and is a statement of 'dispersion in saturated heterogeneous media with no maximum scale' (Philip 1986, Muralidhar and Ramkrishna 1993) because the asymptotic diffusivity grows as an unbounded power function.

5. Dimensions of the Parameters in fPDEs

Once new fPDEs are derived, an important step is to ensure that the dimensions or units of the parameters in the new model are physically meaningful. For simple fPDEs such as the fDWE with one-term fractional derivatives in time or space, fractional dimensions can be easily defined. For more complicated fractional models, new definitions are needed for each parameter in the fPDEs so that the dimensions of the fPDE are physically meaningful.

In defining dimensions of the fractional diffusion coefficient in a time-fractional diffusion equation, Kilbas et al. (2006) suggested that a new parameter may be added to the ordinary diffusion coefficient parameter so that the dimension of the conventional parameter is retained while ensuring a correct dimension in the time-fractional diffusion equation. In their approach, the new fractional diffusion coefficient is given as:

$$D_f = D\tau^{1-\beta} \qquad\qquad (4.87)$$

where β is the order of the time-fractional diffusion equation,

D is the classical diffusion coefficient with the dimension $[L^2 T^{-1}]$, and

τ is the new time parameter which accommodates the new dimensions.

The advantage of Kilbas' method is that the new dimensions of the diffusion coefficient is consistent with the time-fractional diffusion equation. With the new models derived from the CTRW theory or other methods, it is evident that the dimensions of the new diffusion coefficient or diffusivity in the fDWE do not have to follow those of the classical counterparts.

Another method to reconcile the dimension in an fPDE is to adapt Compte's approach (1997) which is outlined in sections 2.16 and 3.3, where the new parameters $A = \dfrac{\tau_a}{\tau^\beta}$ and $D = \dfrac{\sigma^2}{\tau^\beta}$ are incorporated in the formulation.

6. Variable-order Fractional Derivatives and Related fPDEs

6.1 Origins of the Ideas

In addition to fPDEs with a constant order of fractional derivatives, the concepts of variable orders of fractional derivation and integration were first investigated by Unterberger and Bokobza (1965a) and Višik and Èskin (1967).

6.1.1 Time-fractional PDEs of Distributed Order

A special form of the variable-order fractional derivative is the distributed-order time-fractional derivative which, according to Gorenflo and Mainardi (2005), was developed by Caputo (1969). Caputo developed distributed-order differential equations with the order of fractional derivatives as a distribution function, say β. Caputo (1995, 2001) and Bagley and Torvik (2000) further developed these ideas. For example, the distributed-order time-fractional diffusion-wave equation (Gorenflo and Mainardi 2005) can be written as:

$$\int_0^1 b(\beta) D_*^\beta c(x,t) d\beta = D \frac{\partial^2 c(x,t)}{\partial x^2} \tag{4.88}$$

with

$$\int_0^1 b(\beta) d\beta = 1 \tag{4.89}$$

wherein the IC is $C(x, 0) = \delta(x)$ with $\delta(x)$ as the Dirac delta function, and $b(\beta) \geq 0$ is a distribution function which determines the order of fractional derivatives.

The upper limit for integration in Eq. (4.88) can be extended to 2 so that the diffusion-wave equation of different forms can be investigated. The integral in Eq. (4.89) for the distributed-order can be approximated by a discrete form:

$$b(\beta) = \sum_{i=1}^n b_i \delta(\beta - \beta_i), \qquad 0 < \beta_1 < \beta_2 \ldots < \beta_i \leq 1 \tag{4.90}$$

where $\delta(\beta - \beta_i)$ are the Kronecker delta functions, a discrete form of the Dirac delta function.

Any number of terms can be chosen from Eq. (4.90) which will be presented in section 6.2 below. The most widely used form of the time distributed-order fPDE in this book is the two-term time-fractional distributed order of the form:

$$b(\beta) = b_1 \delta(\beta - \beta_1) + b_2 \delta(\beta - \beta_1) \tag{4.91}$$

which corresponds to $0 < \beta_1 < \beta_2 < 1$, $b_1 > 0$, $b_2 > 0$, and $b_1 + b_2 = 1$ for the two-term distributed-order fDWE such that:

$$b(\beta_1)\frac{\partial^{\beta_1} c(x,t)}{\partial t^{\beta_1}} + b(\beta_2)\frac{\partial^{\beta_2} c(x,t)}{\partial t^{\beta_2}} = D\frac{\partial^2 c(x,t)}{\partial x^2} \tag{4.92}$$

Equation (4.92) and its extended forms will be discussed in detail in subsequent chapters, in relation to water flow and solute movement in soils and aquifers with dual porosity.

Since the initiation by Unterberger and Bokobza (1965a) and Višik and Èskin (1967), according to Samko (1995, 2013), variable-order or functional-order fractional integration and differentiation were further investigated by Samko and Ross (1993), and functional orders of fractional derivatives were investigated by Jacob and Leopold (1993). Following the introduction of variable-order fractional derivatives, different forms of variable-order fractional derivation and integration have appeared in literature, which can be categorized into the following types:

1. fractional derivative as a function of the space or time (Višik and Èskin 1967);
2. fractional derivative as a function of the space variable, such as $\beta(x)$ (Jacob and Leopold 1993, Samko and Ross 1993, Chechkin et al. 2005, Samko 2013);
3. fractional derivative as a function of time, such as $\beta(t)$ (Umarov and Steinberg 2009, Valério and da Costa 2011);
4. fractional derivation and integration as a function of time and space variables, such as $\beta(x, t)$ (Lorenzo and Hartley 1998, 2002, Zayernouri and Karniadakis 2015);
5. fractional derivative as a function of the material concentration such as moisture content, $\beta(\theta(x, t))$ (Gerasimov et al. 2010, Sun et al. 2014);
6. multivariable-order fractional integration and differentiation (Lorenzo and Hartley 2002), which involve multiple space and time variables; and
7. random orders (Li et al. 2009, Sun et al. 2010, 2011a), where the order $\beta(x, t)$ is a random function such as $\beta(x, t) = \beta_0 + \varepsilon(x, t)$ with β_0 as the deterministic component and $\varepsilon(x, t)$ as the random or stochastic component.

Category 5 by Gerasimov et al. (2010) is based on experimental evidence that as the moisture ratio of a porous media increases, the flow velocity in the media also increases, resulting in a variable order of fractional derivatives in the diffusion equation, hence leading to an increasing flow rate. This finding is intuitive and consistent with Darcy's law while other categories, though mathematically innovative, await experimental verification. The variable-order $\beta(\theta(x, t))$ by Gerasimov et al. (2010) can embrace space and time because the moisture content in an unsaturated medium is a function of both space and time, where a random component can also be added to account for realistic water movement and solute transport in heterogeneous media.

6.1.2 Space-fractional PDEs of Distributed Order

In literature, the time-fractional PDEs of distributed order are reported more frequently. In fact, the spatial fractional derivatives of distributed order are also important. Umarov and Steinberg (2006) reported the space-fractional PDE of distributed order, expressed as:

$$\frac{\partial c(x,t)}{\partial t} = \int_0^2 a(\lambda) D_0^\lambda c(x,t) d\alpha, \qquad\qquad t > 0, \quad x \in R^N \qquad (4.93)$$

where $a(\lambda)$ is a positively defined integrable function. The difference between the space-fractional PDE in Eq. (4.93) and the time-fractional PDE in Eq. (4.88) is noteworthy, and their similarity lies in the distributed functions for the orders of fractional derivatives.

Similar to Eq. (4.90), the integral in Eq. (4.93) for the space distributed order can be approximated by a discrete form as:

$$a(\lambda) = \sum_{k=1}^M a_k \delta(\lambda - \lambda_k), \qquad\qquad 0 < \lambda_1 < \lambda_2 \ldots < \lambda_k \le 2 \qquad (4.94)$$

where $\delta(\lambda - \lambda_k)$ is the Kronecker delta function, which is a discrete Dirac delta function. Any number of terms can be chosen from Eq. (4.94) for the space distributed order. For example, a multi-term space-fractional PDE presented by Umarov and Steinberg (2006) is of the form:

$$\frac{\partial c(x,t)}{\partial t} = \sum_{k=1}^M a_k D_0^{\lambda_k} c(x,t) \qquad (4.95)$$

6.2 More Generic fPDEs of Functional Order

The fPDEs presented for water movement and solute transport in this book have two or more terms of temporal differentiation, accounting for flow and solute motion in two or more levels of pores. The two-term fPDEs for water flow and solute transport in the mobile-immobile (or large-small) pores are compatible with the concept given by Rubinshtein (1948) and further investigated by Barenblatt et al. (1960) and others, and will be discussed more extensively than other forms of distributed-order fPDEs.

In reality, the order of fractional derivatives can be a function of one or more variables such as time, temperature, concentration of substances, space variables, etc. Lorenzo and Hartley (1998) briefly overviewed the gradual change from constant order to variable order of fractional derivatives. They showed that the dependence of viscoelasticity of certain materials on temperature, which was first observed by Bland (1960), can be expressed as a fractional derivative with an order q which varies from $q \cong 0$ for elastic to $q \cong -1$ for viscoelastic or viscous materials. Smit and de Vries (1970) studied the stress-strain relationships of viscoelastic materials such as textile fibers and showed that the order of the fDE, α, describing the stress-strain relationship varied in the range $0 \le \alpha \le 1$. Bagley's (1991) experiments on viscoelastic stress-relaxation in linear polymers also showed the temperature dependence of the order of fractional derivatives, and a similar temperature dependence of the order of fractional derivatives describing the reaction kinetics of proteins was investigated by Glöckle and Nonnenmacher (1995). Based on these experimental evidences, Lorenzo and Hartley (1998) proposed the functional order $q(t)$ wherein t denotes time. Later on, Lorenzo and Hartley (2002) initiated the use of the functional order $q(c, t)$, where c is the dependence variable such as the concentration of the substance or solute, etc.

Chechkin et al. (2005) showed that when the CTRW theory is applied to investigate transport processes in heterogeneous media, a variable-order space-fractional PDE results, which is of the form (Chechkin et al. 2005):

$$\frac{\partial p}{\partial t} = \frac{\partial^2}{\partial x^2}(K(x)D_t^{1-\lambda(x)}p(x,t)), \qquad p(x,0)=\delta(x) \qquad (4.96)$$

where $D_t^{1-\lambda(x)}$ is the generalization of the Riemann-Liouville derivative of the order $1-\lambda(x)=\mu(x)$ defined as:

$$D_t^{\mu(x)} = \frac{1}{\Gamma[1-\mu(x)]}\frac{\partial}{\partial t}\int_0^t \frac{p(x,\tau)}{(t-\tau)^{\mu(x)}}d\tau \qquad (4.97)$$

with $0<1-\lambda(x)\le 1$ or $0<\mu(x)\le 1$.

Gerasimov et al. (2010) investigated moisture absorption in porous media with a variable-order time-fractional PDE in which the variable order, $\beta(\theta)$, is a function of the moisture content, θ. Ricciuti and Toaldo (2017) demonstrated that with the semi-Markov concept (Doeblin 1940, Levy 1954, Smith 1955, Pyke 1961, Cinlar 1969) and the CTRW theory, for holding times as a power-law decaying density with the exponent depending on the state itself, a variable-order fPDE was derived with a suitable limit, which models anomalous diffusion in heterogeneous media. As an example, one of these models for one-dimensional water movement in non-swelling soils can be written as:

$$b_1\frac{\partial^{\beta_1(x)}\theta}{\partial t^{\beta_1}} + b_2\frac{\partial^{\beta_2(x)}\theta}{\partial t^{\beta_2}} = D\frac{\partial^\lambda\theta}{\partial z^\lambda} - \frac{\partial K}{\partial z} \qquad (4.98)$$

where b_1 and b_2 are the relative porosities in immobile and mobile zones respectively and are given by Eqs. (5.68) and (5.69) respectively.

Straka (2018) recently demonstrated that a functional order fPDE can be derived from the CTRW theory. Sun et al. (2019) reviewed related reports on variable-order and functional order fDEs and fPDEs. In addition, Fedotov et al. (2019) also demonstrated the functional order fPDEs for diffusion and diffusion-advection processes with a variable diffusivity and presented a unified asymptotic solution applicable to both diffusion and diffusion-advection processes for large times when the fractional advection process is negligible.

6.3 *Moments of Variable-order fPDEs*

In section 3.4, the moments of fDWEs were given based on Zaslavsky's RGK method. Here, we briefly discuss the moments of fPDEs of variable order. A special case of Eq. (4.88) is the variable time-fractional diffusion-wave equation of the form:

$$\frac{\partial^{\beta(t)}c(x,t)}{\partial t^{\beta(t)}} = D\frac{\partial^2 c(x,t)}{\partial x^2} \qquad (4.99)$$

$$c(x,t)=\delta(x),\ t=0 \qquad (4.100)$$

and the moment of such an fDWE of functional order is given as (Sun et al. 2010):

$$\left\langle x^2(t) \right\rangle = \frac{2Dt^{\beta(t)}}{\Gamma[1+\beta(t)]} \tag{4.101}$$

which is comparable to Eq. (4.53) for fDWEs of fractional order.

7. Summary

This chapter briefly outlines the transition of mathematical approaches developed for classical diffusion to anomalous diffusion. By revisiting some early ideas, methods and models, the different approaches to non-locality and generalized diffusion as well as fPDEs of various orders have been briefly discussed. The topics cover the transition in theory and methodologies from the integer PDEs to fPDEs and fundamental concepts and methodologies for models for water flow and solute transport in porous media, which can be extended to overland flow and poroelasticity to be discussed in subsequent chapters. Together with Chapters 2 and 3, this chapter concludes the preparations for this book while addressing issues which are important for further discussions and investigations in this book.

Chapter 5

Fractional Partial Differential Equations for Water Movement in Soils

1. Introduction

Water is a central concept for describing processes in hydrology, soil science and geomechanics. Water-related phenomena such as infiltration into and evaporation from soil, and run-off on land play crucial roles in the environment by reshaping material and energy distributions. Water carries nutrients, microbes and other forms of solutes to deeper geological strata or moves them upwards during evaporation resulting in salinity in waterways and on land. Water is also the cause of changes in elasticity of soils which determines the strength and the stability of earth foundations and soil structures, which are important in environmental and civil engineering. Infiltration of water into soils, distribution of water in soils, and runoff overland are major concerns in irrigation practices where water may carry nutrient solutes, pollutants and possibly even microbes, including pathogens, across the agricultural landscape. All these water-related processes require quantitative methods of varying levels of complexity to provide guiding information for relevant practitioners.

Since Darcy's quantitative approach to water flow in porous media in 1856, mathematical models based on integer calculus have played a crucial role and generated significant knowledge for the understanding and management of natural resources and for environmental and engineering practices. Findings in physics and mathematics (as well as stochastics) over the past few decades, as outlined in Chapter 4, have encouraged the introduction of methods based on fractional partial differential equations (fPDEs) into other fields, with extensive applications in hydrology and soil science since the 1990s.

The fPDE-based models presented in this chapter cover various water-related topics which include the governing equations of water movement in both non-swelling and swelling soils, water exchange between pores of different properties (such as mobile and immobile zones or variable geometries), coupled water movement and heat transfer in soils, and issues related to these governing equations based on fPDEs. Many of these fPDE-based models are extensions of integer PDEs, resulting in further levels of information which connect the scaling parameters and fractals to the orders of fractional derivatives in time and space or mass (Zaslavsky 2002, Su 2014).

This chapter also discusses some new ideas relevant to these governing equations of water flow and related processes by considering the orders of fractional derivatives as functions of time, space, water content, solute concentration, temperature and other variables.

Applications of the fPDE-based models outlined in this Chapter are presented in Chapter 6 which is focused on the derivation of solutions of distributed-order fFDEs (or dofPDEs), and further development of equations of infiltration into soils in particular.

2. Integer Calculus-based Models for Water Flow in Soils

2.1 Non-isothermal, Incompressible, Laminar Flow in Non-elastic Soils

Fick's law of diffusion as an extension of Fourier's heat theory plays an important role in the analysis of material diffusion resulting from the concentration gradient that drives the diffusion process. However, particle diffusion is a complex process which is affected by different forces, particularly in porous media. Interest in mass transfer induced by thermal gradient has been a topic in physics for over 160 years, at least since the time of W. Thomson (Lord Kelvin) (1842) when heat transfer, electrical conduction and diffusion were considered as mutual and interfering processes. Ludwig (1856) further investigated coupled thermal diffusion in multi-component mixtures (Dhont et al. 2007).

Some important investigations of lasting impact on concurrent thermal transfer and material diffusion are attributed to Dufour (1872, 1873) and Soret (1879, 1880). Dufour found that the occurrence of a heat flux results from the existence of a chemical potential gradient. This is now known as *the Dufour effect* (Postelnicu 2004). As an opposite phenomenon, Soret found that diffusion of chemicals can be caused by a heat gradient. This is now called *the Soret effect* (Mortimer and Eyring 1980, Platten and Costesèque 2004). The Soret effect and the Dufour effect are reciprocal, implying that a unified mathematical model can connect both these effects. This mathematical model was later proposed by Onsager (1931a, b) and is now called the *Onsager reciprocal relationship* (Mortimer and Eyring 1980). More complicated mechanisms related to fluid flow, mass transport, elasticity, deformation, consolidation and reciprocal effects by energy transfer, etc. have been widely investigated in geophysics and chemical engineering (Biot 1941, 1956a, b, c, 1973, Bird et al. 1960, McTigue 1986).

As an attempt to modify Fick's law of diffusion, Soret (1879) presented the following equation, in modern terminology of calculus, for diffusion of salts in fluids subject to a thermal gradient:

$$\frac{\partial q}{\partial t} = -\frac{\partial}{\partial x}\left(\alpha_0 \frac{\partial q}{\partial x} + \beta_0 \frac{\partial T}{\partial x}\right) \tag{5.1}$$

where q represents salt concentration, T is temperature, x denotes distance, t is time, and α_0 and β_0 are functions of q and T. Soret also defined salt mass fraction u related to salt concentration and density ρ as:

$$q = \rho u \tag{5.2}$$

with

$$\alpha_0 = -D \tag{5.3}$$

$$\beta_0 = -\rho D_T u_0 (1 - u_0) = -\rho D_T \frac{q_0}{\rho}\left(1 - \frac{q_0}{\rho}\right) \tag{5.4}$$

where D is a constant related to the third-order gradient $\left(\dfrac{\partial q}{\partial x}\right)^3$, D_T is thermal diffusivity, and the subscripts for q and u denote their initial values.

Similar to the Onsager reciprocal relationship (Mortimer and Eyring 1980), but with terminologies commonly used in hydrology and soil science, Philip and de Vries (1957) proposed the governing equations for concurrent water movement and heat transfer in porous media. Moreover, they, individually, further fine-tuned the approaches (Philip 1957c, de Vries 1958, 1987) by neglecting elasticity, deformation and consolidation. The equations coupling heat transfer and moisture movement given by Philip and de Vries (1957), and Philip (1957c) are of the form:

$$\frac{\partial \theta}{\partial t} = \nabla \cdot (D(\theta) \nabla \theta) + \nabla \cdot (D(T) \nabla T) \pm \frac{\partial K}{\partial z} \tag{5.5}$$

$$C \frac{\partial T}{\partial t} = \nabla \cdot (\lambda \nabla T) + \rho L \nabla \cdot (D_{\theta vap} \nabla \theta) \tag{5.6}$$

where θ is the volumetric moisture content of soil, T is temperature, $D(\theta)$ is the diffusivity of soil water, $D(T)$ is thermal moisture diffusivity having liquid and vapour components, $D_{\theta vap}$ is the vapor moisture diffusivity of soil, K is the unsaturated hydraulic conductivity of soil, C is the volumetric heat capacity of soil, λ is the thermal conductivity of soil, ρ is the density of liquid water, z is the vertical coordinate where the sign $\pm$ depends on the direction of the flow (positive for downward flow and negative for evaporation), and ∇ is the Laplace operator.

de Vries (1987) further considered the hysteresis of soil moisture curves and presented new formulations of the equations for heat flux, microscopic apparent thermal conductivity and microscopic vapor flux density in gaseous phase. Based on Biot's (1941) work, McTigue (1986) developed a comprehensive framework with a set of equations connecting fluid flow, heat transfer, stress, strain and displacement in porous media. The simplified equations of heat flux and fluid flux given by McTigue (with the gravity term included) are identical to Eqs. (5.5) and (5.6).

In one dimension, the above formulations become:

$$\frac{\partial \theta}{\partial t} = \frac{\partial}{\partial z}\left(D(\theta)\frac{\partial \theta}{\partial z}\right) + \frac{\partial}{\partial z}\left(D(T)\frac{\partial T}{\partial z}\right) \pm \frac{\partial K(\theta)}{\partial z} \tag{5.7}$$

and

$$C\frac{\partial T}{\partial t} = \frac{\partial}{\partial z}\left(\lambda\frac{\partial T}{\partial z}\right) + \rho L\frac{\partial}{\partial z}\left(D_{\theta vap}\frac{\partial \theta}{\partial z}\right) \tag{5.8}$$

Similar to the coupled equations of particle movement and heat transfer (in porous media), which account for the Soret and Dufour effects, the Onsager reciprocal relationship with equal phenomenological coefficients (Mortimer and Eyring 1980) also offers a possibility to measure the thermal diffusion coefficient and the water (or solute) diffusion coefficient in Eqs. (5.7) and (5.8) for concurrent thermal transfer and water (or solute) movement in porous media.

2.2 Isothermal, Incompressible, Laminar Flow in Non-elastic Soils

The equation for the movement of soil water under isothermal conditions is the following partial differential equation given by Richards (1931) and rewritten in a compact form in terms of moisture ratio (or content), by Klute (1952) (Philip 1969a, 1974), as:

$$\frac{\partial \theta}{\partial t} = \nabla \cdot (D\nabla \theta) \pm \frac{dK}{d\theta}\frac{\partial \theta}{\partial z} \tag{5.9}$$

or through a functional relationship of matrix potential $\psi(\theta)$, with water content θ, as:

$$\frac{\partial \psi}{\partial t} = \nabla \cdot (K\nabla \psi) \pm \frac{dK}{d\theta}\frac{\partial \psi}{\partial z} \tag{5.10}$$

where diffusivity D and hydraulic conductivity K are related through the following relationship:

$$D = K\frac{d\psi}{d\theta} \tag{5.11}$$

Notice that the sign $\pm$ is used in Eqs. (5.9) and (5.10): when the vertical coordinate is chosen to point downwards, a negative sign is used. Due to the effect of air, both equations are *not completely equivalent* in some cases such as when ψ exceeds the air-entry value and Eq. (5.10) may apply but Eq. (5.9) does not (Philip 1969a).

The one-dimensional form of Richards equation given as Eq. (5.9), with the coordinate being positive downwards is:

$$\frac{\partial \theta}{\partial t} = \frac{\partial}{\partial z}\left(D\frac{\partial \theta}{\partial z}\right) - \frac{\partial K}{\partial z} \tag{5.12}$$

Both D and K are functions of θ and may also be functions of solute concentrations under an isothermal condition.

The one-dimensional forms of Eqs. (5.9) and (5.10) for the horizontal movement of water are Eqs. (5.13) and (5.14) given as:

$$\frac{\partial \theta}{\partial t} = \frac{\partial}{\partial x}\left(D\frac{\partial \theta}{\partial x}\right) \tag{5.13}$$

and

$$\frac{\partial \psi}{\partial t} = \frac{\partial}{\partial x}\left(K \frac{\partial \psi}{\partial x} \right) \tag{5.14}$$

Equation (5.9) or (5.10) is known as Richards equation even though, to a large extent, it is identical to the formulation proposed by Gardner and Widstoe (1921) a decade earlier.

2.3 Isothermal and Laminar Flow in Heterogeneous Soils

Departing from Miller-Miller's similitude concept, Philip (1967) proposed the concept of scale-heterogeneous media with the following relationships for one-dimensional scale-heterogeneous medium:

$$K_a(\theta) = [\lambda(x)]^2 K_*(\theta) \tag{5.15}$$

and

$$\Psi_a(\theta, x) = \frac{\Psi_*(\theta, x)}{\lambda(x)} \tag{5.16}$$

where the spatial variation of medium properties such as hydraulic conductivity $K_a(\theta)$ and capillary potential $\Psi_a(\theta, x)$ are embodied wholly in the spatial variability of $\lambda(x)$. The parameters $K_*(\theta)$ and $\Psi_*(\theta, x)$ are the reduced hydraulic conductivity and capillary potential respectively or the values of $K_a(\theta)$ and $\Psi_a(\theta, x)$ at the surface of the soil as references (Broadbridge 1987, 1988).

For isothermal transport of moisture in heterogeneous media, a one-dimensional Fokker-Planck equation for scale-heterogeneous media, proposed by Philip (1967), is of the form:

$$\frac{\partial \theta}{\partial t} = \frac{\partial}{\partial z}\left[K(\theta, z)\frac{\partial \Psi(\theta, z)}{\partial z} \right] - \frac{\partial}{\partial z} K(\theta, z) \tag{5.17}$$

where $\Psi(\theta, z)$ is the capillary potential as a function of scale and moisture content, and $K(\theta, z)$ is the unsaturated hydraulic conductivity.

2.4 Isothermal, Incompressible, Laminar Flow in Soils Affected by the Flow of Air

Green and Ampt (1911) were best known for their first model of infiltration, but their work on models of air permeability P_a, liquid permeability P_w to soil, and their investigation of swelling properties of soil are not well acknowledged in literature.

Based on Poiseuille's law of flow in capillaries, Green and Ampt (1911) derived air permeability P_a and liquid permeability P_w to soil, as follows:

$$P_a = \frac{l}{Ab(t_2 - t_1)}\ln\left(\frac{h_1}{h_2}\right) = \frac{2.303l}{Ab(t_2 - t_1)}\log_{10}\left(\frac{h_1}{h_2}\right) \tag{5.18}$$

$$P_w = \frac{Vl}{Aht} \tag{5.19}$$

and defined the air-water permeability ratio, which can be called the *Green-Ampt conductivity ratio* (Green and Ampt 1911) as:

$$K_{GA} = \frac{\eta_a P_a}{\eta_w P_w} \tag{5.20}$$

where η_a and η_w are the viscosities of air and water respectively.

Comparing with the hydraulic conductivity in Chapter 3, the Green-Ampt conductivity ratio K_{GA} measures the relative magnitude of conductivity of air and water in a medium, ranging from $K_{GA} = 1$ for sands to 14 or higher for a packed clay soil. To this end, K_{GA} is an indicator of swelling-shrinking properties of soil.

Two-phase flow of water and air in unsaturated porous media is a process that occurs in many environments. Typical examples of two-phase flow include water-oil or oil-gas displacement in natural petroleum reservoirs (Morel-Seytoux 1973, Barenblatt et al. 1990); material transport in the environment (Morel-Seytoux 1973, Miller et al. 1998); water and air flow in natural unsaturated soils (Green and Ampt 1911); water and air flow in soils during conventional flood irrigation (Dixon and Linden 1972); and water and air (or oxygen) movement in soils during aerated subsurface drip irrigation or oxygation (Su and Midmore 2005, Bhattarai et al. 2005, 2006). Examples of this type of flow patterns common in engineering practices include oil and gas flow from or to point or line sources/sinks, soil water flow during irrigation, aquifer tracer testing and well hydraulics (Chrysikopoulos et al. 1990), and parameter determination (Pickens et al. 1981).

Since the pioneering work by Green and Ampt (1911), the mechanics of two-phase flow was identified as an important topic and subsequent experimental (Muskat and Meres 1936, Wyckoff and Botset 1936) and theoretical (Muskat and Meres 1936, Leverett 1939) investigations followed. Multi-phase flow was also identified as an important issue.

According to Chen (1988), the first classic model of multi-phase flow was proposed by Leverett (1941) and its simplification was made by Buckley and Leverett (1942). The reader is referred to McWhorter (1971), Morel-Seytoux (1973), Chen (1988), Barenblatt et al. (1990), McWhorter and Sunada (1990, 1992) and Miller et al. (1998) for background reviews on studies of generic two-phase and multi-phase flow. A brief review of the examples of analytic investigations of two-phase flow for concurrent flow of water and air in porous media can be found in the works of Su (2009b).

The equations governing two-phase flow in porous media were written as follows by McWhorter (1971), Morel-Seytoux (1973), Chen (1988), and McWhorter and Sunada (1990):

$$q_a = -\frac{kk_{ra}}{\mu_a}(\nabla p_a + \rho_a g) \tag{5.21}$$

$$q_w = -\frac{kk_{rw}}{\mu_w}(\nabla p_w + \rho_w g) \tag{5.22}$$

$$\phi \frac{\partial S}{\partial t} = -\nabla q_w \tag{5.23}$$

and

$$\phi \frac{\partial}{\partial t}(\rho_a S) = \nabla \cdot (\rho_a q_a) \tag{5.24}$$

where

S is liquid saturation defined as:

$$S = \frac{\theta - S_r}{S_s - S_r} \tag{5.25}$$

with S_s and S_r being the saturated and the residual values of S; θ is the liquid content; q_a and q_w are the volumetric flow rates (per unit of area) of the gas and the liquid; k is the saturated permeability; k_{ra} and k_{rw} are the relative permeabilities (the ratio of effective permeability to saturated permeability) of the gas and the liquid; p_a and p_w are partial pressures of the gas and the liquid; ρ_a and ρ_w are the densities of the gas and the liquid; μ_a and μ_w are the dynamic viscosities of the gas and the liquid; ϕ is the porosity of the medium; g is the acceleration due to gravity, and t denotes time.

Sander et al. (1988) rewrote McWhorter's (1971) formulation of the one-dimensional equation of flow of water affected by air flow, which, with notation more familiar in hydrology literature, is of the form:

$$\frac{\partial \theta}{\partial t} + \frac{\partial q}{\partial z} = 0 \tag{5.26}$$

where

$$q = -D(\theta, r)\frac{\partial \theta}{\partial z} + CK(\theta) + q_t f_w(\theta) \tag{5.27}$$

with

$$q_t = q_w + q_a \tag{5.28}$$

$$C = 1 - \rho_a/\rho_w \tag{5.29}$$

Here, $D(\theta)$ is the two-phase diffusivity and $K(\theta)$ is the two-phase unsaturated hydraulic conductivity; $D(\theta)$ and $K(\theta)$ are related to the one-phase diffusivity D_w (water only) and unsaturated hydraulic conductivity K_w by the following relationships:

$$D(\theta) = D_w(1 - f_w) \tag{5.30}$$

$$K(\theta) = K_w(1 - f_w) \tag{5.31}$$

where f_w is the fractional flow function defined as (McWhorter 1971):

$$f_w = \left(1 + \frac{k_{ra}\mu_w}{k_{rw}\mu_a}\right)^{-1} \tag{5.32}$$

or

$$f_w = \frac{K_w}{K_w + K_a} \tag{5.33}$$

where K_w and K_a are the unsaturated conductivities of water and air respectively. For concurrent water and air flow, $C \approx 1$ in Eqs. (5.27) and (5.29). Hence, Eqs. (5.26) and (5.27) can be written as a usual convection-dispersion equation:

$$\frac{\partial \theta}{\partial t} = \frac{\partial}{\partial z}\left[D(\theta)\frac{\partial \theta}{\partial z} \right] - \frac{\partial}{\partial z}[K(\theta) + q_t f_w(\theta)] \tag{5.34}$$

Equation (5.34) in Cartesian coordinates has been extensively investigated in the context of concurrent flow of water and air (see Su 2009b, Su and Midmore 2005 for brief reviews). Its counterpart in radial coordinates, governing the two-phase radial flow of two immiscible, incompressible fluids in porous media, is given as follows (Chen 1988, Sander et al. 2005):

$$\phi\frac{\partial S}{\partial t} = \frac{\partial}{r\partial r}\left[D(S)r\frac{\partial S}{\partial r} - q_t f_w \right] \tag{5.35}$$

where

$$D(S) = \frac{k_{ra}f_w}{\mu_a}\frac{dP_c}{dS} \tag{5.36}$$

with P_c denoting capillary pressure and r being radial distance. Weeks et al. (2003) analyzed Eq. (5.35) in the context of two-phase flow by choosing the two functions in Eq. (5.35) to be:

$$D(S) = D_1 S^a \tag{5.37}$$

and

$$f_w = S^{na/2+1} \tag{5.38}$$

where D_1 and a are constants determined from flow properties.

2.5 Water Flow Affected by Solute and Energy Gradients in Soils

The effect of solutes on hydraulic parameters in models for water movement in soils is an important issue and was investigated a few decades ago. Significant reductions in saturated hydraulic conductivity K_s, resulting from variations in ions or solutes in soil, have also been reported in literature. For example, reductions in the saturated hydraulic conductivity up to 50 percent (or even more) were detected (Suarez et al. 1984, Suarez 1985) which were also analyzed by Nielsen et al. (1986), who envisaged significant potential in the understanding of relevant physical and chemical processes which affect hydraulic processes.

Bear (1972) further expanded the idea of including the solute effect on moisture diffusivity, thermal diffusivity and hydraulic conductivity in the models by Philip and de Vries (1957) and Philip (1957c) (see section 2.1 in this chapter). By extending Bear's suggestion to the flow of moisture, the following expression result:

$$\frac{\partial \theta}{\partial t} = \nabla \cdot (D_{(\theta,T,C)}\nabla \theta) + \nabla \cdot (D_{(\theta,T,C)}\nabla T) \pm \frac{\partial}{\partial z}K_{(\theta,T,C)} \tag{5.39}$$

for water flow subject to solute and thermal effects, and the following:

$$C\frac{\partial T}{\partial t} = \nabla \cdot (\lambda_{(\theta,T,C)}\nabla T) + \rho L \nabla \cdot (D_{\theta vap(\theta,T,C)}\nabla \theta) \qquad (5.40)$$

for heat transfer subject to water and solute effects.

A set of PDEs presented by Miller et al. (1998) can be used to describe the interdependence of the concerned quantities and parameters in multi-phase flow in porous media.

2.6 Equations Governing Water Movement in Swelling Soils

In literature, it has been generally acknowledged that the concept of flow in swelling soils was formally conceived by Terzaghi (1923) and the shrinkage curve concept was developed by Haines (1923) to characterize swelling agricultural soils. As was seen in section 2.4 of this Chapter, over a decade earlier, Green and Ampt (1911) defined conductivity ratio K_{GA} which characterizes the swelling properties of porous media and varies from 1 for sands and 14 or higher for clays. Before the quantitative analysis by Green and Ampt, measurement of volumetric change was attempted in the early 19th century by Schübler, as was reported by Tempany (1917), which indicates that the swelling properties of soils were already being appreciated by that time (Smiles and Raats 2005).

Terzaghi (1923) developed the concept of effective stress and material coordinates defined in terms of distribution of solid, which was discarded in his later work (Terzaghi 1956) due to difficulties associated with its understanding for application purposes (Smiles 2000c). The equations governing the flow of liquids in swelling media, formulated in a material coordinate, were developed in the 1960s (Raats 1965, Raats and Klute 1968), and in particular, a form of flow equations was presented in the material coordinate which was based on an integral transform of the moisture ratio (Smiles and Rosenthal 1968, Philip and Smiles 1969, Philip 1969b). This initiation was followed by a short period of intensive studies on the topic of flow in swelling porous media (Philip 1969c, 1969d, 1970a, b, c, 1972, 1992a, Philip and Smiles 1969, Smiles 1974, 2000a, b, Talsma and van der Lelij 1976, Broadbridge 1990, Karalis 1992, Smiles and Raats 2005).

Since the 1960s, after the equations of flow in swelling porous media were presented, methods based on modified theories of flow in rigid media, and their applications to swelling media continued being reported (Giráldez and Sposito 1978, 1985, Giráldez et al. 1983, Baveye et al. 1989, Garnier et al. 1997, 1998, Camporese et al. 2006, Romkens and Prasad 2006). For more complete literature reviews of topics related to flow in swelling soils, the reader is referred to Smiles (2000c) and Smiles and Raats (2005).

Green and Ampt (1911) measured and quantified the permeability of air and water in swelling soils as the first of such studies in soil science and hydrology. They investigated the concurrent flow of air and water in different soils and derived expressions for air permeability P_a, liquid permeability P_w to the soil, and their ratio K_{GA} for quantitative description of flow of air and water in different soils with varying textures. To further the quantitative analysis of flow in swelling-shrinking media, the material coordinate m is defined by the following integral (Smiles and Rosenthal 1968, Philip 1969b):

$$m = \int_0^z (1+e)^{-1} dz_1 \tag{5.41}$$

where z is the conventional space coordinate in the vertical direction and e is the void ratio given by $e = \theta/(1 - \theta)$ with θ being the moisture ratio defined as $\theta = \theta_l/\theta_s$ wherein θ_l and θ_s are the volume fractions of liquid and solid respectively.

With the material coordinate, the Richards equation in one dimension for water movement in rigid soils is converted into the following equation for water movement in swelling soils, which is credited to several investigators (Smiles and Rosenthal 1968, Philip 1969b, Smiles and Raats 2005):

$$\frac{\partial \theta}{\partial t} = \frac{\partial}{\partial m}\left(D_m(\theta) \frac{\partial \theta}{\partial m} \right) - \frac{\partial K_m(\theta)}{\partial m} \tag{5.42}$$

where

$$D_m(\theta) = \frac{K_m(\theta)}{1+\theta} \frac{d\Phi}{d\theta} \tag{5.43}$$

with Φ as the unloaded matrix potential and $K_m(\theta)$ being the unsaturated material hydraulic conductivity defined as (Smiles and Raats 2005):

$$K_m(\theta) = \frac{k_m}{\theta_s}(\gamma_n \alpha - 1) \tag{5.44}$$

Here, k_m is the saturated material hydraulic conductivity expressed as:

$$k_m = K(\theta)\theta_s \tag{5.45}$$

with $K(\theta)$ as the conventional unsaturated hydraulic conductivity.

3. Fractional Calculus-based Models for Water Movement in Soils

Based on the background on water movement in soils in section 2 and published reports on several forms of fPDEs for water movement in soils (Su 2010, 2012, 2014), this section further discusses fPDEs for water movement in soils.

3.1 Fractional Models for Water Movement in Soils Resulting from Stochastic Processes

For anomalous flow in soils, different approaches can be used to derive the associated governing equations. One approach is based on the equations for conservation of mass and momentum. The examples of this approach are Eq. (5.9) and Eq. (5.12). The second approach is based on the stochastic process of water movement or particle motion in porous media, which is a consequence of continuous-time random walks (CTRW) of particles or water parcels in soil as described by Su (2014) and outlined in Chapter 4. For analyzing anomalous flow, other reports introduce an empirical exponent in the diffusion equation formulated for rigid media (Küntz and Lavallee 2001, Lockington and Parlange 2003, Gerolymatou et al. 2006, Valdes-Parada et al. 2007).

Background information on fPDEs, the CTRW theory, and their applications in hydrology and soil science can be found in literature (Zhang et al. 2009, Su 2014, Benson et al. 2013, Su 2017a). This section is dedicated to fPDEs for water movement in soils, which can incorporate a number of important issues such as flow in soils with swelling properties, effects of thermal and solute concentration, heterogeneity, coupled transport of water and solute, coupled water movement and heat transfer. The new formulations explain different anomalous mechanisms of fluid dynamics in natural porous media by including space-time and mass-time fractional derivatives in PDEs.

3.2 Mass-time and Space-time fPDEs for Anomalous Water Flow in Soils

3.2.1 Fractional Diffusion-wave Equation for Water Movement in Non-swelling Soils

In Chapter 4, the CTRW theory was applied to derive the fractional diffusion-wave equation (fDWE) for water movement in non-swelling soils. The equation was of the following form:

$$\frac{\partial^\beta \theta(x,t)}{\partial t^\beta} = D \frac{\partial^\lambda \theta(x,t)}{\partial x^\lambda} \tag{5.46}$$

where D denotes diffusivity. Zaslasvky (2002) established the following relationship which connects the orders of spatio-temporal fractional derivatives, λ and β, and spatio-temporal scaling parameters, L_s and L_t, as:

$$\frac{\beta}{\lambda} = \frac{\ln L_s}{\ln L_t} = \frac{\mu}{2} \tag{5.47}$$

which also implies that the transport exponent, μ, expressed as:

$$\mu = \frac{2 \ln L_s}{\ln L_t} \tag{5.48}$$

or

$$\mu = \frac{2\beta}{\lambda} \tag{5.49}$$

can be used to identify the types of diffusion: $\mu < 1$ for sub-diffusion, $\mu = 1$ for classical diffusion, and $\mu > 1$ for super-diffusion (Zaslavsky 2002).

The moment equation for Eq. (5.46), with D as a constant, is given as (Zaslavsky 2002):

$$\left\langle x^\lambda \right\rangle = \frac{\Gamma(1+\lambda)}{\Gamma(1+\beta)} D t^\beta \tag{5.50}$$

which represents ensemble averaging of distance or displacement, x, over repeated observations. The ensemble mean in Eq. (5.50) is a very important result derived from an instantaneous input or a Dirac delta function input (Saichev and Zaslavsky 1997) and can be used to estimate the model parameters and measure the distance between the initial reference and the diffusion front.

For a more generic anomalous diffusivity in different dimensions, Uchaikin and Saenko (2003) showed that CTRW in the case of anomalous diffusion for n dimensions yields Eq. (4.33), that is:

$$D = \frac{\Gamma(n/2)\Gamma(1-\lambda/2)A}{2^{\lambda}\Gamma[(\lambda+n)/2]\Gamma(1-\beta)B}, \qquad n = 1,2,3 \tag{5.51}$$

which, in one dimension, becomes Eq. (4.40):

$$D = \frac{\pi^{1/2}\Gamma(1-\lambda/2)A}{2^{\lambda}\Gamma[(\lambda+1)/2]\Gamma(1-\beta)B} \tag{5.52}$$

and D in two ($n = 2$) and three ($n = 3$) dimensions are given by Eqs. (4.41) and (4.42) respectively.

3.2.2 fPDEs for Water Movement with Diffusion and Convection in Non-swelling Soils

The approach with the CTRW theory can be also applied to derive an fPDE for water movement subject to convection, with a shift jump size distribution $(f^{*}(x - V_t)) = f(x)$ in one dimension, which results in the following fPDE:

$$\frac{\partial^{\beta}\theta}{\partial t^{\beta}} = D\frac{\partial^{\lambda}\theta}{\partial x^{\lambda}} - \frac{\partial K(\theta)}{\partial x} \tag{5.53}$$

For water movement in higher dimensions, Eq. (5.53) is extended with $(f^{*}(\mathbf{x} - V_t)) = f(\mathbf{x})$ in multiple dimensions as (Zhang et al. 2009):

$$\frac{\partial^{\beta}\theta}{\partial t^{\beta}} = \frac{\partial}{\partial x_i}\left(\mathbf{D}_0\frac{\partial^{\eta}\theta}{\partial x_i^{\eta}}\right) - \frac{\partial K(z,\theta)}{\partial z} \tag{5.54}$$

3.2.3 fPDEs for Water Movement in Swelling Soils

fPDEs incorporating both swelling properties and anomalous flow in soils have been proposed a few years ago (Su 2009a, 2010, 2012, 2014), with the aid of material coordinates (Smiles and Rosenthal 1968, Philip 1969b) for fractional-order models. One of the simplest fPDEs for water movement in swelling soils is written as:

$$\frac{\partial^{\beta}\theta}{\partial t^{\beta}} = D_m\frac{\partial^{\lambda}\theta}{\partial m^{\lambda}} \tag{5.55}$$

which results from an analogue of the long-time result of CTRWs without convection in soils. The introduction of material coordinates accounts for anomalous flows in swelling soils, which is not accommodated by classical CTRW models.

For flow in swelling soils, a relationship similar to Eq. (5.47) holds as follows (Su 2014):

$$\frac{\beta}{\lambda} = \frac{\ln L_m}{\ln L_t} = \frac{\mu}{2} \tag{5.56}$$

where L_m is the mass scaling parameter, and the transport exponent is expressed as:

$$\mu = \frac{2\beta}{\lambda} \tag{5.57}$$

which establishes the connection among β, λ and the scaling parameters for swelling soils. Equation (5.56) yields another identity with the transport exponent and the mass and time scaling parameters as:

$$\mu = \frac{2\ln L_m}{\ln L_t} \tag{5.58}$$

The moment equation for flow in swelling media is given as:

$$\left\langle m^\lambda \right\rangle = \frac{\Gamma(1+\lambda)D}{\Gamma(1+\beta)} t^\beta \tag{5.59}$$

which is the mass ensemble mean. The form of diffusivity in Eq. (5.59) is identical to that in Eq. (5.50). However, their values determined by the parameters A and B in Eq. (5.52) and the orders of fractional derivatives in time and mass are specific for swelling media.

3.2.4 fPDEs for Water Movement in Swelling Soils with Diffusion and Convection

Similar to the idea of extending Eq. (5.46) to the form in Eq. (5.53) to account for convection, Eq. (5.55) is based on the CTRW theory for flow in swelling soils. Furthermore, when convection is present, Eq. (5.55) is extended to form the fPDE of the form:

$$\frac{\partial^\beta \theta}{\partial t^\beta} = D_m \frac{\partial^\lambda \theta}{\partial m^\lambda} - \frac{\partial K_m(\theta)}{\partial m} \tag{5.60}$$

with a shift jump size distribution $(f^*(m - V_t)) = f(m)$ in one dimension.

Unlike Eq. (5.54), which is in three dimensions for non-swelling soils, Eq. (5.60) considers swelling only in the vertical direction so that no extension into higher dimensions is permitted.

3.3 fPDEs for Anomalous Water Diffusion-Convection in Heterogeneous Soils

For water flow in non-swelling soils with heterogeneity, the one-dimensional form of Eq. (5.54) may be modified as follows:

$$\frac{\partial^\beta \theta}{\partial t^\beta} = \frac{\partial^\lambda}{\partial z^\lambda}(D(z,\theta)\theta) - \frac{\partial K(z,\theta)}{\partial z} \tag{5.61}$$

which can be further extended into three dimensions as (Su 2014):

$$\frac{\partial^\beta \theta}{\partial t^\beta} = \frac{\partial^\lambda}{\partial x_i^\lambda}(\mathbf{D_0}(x_i, \theta)\theta) - \frac{\partial K(z, \theta)}{\partial z} \tag{5.62}$$

When diffusivity and hydraulic conductivity are functions of moisture content only, Eq. (5.61) can be written as:

$$\frac{\partial^\beta \theta}{\partial t^\beta} = \frac{\partial^\lambda}{\partial z^\lambda}(D(\theta)\theta) - \frac{dK(\theta)}{d\theta}\frac{\partial \theta}{\partial z} \tag{5.63}$$

where the term $\dfrac{dK(\theta)}{d\theta} = V$ is convection velocity. When the function $K(\theta)$ is known, velocity is readily known as $V = \dfrac{dK(\theta)}{d\theta}$.

For water movement in swelling soils, the governing equation similar to Eq. (5.61) is:

$$\frac{\partial^\beta \theta}{\partial t^\beta} = \frac{\partial^\lambda}{\partial m^\lambda}(D_m(m, \theta)\theta) - \frac{\partial K(m, \theta)}{\partial m} \tag{5.64}$$

and when diffusivity and hydraulic conductivity are functions of moisture ratio only, Eq. (5.64) is simplified as:

$$\frac{\partial^\beta \theta}{\partial t^\beta} = \frac{\partial^\lambda}{\partial m^\lambda}(D_m(\theta)\theta) - \frac{\partial K(\theta)}{\partial m} \tag{5.65}$$

3.4 Distributed-order fPDEs for Water Flow in Soils with Mobile and Immobile Zones

The fPDEs discussed so far regard the movement of water in soils as flow in a uniform medium, without distinguishing among pores of different sizes. In fact, pore sizes in soils vary with different size distributions, depending upon the type of soil. In this section, distributed-order fPDE (dofPDEs) are discussed for water movement in soils to account for water flow at different levels of pore size.

Here, both non-swelling and swelling soils are considered to have properties of dual porosity by virtue of which the fluid moves faster in larger mobile pores and slowly (or is stagnant) in relatively immobile pores. dofPDEs are discussed as more general equations for the movement of water in soils. This approach provides a mathematical representation of anomalous diffusion at two scales in soils with mobile-immobile zones or large-small pores.

3.4.1 Two-term Distributed-order fPDEs for Flow in Non-swelling Soils

These dofPDEs are based on the concept of water flow and transport in dual porous media with mobile and immobile zones. The concept of dual porous media and related mathematical models were first proposed Rubinshtein (1948) over seven decades ago and investigated by others subsequently (Barenblatt et al. 1960, Nielsen and Biggar 1961, Deans 1963, Coats and Smith 1964, Villermaux and van Swaay

1969, Passioura 1971, van Genuchten and Wierenga 1976, Gaudet et al. 1977, de Smedt and Wierenga 1979, Nielsen et al. 1986, Bond and Wierenga 1990, Gerke and van Genuchten 1993, Griffioen et al. 1998, Haggerty and Gorelick 1995, Haggerty et al. 2004, Gao et al. 2009, 2010, Khuzhayorov et al. 2010).

The mobile-immobile models have been investigated for both single-rate transfer (most of the studies) and multiple-rate mass transfer (Haggerty and Gorelick 1995, Haggerty et al. 2004) between the two regions. Gao et al. (2010) studied scale-dependent reactive solute transport in porous media with mobile-immobile zones.

The mobile-immobile concept has been applied to both unsaturated and saturated media (Culkin et al. 2008, Khuzhayorov et al. 2010) and tested using outflow methods or non-destructive methods such as nuclear magnetic resonance (Culligan et al. 2001). This concept has also been used to describe solute transport in different forms of porous media or under different flow conditions such as density-driven flow (Starr and Parlange 1976), flow in macropores (Seyfried and Rao 1987), flow in randomly heterogeneous media (Barry and Sposito 1989) and in stratified porous media (Li et al. 1994).

The interesting properties and the preliminary success of dofPDEs in modelling solute transport and water flow in porous media (Schumer et al. 2003b, Su 2014, 2017a) justify the further investigation of dofPDEs in water flow and solute transport in porous media. In this section, only two terms in the multi-term fPDEs are retained, that is, $\varphi\left(D_t^\beta + \sum_{i=1}^{s} a_i D_t^{\beta_i}\right)(x,t)$ in the general form of the fPDE for water movement.

In this case, for water movement in no-swelling soils with mobile-immobile zones, Eq. (5.63) is extended to the following fPDE based on fractional gradient (Su 2014):

$$b_1 \frac{\partial^{\beta_1}\theta}{\partial t^{\beta_1}} + b_2 \frac{\partial^{\beta_2}\theta}{\partial t^{\beta_2}} = \frac{\partial}{\partial z}\left[D\frac{\partial^\eta\theta}{\partial z^\eta}\right] - \frac{dK(\theta)}{d\theta}\frac{\partial\theta}{\partial z} \tag{5.66}$$

or the following equation, based on the fractional derivatives of the gradient:

$$b_1 \frac{\partial^{\beta_1}\theta}{\partial t^{\beta_1}} + b_2 \frac{\partial^{\beta_2}\theta}{\partial t^{\beta_2}} = \frac{\partial^\eta}{\partial z^\eta}\left(D(\theta)\frac{\partial\theta}{\partial z}\right) - \frac{dK(\theta)}{d\theta}\frac{\partial\theta}{\partial z} \tag{5.67}$$

Here, b_1 and b_2 are the relative porosities in immobile and mobile zones respectively, that is, $b_1 = \frac{\phi_{im}}{\phi}$ and $b_2 = \frac{\phi_m}{\phi}$, with ϕ_{im}, ϕ_m and ϕ being porosities in the immobile zone, the mobile zone, and total porosity respectively. With the new dimensions introduced as a result of fractional differentiation, b_1 and b_2 can be written as:

$$b_1 = \frac{\phi_{im}}{\phi}\tau^{\beta_1 - 1} \tag{5.68}$$

$$b_2 = \frac{\phi_m}{\phi}\tau^{\beta_2 - 1} \tag{5.69}$$

The parameters are set to be

$$
\left.\begin{array}{ll}
0 < \beta_1 \le 2; & 0 < \beta_2 \le 2 \\
b_1 + b_2 = 1; & 0 < \eta \le 1
\end{array}\right\}
\tag{5.70}
$$

3.4.2 The Moment Equation, Ensemble Mean, Scaling and Asymptotic Properties of the Moment

Based on the scaling parameters and moment equations for diffusion and convection in previous sections, and the works of Chechkin et al. (2002) and Sandev et al. (2015), as a special case with $\eta = 1$ (without convection), the moment equation for the distributed-order fPDE in Eq. (5.66), with constant D, becomes:

$$
\left\langle x^2 \right\rangle = \frac{2D}{b_2} t^{\beta_2} E_{\beta_2 - \beta_1, \beta_2 + 1}\left(-\frac{b_1}{b_2} t^{\beta_2 - \beta_1} \right)
\tag{5.71}
$$

where $E_{\beta_2 - \beta_1, \beta_2 + 1}\left(-\dfrac{b_1}{b_2} t^{\beta_2 - \beta_1} \right)$ is the Mittag-Leffler function.

The moment equation in Eq. (5.71) for the distributed-order fPDE has the following asymptotic properties for $t \to 0$ and $t \to \infty$, through an extension of the analysis by Chechkin et al. (2002):

$$
\left\langle x^2 \right\rangle \approx \frac{2D}{\Gamma(1 + \beta_2) b_2} t^{\beta_2} \propto t^{\beta_2}
\qquad \text{for } t \to 0
\tag{5.72}
$$

and

$$
\left\langle x^2 \right\rangle \approx \frac{2D}{\Gamma(1 + \beta_1) b_1} t^{\beta_1} \propto t^{\beta_1}
\qquad \text{for } t \to \infty
\tag{5.73}
$$

The above asymptotic solutions imply that large pores are more important at the early stage of water flow, whereas small pores are more important at large times.

For uniform pores with no distinction of mobile and immobile zones, Eq. (5.71) has the following asymptotic properties (Chechkin et al. 2002):

$$
\left\langle x^2 \right\rangle \approx 2Dt \ln\frac{1}{t}
\qquad \text{for } t \to 0
\tag{5.74}
$$

and

$$
\left\langle x^2 \right\rangle \approx 2D \ln t
\qquad \text{for } t \to \infty
\tag{5.75}
$$

where ln is natural logarithm. These asymptotic results are useful for examining the trend of flow in soils and for evaluating the model parameters, given observation data on the positions over time.

3.4.3 Two-term Distributed-order fPDE for Water Flow in Swelling Soils with Mobile and Immobile Zones

In this case, the governing equation for flow in swelling soils, in terms of fractional gradient, is of the form (Su 2014):

$$b_1 \frac{\partial^{\beta_1}\theta}{\partial t^{\beta_1}} + b_2 \frac{\partial^{\beta_2}\theta}{\partial t^{\beta_2}} = \frac{\partial}{\partial m}\left[D_m(\theta)\frac{\partial^{\eta}\theta}{\partial m^{\eta}} \right] - (\gamma_n\alpha - 1)\frac{\partial K_m(\theta)}{\partial m} \tag{5.76}$$

and in terms of the fractional derivatives of the gradient, it is

$$b_1 \frac{\partial^{\beta_1}\theta}{\partial t^{\beta_1}} + b_2 \frac{\partial^{\beta_2}\theta}{\partial t^{\beta_2}} = \frac{\partial^{\eta}}{\partial m^{\eta}}\left[D_m(\theta)\frac{\partial\theta}{\partial m} \right] - (\gamma_n\alpha - 1)\frac{\partial K_m(\theta)}{\partial m} \tag{5.77}$$

where

$$D_m(\theta) = \frac{K_m(\theta)}{1+\theta}\frac{d\Phi}{d\theta} \tag{5.78}$$

with Φ being the unloaded matrix potential and $K_m(\theta)$ being the unsaturated material hydraulic conductivity defined as (Smiles and Raats 2005):

$$K_m(\theta) = \frac{k_m}{\theta_s}(\gamma_n\alpha - 1) \tag{5.79}$$

where k_m is the saturated material hydraulic conductivity (Smiles and Raats, 2005) given as:

$$k_m = K(\theta)\theta_s \tag{5.80}$$

The moment equation or the assemble mean for flow in swelling soils for the special case without convection is expressed as:

$$\langle m^2 \rangle = \frac{\Gamma(1+\lambda)D_m}{b_2}t^{\beta_2}E_{\beta_2-\beta_1,\beta_2+1}\left(-\frac{b_1}{b_2}t^{\beta_2-\beta_1} \right) \tag{5.81}$$

The special cases and asymptotic properties of Eq. (5.71) discussed for non-swelling soils, with different forms of λ, β_1 and β_2, apply to swelling soils as well when x is replaced with m and D with D_m.

3.4.4 Special Cases of fPDEs with $D_m(\theta)$ and $K_m(\theta)$ as Power Functions

Whereas diffusivity and hydraulic conductivity have been defined above, in practice, empirical formulae are often used to relate diffusivity and hydraulic conductivity with moisture ratio. In particular, power functions are often used for expressing diffusivity and hydraulic conductivity (Philip 1992b) with diffusivity being of the form:

$$D_m(\theta) = D_0\theta^b \tag{5.82}$$

and hydraulic conductivity expressed as:

$$K_m(\theta) = K_0\theta^k \tag{5.83}$$

where D_0, b, K_0 and k are constants which need to be determined experimentally.

As an example of the application of Eqs. (5.82) and (5.83), Eq. (5.77) based on fractional derivatives of moisture gradient is written as:

$$b_1 \frac{\partial^{\beta_1}\theta}{\partial t^{\beta_1}} + b_2 \frac{\partial^{\beta_2}\theta}{\partial t^{\beta_2}} = \left(\frac{D_0}{b+1}\right)\frac{\partial^{\eta}}{\partial m^{\eta}}\left[\frac{\partial\theta^{b+1}}{\partial m} \right] - (\gamma_n\alpha - 1)K_0k\theta^{k-1}\frac{\partial\theta}{\partial m} \tag{5.84}$$

Equation (5.84) can also be derived from the CTRW theory with distributed-order time fractional derivatives resulting from the two time-scaling property and convection due to a shift jump size distribution (Zhang et al. 2009). The relationship between CTRW and the distributed-order fDWE was established by Chechkin et al. (2002), justifying the applicability of Eq. (5.84) as an extension of the CTRW theory and fPDE to water movement in soils. Similar principles apply to both swelling and non-swelling soils.

Note that different terminologies appear in literature for flow patterns when the orders of fractional derivatives, β_1 and β_2, take different values. Our case here with $0 < \beta_1 \leq 2$ and $0 < \beta_2 \leq 2$ is similar to the concept of a decoupled CTRW (Meerschaert et al. 2010) in the two time-scaling case. It is also similar to the alternative derivation by Zaslavsky (2002), where β_1 and β_2 were not restricted. In the above formulation, the relationships $b_1 < b_2$ and $\beta_1 < \beta_2$ hold as there is no need to distinguish between mobile and immobile zones for $\beta_1 = \beta_2$. For $0 < \beta_1 \leq 2$ and $0 < \beta_2 \leq 2$, the fPDE can be solved using integral transform methods (Debnath 2003a, Mainardi et al. 2001) in fluid mechanics (in addition to numerical methods).

Two-term distributed-order fPDEs are special forms of variable-order fPDEs. According to Gorenflo and Mainardi (2005), it was Caputo (1969) who developed the idea of distributed-order differential equations with a distribution function being used for β. Algebraic functions have also been reported for β (Jacob and Leopold 1993, Lorenzo and Hartley 2002). Here, the two-term distributed-order fPDE (Gorenflo and Mainardi 2005) is used to model different flow patterns in two levels of pores. An alternative to this approach is to use different values for the fractional orders (Hahn and Umarov 2011) at different stages of changing diffusion patterns, with one stage described by β_1 and the next by β_2.

With respect to the forms for diffusivity and hydraulic conductivity, Philip (1960a, b) reported a very large set of diffusivity functions that yield exact solutions of a concentration-dependent diffusion equation. With those different functions for diffusivity, different solutions can be derived for Eq. (5.84) for swelling soils and Eq. (5.66) for non-swelling soils. For example, Gerolymatou et al. (2006) used an exponential function for diffusivity in the fractional diffusion equation to model water absorption.

Equation (5.84) is specifically discussed here because diffusivity as a power function bridges the diffusion equation and the non-linear porous media equation (Vazquez 2007) characterized by the term $\dfrac{\partial}{\partial m}\left[\dfrac{\partial \theta^{b+1}}{\partial m}\right]$. Due to the functional relationship between diffusivity and hydraulic conductivity, hydraulic conductivity is also a power function when Eq. (5.84) is used for diffusivity.

As a counterpart of the space fractional derivative in terms of Riemann-Liouville fractional derivatives (Voller 2014), the following identity for fractional mass derivatives (Podlubny 1999):

$$\frac{\partial^{\eta}}{\partial m^{\eta}}\left[\frac{\partial \theta^{b+1}}{\partial m}\right] = \frac{\partial^{\eta+1}\theta^{b+1}}{\partial m^{\eta+1}} \tag{5.85}$$

enables Eq. (5.84) to be written as:

$$b_1 \frac{\partial^{\beta_1}\theta}{\partial t^{\beta_1}} + b_2 \frac{\partial^{\beta_2}\theta}{\partial t^{\beta_2}} = \frac{D_0}{b+1}\frac{\partial^{\lambda}\theta^{b+1}}{\partial m^{\lambda}} - (\gamma_n\alpha - 1)K_0 k\theta^{k-1}\frac{\partial\theta}{\partial m} \tag{5.86}$$

with

$$\left.\begin{array}{l} 0 < \beta_1 \le 2; \quad 0 < \beta_2 \le 2 \\ 0 < \lambda \le 2; \quad \lambda = \eta + 1 \\ b_1 + b_2 = 1 \end{array}\right\} \tag{5.87}$$

Without considering immobile zones in soils, that is, $b_1 = 0$ (as $b_1 + b_2 = 1$) and $\beta_1 = 0$, and by writing $\beta_2 = \beta$, Eq. (5.86) is simplified as:

$$\frac{\partial^{\beta}\theta}{\partial t^{\beta}} = \frac{D_0}{b+1}\frac{\partial^{\lambda}\theta^{b+1}}{\partial m^{\lambda}} - (\gamma_n\alpha - 1)K_0 \frac{\partial\theta^k}{\partial m} \tag{5.88}$$

with

$$0 < \beta \le 2; 0 < \lambda \le 2 \tag{5.89}$$

With the formulation in Eq. (5.88), it is believed that the parameters and the material coordinates characterize the different flow patterns in swelling soils with mobile and immobile zones. The parameters in functions for the material diffusivity, $D_m(\theta)$, and hydraulic conductivity, $K_m(\theta)$, may also be functions of mass or direction when the soils are heterogeneous and/or anisotropic.

3.4.5 Special Forms of Space-time fPDEs for Water Movement in Non-swelling Soils

With power functions for diffusivity $D(\theta)$ and hydraulic conductivity $K(\theta)$ in Eqs. (5.82) and (5.83), the fPDE in Eq. (5.67) for water movement in non-swelling soils with mobile-immobile zones can be written as:

$$b_1 \frac{\partial^{\beta_1}\theta}{\partial t^{\beta_1}} + b_2 \frac{\partial^{\beta_2}\theta}{\partial t^{\beta_2}} = \frac{D_0}{b+1}\frac{\partial^{\lambda}\theta^{b+1}}{\partial z^{\lambda}} - K_0 \frac{\partial\theta^k}{\partial z} \tag{5.90}$$

$$\left.\begin{array}{l} 0 < \beta_1 \le 2; \quad 0 < \beta_2 \le 2 \\ b_1 + b_2 = 1; \quad 0 < \lambda \le 2 \end{array}\right\} \tag{5.91}$$

where z is the usual vertical physical coordinate. Based on Eq. (5.66), Eq. (5.90) can be extended to two and three dimensions for non-swelling soils as:

$$b_1 \frac{\partial^{\beta_1}\theta}{\partial t^{\beta_1}} + b_2 \frac{\partial^{\beta_2}\theta}{\partial t^{\beta_2}} = \frac{\partial}{\partial x_i}\left(D_0\theta^b \frac{\partial^{\eta}\theta}{\partial x_i^{\eta}}\right) - K_0 \frac{\partial\theta^k}{\partial z} \tag{5.92}$$

with

$$\left.\begin{array}{l} 0 < \beta_1 \le 2; \quad 0 < \beta_2 \le 2 \\ b_1 + b_2 = 1; \quad 0 < \eta \le 1 \end{array}\right\} \tag{5.93}$$

where $i = 1,2,3$ is the number of dimensions, and $\mathbf{D}_0$, b, K_0 and k may vary with space or direction in different dimensions x_i.

For soils without immobile zones, Eq. (5.92), with $b_1 = 0$ and $\beta_2 = \beta$, can be simplified as:

$$\frac{\partial^\beta \theta}{\partial t^\beta} = \frac{\partial}{\partial x_i} \left(D_0 \theta^b \frac{\partial^\eta \theta}{\partial x_i^\eta} \right) - K_0 \frac{\partial \theta^k}{\partial z} \tag{5.94}$$

with

$$0 < \beta \leq 2; 0 < \eta \leq 1 \tag{5.95}$$

As space- and time-fractional models are shown to better represent first and second moments of solute transport at early and later times of tracer plumes in contrast to the classical advection-diffusion equation (ADE) (Zhang et al. 2009), the fractional models are then expected to perform better than their classical counterpart for water flow in soils as water is the carrier for solutes. An example of improved accuracy with fPDEs for water flow is infiltration into soil, which has been demonstrated earlier (Su 2010, 2012, 2014).

3.5 *fPDEs for Water Movement in Soils on Hillslopes*

In literature associated with hydrology and soil physics, the majority of publications with theoretical analyses on water flow in soils are concerned with flow on flat surfaces, with only a small number directed to flow in soils on slopes. With this disparity, most of the current theory and methodologies on water movement in soils disregard the slope effects. In reality, however, natural surfaces either have diverse variability with different slope gradients such as in a catchment, or a certain level of local slope gradients on a plain. Rainfall on slopes can generate different quantities of infiltration and runoff under different conditions and the mechanisms of infiltration and runoff generation on hillslopes are important for practical reasons.

3.5.1 *Planar Slope*

The geometries of natural hillslopes are very complex and they can change in three dimensions (O'Loughlin 1986). To classify the geometries of hillslopes, Agnese et al. (2007) postulated nine types of hillslopes as a matrix of *convergent, divergent* and *planar* geometries corresponding to *concave, straight* and *convex downslope* shapes. In Chapter 7, overland flow on hillslopes with different geometries will be discussed, which can be combined with water movement in soils for further integrated analysis of more complex processes such as water flow in soils on an eroding hillslope (Su 2002).

Theoretical analyses of water flow in soils affected by slopes with different geometries can be found in the works of Philip (1991a, b, c) and Philip and Knight (1997). This section only briefly discusses the modifications and applications of fPDEs for water flow on planar hillslopes.

First, refer to Fig. 5.1 and consider water flow in soils on a slope inclined with an angle λ_s.

Similar to the analysis of water movement in soils on a slope with Richards equation, by Philip (1991a), the coordinates (x_*, z_*) is rotated by an angle λ_s:

$$x_* = x \cos \lambda_s + z \sin \lambda_s \tag{5.96}$$

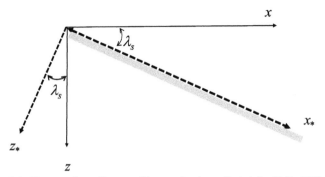

Fig. 5.1. The rotated coordinates to illustrate the slope effects (after Philip 1991a).

$$z_* = -x \sin \lambda_s + z \cos \lambda_s \tag{5.97}$$

With the saturated moisture content or ratio θ_s on a surface with excess water available for runoff, and an initial moisture content θ_i, Philip (1991a) showed that Richards equation can be modified to account for the slope effect by considering $\cos \lambda_s$ in the gravity term, with the term $x \cos \lambda_s$ essentially negligible. Following this approach, Eqs. (5.66) and (5.67) for non-swelling soils can be modified as follows, based on the fractional gradient:

$$b_1 \frac{\partial^{\beta_1}\theta}{\partial t^{\beta_1}} + b_2 \frac{\partial^{\beta_2}\theta}{\partial t^{\beta_2}} = \frac{\partial}{\partial z_*}\left[D(\theta)\frac{\partial^{\eta}\theta}{\partial z_*^{\eta}}\right] - \cos \lambda_s \frac{dK(\theta)}{d\theta}\frac{\partial\theta}{\partial z_*} \tag{5.98}$$

and based on the fractional derivatives of the gradient as:

$$b_1 \frac{\partial^{\beta_1}\theta}{\partial t^{\beta_1}} + b_2 \frac{\partial^{\beta_2}\theta}{\partial t^{\beta_2}} = \frac{\partial^{\eta}}{\partial z_*^{\eta}}\left(D(\theta)\frac{\partial\theta}{\partial z_*}\right) - \cos \lambda_s \frac{dK(\theta)}{d\theta}\frac{\partial\theta}{\partial z_*} \tag{5.99}$$

For swelling-shrinking soils, a similar approach can be applied by including $x \cos \lambda_s$ in the gravity term to yield, based on Eq. (5.76) with the fractional flux, to yield the following fPDE:

$$b_1 \frac{\partial^{\beta_1}\theta}{\partial t^{\beta_1}} + b_2 \frac{\partial^{\beta_2}\theta}{\partial t^{\beta_2}} = \frac{\partial}{\partial m}\left[D(\theta)\frac{\partial^{\eta}\theta}{\partial m^{\eta}}\right] - \cos \lambda_s \frac{dK(\theta)}{d\theta}\frac{\partial\theta}{\partial m} \tag{5.100}$$

and based on Eq. (5.77) with fractional derivatives of the flux to yield the following fPDE:

$$b_1 \frac{\partial^{\beta_1}\theta}{\partial t^{\beta_1}} + b_2 \frac{\partial^{\beta_2}\theta}{\partial t^{\beta_2}} = \frac{\partial^{\eta}}{\partial m^{\eta}}\left(D(\theta)\frac{\partial\theta}{\partial m}\right) - \cos \lambda_s \frac{dK(\theta)}{d\theta}\frac{\partial\theta}{\partial m} \tag{5.101}$$

3.5.2 Convergent and Divergent Slopes

For soil water movement on convergent and divergent slopes, Philip (1991b) introduced the following coordinate transformation:

$$r_* = r \cos \lambda_s + z \sin \lambda_s \tag{5.102}$$

$$z_* = -r \sin \lambda_s + z \cos \lambda_s \tag{5.103}$$

where $\lambda_s < 0$ holds for convergent slopes and $\lambda_s > 0$ for divergent slopes. These transformations, with the aid of perturbation analysis, transform Richards equation into the following (Philip 1991b):

$$\frac{\partial \theta}{\partial t} = \frac{\partial}{\partial z_*} \left(D \frac{\partial \theta}{\partial z_*} \right) - \left(\sin \lambda_s \frac{D}{r_0} + \cos \lambda_s \frac{dK}{d\theta} \right) \frac{\partial \theta}{\partial z_*} \tag{5.104}$$

where r_0 is the value of r at the surface $(r_*, z_*) = (r_0 \sec \lambda_s, 0)$ and both D and K are functions of θ.

With the transformations in Eqs. (5.102) and (5.103), similar to fPDEs for planar slopes, fPDEs for water movement on convergent and divergent slopes can be written as follows, based on fractional flux:

$$b_1 \frac{\partial^{\beta_1} \theta}{\partial t^{\beta_1}} + b_2 \frac{\partial^{\beta_2} \theta}{\partial t^{\beta_2}} = \frac{\partial}{\partial z_*} \left(D \frac{\partial^\eta \theta}{\partial z_*^\eta} \right) - \left(\sin \lambda_s \frac{D}{r_0} + \cos \lambda_s \frac{dK}{d\theta} \right) \frac{\partial \theta}{\partial z_*} \tag{5.105}$$

and as follows, based on the fractional derivatives of flux:

$$b_1 \frac{\partial^{\beta_1} \theta}{\partial t^{\beta_1}} + b_2 \frac{\partial^{\beta_2} \theta}{\partial t^{\beta_2}} = \frac{\partial^\eta}{\partial z_*^\eta} \left(D \frac{\partial \theta}{\partial z_*} \right) - \left(\sin \lambda_s \frac{D}{r_0} + \cos \lambda_s \frac{dK}{d\theta} \right) \frac{\partial \theta}{\partial z_*} \tag{5.106}$$

Similar changes can also be introduced to fPDEs for swelling-shrinking soils, based on fractional flux as:

$$b_1 \frac{\partial^{\beta_1} \theta}{\partial t^{\beta_1}} + b_2 \frac{\partial^{\beta_2} \theta}{\partial t^{\beta_2}} = \frac{\partial}{\partial m_*} \left(D \frac{\partial^\eta \theta}{\partial m_*^\eta} \right) - (\gamma_n \alpha - 1) \left(\sin \lambda_s \frac{D}{r_0} + \cos \lambda_s \frac{dK}{d\theta} \right) \frac{\partial \theta}{\partial m_*} \tag{5.107}$$

and the following, based on the fractional derivatives of flux:

$$b_1 \frac{\partial^{\beta_1} \theta}{\partial t^{\beta_1}} + b_2 \frac{\partial^{\beta_2} \theta}{\partial t^{\beta_2}} = \frac{\partial^\eta}{\partial m_*^\eta} \left(D \frac{\partial \theta}{\partial m_*} \right) - (\gamma_n \alpha - 1) \left(\sin \lambda_s \frac{D}{r_0} + \cos \lambda_s \frac{dK}{d\theta} \right) \frac{\partial \theta}{\partial m_*} \tag{5.108}$$

3.6 Multi-term fPDEs for Water Movement in Multifractal Soils

The equations for water movement in soils presented earlier have been investigated for both one and two temporal terms to account for uniform (Su 2010) and mobile-immobile pores (Su 2012, 2014). The pores in soil particles and between particles as well as aggregates essentially have multiple levels of structures which can be interpreted with multifractals (Martin et al. 2009). Multifractals are geometrical patterns generated by continuously distributed variables (such as pore size, porosity, etc.) and the properties of which exhibit a multitude of power-law scaling relations (Perfect et al. 2009). Hence, a multi-term fPDE in temporal fractional derivatives is a more reasonable representation of the flow process in multifractal porous media (Su 2017a).

As explained in Chapter 2, the signs for SFDs $\dfrac{\partial^\lambda \theta}{\partial |x|^\lambda}$ and $D_0^\lambda \theta$ used here being

identical, $D_0^\lambda \theta = \dfrac{\partial^\lambda \theta}{\partial x^\lambda}$ is used throughout the text as synonymous with symmetric

fractional derivatives.

With SFDs, Eq. (5.66) for non-swelling soils can be expanded to multiple dimensions with multi-term fractional derivatives in time as:

$$\sum_{j=1}^{s} a_j D_t^{\beta_j} \theta = \frac{\partial}{\partial x_i}\left(\mathbf{D}(x_i, \theta)\frac{\partial^\eta \theta}{\partial x_i^\eta} \right) - \frac{\partial K(z, \theta)}{\partial z}, \qquad i = 1, 2, 3, \tag{5.109}$$

which is expressed in terms of the fractional gradient;

$$\sum_{j=1}^{s} a_j D_t^{\beta_j} \theta(z, t) = \frac{\partial^\eta}{\partial x_i^\eta}\left(\mathbf{D}(x_i, \theta)\frac{\partial \theta}{\partial x_i} \right) - \frac{\partial K(z, \theta)}{\partial z}, \qquad i = 1, 2, 3, \tag{5.110}$$

which is expressed as fractional derivatives of the gradient. A mixed fractional

derivatives can also be used such as $\dfrac{\partial^{\eta_1}}{\partial x_i^{\eta_1}}\left(\mathbf{D}(x_i, \theta)\dfrac{\partial^{\eta_2} \theta}{\partial x_i^{\eta_2}} \right)$ which appeared in an

earlier report by He (1998).

For swelling soils, the above models are equivalent to the following forms:

$$\sum_{j=1}^{s} a_j D_t^{\beta_j} \theta(m, t) = \frac{\partial}{\partial m}\left(D(m, \theta)\frac{\partial^\eta \theta}{\partial m^\eta} \right) - \frac{\partial K(m, \theta)}{\partial m} \tag{5.111}$$

in terms of fractional gradient, and

$$\sum_{j=1}^{s} a_j D_t^{\beta_j} \theta(m, t) = \frac{\partial^\eta}{\partial m^\eta}\left(D(m, \theta)\frac{\partial \theta}{\partial m} \right) - \frac{\partial K(m, \theta)}{\partial m} \tag{5.112}$$

in terms of fractional derivatives of gradient.

Distributed-order fPDEs and multi-fractality were also discussed by Sandev et al. (2015), whereas the analytical solutions of multi-term and two-term fPDEs were given by Jiang et al. (2012a, b) and Su (2017b) respectively.

As mentioned in Chapter 4, the orders of fractional derivatives can be functions of time, space, solute concentration, temperature, or other variables. Hence, further generalizations of these models would include orders of fractional derivatives in the form of $\beta(x, c, t)$, $\lambda(x, c, t)$ and $\eta(x, c, t)$.

3.7 *Dimensions of the Parameters in fPDEs for Water Movement in Soils*

Following Kilbas et al. (2006), we modify the parameters in the two-term fPDEs in sections 3.4 to 3.6 as follows:

$$b_{im} = \frac{\theta_{im}}{\theta}\tau^{\beta_1 - 1} \tag{5.113}$$

$$b_m = \frac{\theta_m}{\theta}\tau^{\beta_2 - 1} \tag{5.114}$$

As an example, the new fractional diffusion coefficient in Eqs. (5.82) and (5.90) is given by:

$$D_f = \left(\frac{D_0}{b+1}\right)\tau_d^{\lambda-2} \tag{5.115}$$

where τ and τ_d are parameters for dimension amendments, and b_1, b_2, θ_{im}, θ_m and θ have already been defined earlier. Similar modifications can also be made to other forms of fPDEs.

When multi-term fPDEs are used, the amended dimensions for the multi-term fPDEs in Eqs. (5.99) to (5.112) are modified as follows:

$$\left.\begin{aligned} b_1 &= \frac{\theta_1}{\theta}\tau^{\beta_1-1} \\ b_2 &= \frac{\theta_2}{\theta}\tau^{\beta_2-1} \\ &\ldots\ldots \\ b_n &= \frac{\theta_n}{\theta}\tau^{\beta_n-1} \end{aligned}\right\} \tag{5.116}$$

where $\theta_1, \theta_2,\ldots,\theta_n$ are moisture ratios in $1,2,\ldots, n$ levels of pores.

As discussed in sections 2.16, 3.3 and 5 in Chapter 4, Compte (1997) introduced new parameters which ensured the physical explanation of the fPDEs, where the new parameters $A = \dfrac{\tau_a}{\tau^\beta}$ and $D = \dfrac{\sigma^2}{\tau^\beta}$ are incorporated in the news fPDEs.

4. Conservation of Mass in the Context of fPDEs

There are a variety of fPDEs reported for liquid flow and particle movement in unsaturated and saturated porous media. Conservation of mass, energy and momentum are relevant to the physical interpretation of the models. Wheatcraft and Meerschaert (2008) presented several forms of equations of conservation of mass for different situations depending on the compressibility of the porous medium and the liquid.

For a medium with compressibility varying in the vertical direction only, the conservation of mass presented by Wheatcraft and Meerschaert (2008) is of the form:

$$\frac{\Gamma(\lambda+1)}{(\Delta x)^\lambda}\frac{\partial}{\partial t}(\Delta x_3 \phi \rho) = -\frac{\partial^\lambda}{\partial x_i^\lambda}(\rho q_i) \tag{5.117}$$

or

$$\Gamma(\lambda+1)\Delta x^{1-\lambda}\rho(\beta_s + \phi\beta_w)\frac{\partial p}{\partial t} = -\rho\frac{\partial^\lambda q_i}{\partial x_i^\lambda} \tag{5.118}$$

for constant ρ and a fixed element size, Δx.

The corresponding three-dimensional conservation of mass can also be expressed as:

$$\Gamma(\lambda+1)\Delta x^{1-\lambda}\rho(\beta_s + \phi\beta_w)\frac{\partial p}{\partial t} = -\rho(\nabla^\lambda \mathbf{q}) \tag{5.119}$$

where ρ is the density of the fluid, q_i is specific discharge (with $i = 1,2,3$), ϕ denotes porosity, β_s and β_w represent the compressibility of the porous medium and that of water, and Δx is the increment of the control volume, which represents the scale of heterogeneity as the level of homogeneity increases as $\lambda \to 1$ ($\Delta x^{1-\lambda} \to \Delta x^0 \to 1$).

Olsen et al. (2016, Eq. (20)) presented the conservation of mass based on two-sided fractional derivatives or symmetrical fractional derivatives. For a vertically compressible medium, which has a variable Δx_3, it is of the form:

$$\frac{2^\lambda \Gamma(\lambda+1)}{(\Delta x)^\lambda} \frac{\partial}{\partial t}(\Delta x_3 \phi \rho) = {}^R D_{x_i}^\lambda (\rho q_i) - {}^L D_{x_i}^\lambda (\rho q_i) \tag{5.120}$$

Following Wheatcraft and Meerschaert's (2008) approach to the dependence of Δx_3, ϕ and ρ on liquid pressure p, the conservation of mass presented by Olsen et al. (2016) can be written as:

$$2^\lambda \Gamma(\lambda+1) \Delta x^{1-\lambda} \rho(\beta_s + \phi\beta_w)\frac{\partial p}{\partial t} = \rho {}^R D_{x_i}^\lambda q_i - \rho {}^L D_{x_i}^\lambda q_i \tag{5.121}$$

for constant ρ and equal element sizes, $\Delta x = \Delta x_3$.

The above equations for conservation of mass do not involve fractional derivatives in time. When fractional derivatives in time are introduced, Eqs. (5.117) and (5.120) become, respectively,

$$\frac{\Gamma(\lambda+1)}{(\Delta x)^\lambda} \frac{\partial^\beta}{\partial t^\beta}(\Delta x_3 \phi \rho) = -\frac{\partial^\lambda}{\partial x_i^\lambda}(\rho q_i) \tag{5.122}$$

and

$$\frac{2^\lambda \Gamma(\lambda+1)}{(\Delta x)^\lambda} \frac{\partial^\beta}{\partial t^\beta}(\Delta x_3 \phi \rho) = {}^R D_{x_i}^\lambda (\rho q_i) - {}^L D_{x_i}^\lambda (\rho q_i) \tag{5.123}$$

Note that the expansion of the term $\Delta x_3 \phi \rho$ in fractional derivatives involves the fractional Leibniz rule as a finite series (Podlubny 1999) and is much more complex than its integer counterpart.

Of the two forms of equation of conservation above derived from the mass balance method, the two-sided formulation is consistent with the stochastic approach with which fPDEs are derived from the CTRW theory that incorporates both forward and backward motions of particles or water parcels.

5. fPDEs for Coupled Water Movement, Energy Transfer, Gas Flow and Solute Transport in Porous Media

5.1 Coupled Water Movement and Gas Movement in Soils

Two-phase flow is an important issue in many fields which include hydrology, soil physics, petroleum engineering, etc. The issue of two-phase flow was first quantified by Green and Ampt (1911), who were better known for their first model of infiltration than for the quantitative analysis of hydraulic conductivity of water in soils affected by air.

Section 2.4 in this Chapter briefly reviews related developments in two-phase flow of water and air. Here, the fractional extension of the models for two-phase flow in porous media as reported earlier (Su 2009b) is outlined, which is a time-fractional version of Eq. (5.35):

$$\frac{\partial^{\beta} S}{\partial t^{\beta}} = \frac{1}{r^{j-1}} \frac{\partial}{\partial r} \left(D_0 r^{j-1} \frac{\partial S}{\partial r} \right) - \frac{q_{te}}{r^{j-1}} \frac{\partial S}{\partial r} \tag{5.124}$$

with

$$D_0 = \frac{D_1}{\phi} \tag{5.125}$$

$$q_{te} = \frac{q_t}{\phi} \tag{5.126}$$

The definitions of other parameters appear in section 2.4. Of course, fractional derivatives in space can be also considered in Eq. (5.124) and distributed orders introduced to replace $\dfrac{\partial^{\beta} S}{\partial t^{\beta}}$. The dimensions of Eq. (5.124) are also important which need to be addressed (see section 3.7).

5.2 Coupled Water Movement and Heat Transfer

The background of concurrent mass movement and heat transfer was briefly reviewed in section 2.1. Therein, the equations of concurrent water movement and heat transfer presented by Philip and de Vries (1957) were also outlined. In this section, we attempt at extending the above fractional formulations to coupled PDEs presented by Philip and de Vries (1957) and Philip (1957c).

The success of distributed-order fPDEs for water movement and solute transport in mobile-immobile zones provides justification for their extension to heat transfer in porous media. This extension is equally comprehensible because different zones with variable porosities and densities have different thermal conductivities which lead to different heat fluxes. Ezzat et al. (2012, 2015) derived distributed-order fPDEs for heat transfer in materials with heterogeneous microstructures, based on the two-temperature concept by Tzou (1995). The fPDEs by Ezzat et al. (2012, 2015) indirectly support the coupled dofPDEs for the analysis of water movement and heat transfer in porous media.

The fPDEs based on Philip-de Vries equations (1957) for coupled moisture movement and heat transfer in porous media such as soils can be written in one dimension as:

$$\frac{\partial^{\beta} \theta}{\partial t^{\beta}} = \frac{\partial^{\lambda_1}}{\partial z^{\lambda_1}} (D(\theta)\theta) + \frac{\partial^{\lambda_2}}{\partial z^{\lambda_2}} (D(T)T) \pm \frac{\partial K}{\partial z} \tag{5.127}$$

and

$$C \frac{\partial^{\beta} T}{\partial t^{\beta}} = \lambda \frac{\partial^{\lambda_3} T}{\partial z^{\lambda_3}} + \rho L \frac{\partial^{\lambda_4}}{\partial z^{\lambda_4}} (D_{\theta vap} \theta) \tag{5.128}$$

where $\lambda_1, \lambda_2, \lambda_3$ and λ_4 are the orders of space fractional derivatives. The dofPDEs for concurrent flow of moisture and heat transfer can be written as:

$$b_1 \frac{\partial^{\beta_1} \theta}{\partial t^{\beta_1}} + b_2 \frac{\partial^{\beta_2} \theta}{\partial t^{\beta_2}} = \frac{\partial^{\lambda_1}}{\partial z^{\lambda_1}} (D(\theta)\theta) + \frac{\partial^{\lambda_2}}{\partial z^{\lambda_2}} (D(T)T) \pm \frac{\partial K}{\partial z} \qquad (5.129)$$

and

$$C \left(b_1 \frac{\partial^{\beta_1} T}{\partial t^{\beta_1}} + b_2 \frac{\partial^{\beta_2} T}{\partial t^{\beta_2}} \right) = \lambda \frac{\partial^{\lambda_3} T}{\partial z^{\lambda_3}} + \rho L \frac{\partial^{\lambda_4}}{\partial z^{\lambda_4}} (D_{\theta vap} \theta) \qquad (5.130)$$

These fPDEs can be extended to two and three dimensions with multiple terms to account for concurrent flow of moisture and heat transfer in multifractal media as:

$$\sum_{j=1}^{s} a_j D_t^{\beta_j} \theta = \nabla^{\lambda_1} \cdot (D(\theta)\theta) + \nabla^{\lambda_2} \cdot (D(T)T) \pm \frac{\partial K(z,\theta)}{\partial z} \qquad (5.131)$$

and

$$\sum_{j=1}^{s} a_j D_t^{\beta_j} CT = \nabla^{\lambda_3} \cdot (\lambda T) + \rho L \nabla^{\lambda_4} \cdot (D_{\theta vap} \theta) \qquad (5.132)$$

6. Functional-order Fractional Partial Differential Equations

The orders of the fPDEs presented in this section and earlier sections are fractional constants. In reality, however, the order of fractional derivatives can vary as a function of one or more variables such as time, temperature, concentration of substances or space variables.

Experimentally, Lorenzo and Hartley (1998) overviewed the gradual change from constant order to variable order of fractional derivatives which showed that the dependence of the viscoelasticity of certain materials on temperature, first observed by Bland (1960), can be expressed as a function of the order of fractional derivatives, q, which varies from $q \cong 0$ for elastic to $q \cong -1$ for viscoelastic or viscous materials. Similar studies were carried out by Smit and de Vries (1970) which showed the stress-strain relationships of viscoelastic materials such as textile fibers had the order of the fDE, α, describing the stress-strain relationship varied in the range of $0 \leq \alpha \leq 1$.

Bagley's (1991) experiments on viscoelastic stress-relaxation in linear polymers also showed temperature dependence of the order of fractional derivatives. A similar temperature-dependence of the order of fractional derivatives describing the reaction kinetics of proteins was investigated by Glöckle and Nonnenmacher (1995). Based on these experimental evidence, Lorenzo and Hartley (1998) proposed the functional order $q(t)$, where t represents time. Later, Lorenzo and Hartley (2002) initiated the use of the functional order $q(t, c)$, where c was the dependence variable (such as the concentration of the substance or solute, etc.).

Theoretically, the initiation of variable-order fractional calculus is attributed to Unterberger and Bokobza (1965a) and Višik and Èskin (1967). Their initiation was followed by Leopold (1991), Jacob and Leopold (1993), Negoro (1994), Kikuchi and Negoro (1995), Zayernouri and Karniadakis (2015) and Tavares et al. (2016). Sun et al. (2019) presented an extensive review of variable-order fDEs and their applications.

In the content of soil water movement, Gerasimov et al. (2010) demonstrated that the order of fractional derivative is a function of moisture content $\beta(\theta(x, t))$. Ricciuti and Toaldo (2017) demonstrated that with the semi-Markovian concept (Doeblin 1940, Levy 1954, Smith 1955, Pyke 1961) and the CTRW theory, for holding times as a power-law decaying density with the exponent depending on the state itself, a variable-order fPDE can be derived with a suitable limit, which can be used to model anomalous diffusion in heterogeneous media. This means that when water parcels move following the semi-Markovian chains, the fPDEs for water movement in non-swelling soils (Su 2014), which were outlined in section 3.3, can be extended as:

$$b_1 \frac{\partial^{\beta_1(\theta,z,t)}\theta}{\partial t^{\beta_1\beta_1(\theta,z,t)}} + b_2 \frac{\partial^{\beta_2(\theta,z,t)}\theta}{\partial t^{\beta_2(\theta,z,t)}} = \frac{\partial^{\eta(\theta,z,t)}}{\partial x_i^{\eta(\theta,z,t)}}\left(\mathbf{D}_0 \frac{\partial \theta}{\partial x_i}\right) - \frac{\partial K(z,\theta)}{\partial z} \tag{5.133}$$

or

$$b_1 \frac{\partial^{\beta_1(\theta,z,t)}\theta}{\partial t^{\beta_1\beta_1(\theta,z,t)}} + b_2 \frac{\partial^{\beta_2(\theta,z,t)}\theta}{\partial t^{\beta_2(\theta,z,t)}} = \frac{\partial}{\partial x_i}\left(\mathbf{D}_0 \frac{\partial^{\eta(\theta,z,t)}}{\partial x_i^{\eta(\theta,z,t)}}\right) - \frac{\partial K(z,\theta)}{\partial z} \tag{5.134}$$

Similar to the above formulation, mixed fractional derivatives such as $\dfrac{\partial^{\eta_1}}{\partial x_i^{\eta_1}}\left(\mathbf{D}(x_i,\theta)\dfrac{\partial^{\eta_2}\theta}{\partial x_i^{\eta_2}}\right)$ can also be used. The above formulations were supported by the data given by Gerasimov et al. (2010).

For swelling soils, the above fPDE can be written as:

$$b_1 \frac{\partial^{\beta_1(m,t)}\theta}{\partial t^{\beta_1}} + b_2 \frac{\partial^{\beta_2(m,t)}\theta}{\partial t^{\beta_2}} = \frac{\partial^{\eta(m,t)}}{\partial m^{\eta(m,t)}}\left(D(\theta)\frac{\partial \theta}{\partial m}\right) - \frac{\partial K}{\partial m} \tag{5.135}$$

or

$$b_1 \frac{\partial^{\beta_1(m,t)}\theta}{\partial t^{\beta_1}} + b_2 \frac{\partial^{\beta_2(m,t)}\theta}{\partial t^{\beta_2}} = \frac{\partial}{\partial m}\left(D(\theta)\frac{\partial^{\eta(m,t)}\theta}{\partial m^{\eta(m,t)}}\right) - \frac{\partial K}{\partial m} \tag{5.136}$$

In literature, different functional orders of fractional derivatives have been reported in various forms:

1. as a function of *time* (Coimbra 2003, Sun et al. 2009, 2010, Ludu 2016);
2. as a function of *time* and *space* (Sun et al. 2014, Zayernouri and Karniadakis 2015, Tavares et al. 2016);
3. as a function of *concentration* (Sun et al. 2009); and
4. as a *random order* as a function of time (Li et al. 2009, Sun et al. 2010, 2011b).

Hence, the orders of fractional derivatives (β_1, β_2 and λ) can be functions of time, space, water content, solute concentration, temperature, etc., such as in the form of $\beta_1 = f(\theta, z, T, C, t)$, $\beta_2 = f(\theta, z, T, C, t)$ and $\lambda = f(\theta, z, T, C, t)$.

The above functional-order fPDEs can be further extended to multiple terms to account for the multifractality of porous media. For example, the multi-term one-dimensional form of Eq. (5.54) for non-swelling soils, in terms of fractional gradient, is:

$$\sum_{j=1}^{s} a_j D_t^{\beta_j(\theta(x,t))} \theta = \frac{\partial}{\partial x}\left(D(x,\theta)\frac{\partial^{\eta(\theta(x,t))}\theta}{\partial x^{\eta(\theta(x,t))}} \right) - \frac{\partial K(z,\theta)}{\partial z} \qquad (5.137)$$

The above equation can be written in terms of fractional derivatives of the gradient as:

$$\sum_{j=1}^{s} a_j D_t^{\beta_j(\theta(x,t))} \theta = \frac{\partial^{\eta(\theta(x,t))}}{\partial x^{\eta(\theta(x,t))}}\left(D(x,\theta)\frac{\partial \theta}{\partial x} \right) - \frac{\partial K(z,\theta)}{\partial z} \qquad (5.138)$$

For swelling soils, their equivalents in terms of fractional flux and fractional derivatives of flux are respectively:

$$\sum_{j=1}^{s} a_j D_t^{\beta_j(\theta(m,t))} \theta = \frac{\partial}{\partial m}\left(D(m,\theta)\frac{\partial^{\eta(\theta(m,t))}\theta}{\partial m^{\eta(\theta(m,t))}} \right) - \frac{\partial K(m,\theta)}{\partial m} \qquad (5.139)$$

and

$$\sum_{j=1}^{s} a_j D_t^{\beta_j(\theta(m,t))} \theta(x,t) = \frac{\partial^{\eta(\theta(m,t))}}{\partial x^{\eta(\theta(m,t))}}\left(D(m,\theta)\frac{\partial \theta}{\partial m} \right) - \frac{\partial K(m,\theta)}{\partial m} \qquad (5.140)$$

The challenge with the large number of fractional orders in the formulation is the increasing difficulty of evaluating the parameters β_j with real data.

7. Exchange of Water between Mobile and Immobile Zones

In addition to the equations for water movement at two levels of pores and in uniform pores in soils, fDEs for exchange of water between mobile and immobile zones in soil have also been investigated (Su 2012, 2014).

Based on the dual porosity model, the differential equation for exchange of water between large and small pores (Nielsen et al. 1986) has been modified (Su 2012, 2014) to account for the anomalous exchange of water between the two zones. With the order of fractional derivative, β_3, which incorporates the variable-order fractional derivatives, the model is rewritten as:

$$\frac{\partial^{\beta_3(\theta(x,t))}\theta_{im}}{\partial t^{\beta_3}} = \lambda\theta_{im} + h \qquad (5.141)$$

where

$$\lambda = -\frac{\omega}{b_{im}} \qquad (5.142)$$

$$h = \frac{\omega}{b_{im}}\theta_m \qquad (5.143)$$

The water mass exchange rate, ω, now accommodates new dimensions due to the introduction of β_3. Here, the order of the fractional derivative, β_3, for moisture exchanges between the mobile and immobile zones could have similar values as β_1. However, we leave this issue for further experimental verification.

For a constant β_3, Eq. (5.141) has been analyzed for water exchange between mobile and immobile zones, subject to two types of boundary conditions at the interface (Su 2012). The asymptotic results associated with Eq. (5.141 are summarized in Chapter 6.

8. Summary

This chapter outlines the fPDE-based models for water movement in soils. Some of the fPDEs presented here will be solved in Chapter 6 with infiltration into soils and water flow in soils as examples while some of the fPDEs are suggested here only. The fPDEs for water movement in unsaturated media, compared to the published reports on fPDEs for solute transport in aquifers, require further analysis with data from laboratory and the field.

Few reports are available in literature on coupled water movement, heat transport and solute transport in soils, and further research and verification using data are needed.

Chapter 6

Applications of Fractional Partial Differential Equations to Infiltration and Water Movement in Soils

1. Introduction

This chapter demonstrates the applications of some of the fPDEs presented in Chapter 5. The two major issues discussed in this chapter are infiltration of water into soils and water exchange between large and small pores. In addition, analytical solutions are also presented and briefly discussed for water flow within a soil of finite depth.

Infiltration is the process by which water enters the soil surface. It is one of the most important processes in hydrology and soil science as it connects surface and subsurface flow processes in the hydrological cycle. The interfacing processes driven by infiltration or feedback through coupling with heat transfer involve interactions between surface and subsurface waters, exchange of entrained solutes. Hence, infiltration is one of the most intensively investigated topics in hydrology and soil physics and its quantification has been benefitted by the first infiltration model proposed by Green and Ampt (1911). Since Green and Ampt's pioneering work, the vast majority of studies on infiltration equations reported so far have been derived for infiltration into rigid soils, with a relatively small number of reports on swelling soils.

The approach taken by Green and Ampt (1911) is not only the first hydraulic model for infiltration with rigorous mathematical formulation, but also important for its inception as the first model for two-phase flow in porous media. Many formulations for infiltration have appeared since 1911 and the most widely used infiltration models at present are the Green-Ampt model, Philip's infiltration equation (1954a, 1957a, 1969a), and the empirical models by Kostiakov (1932) and Horton (1939).

Two types of boundary conditions are physically important for infiltration processes. The first type is water content or water ratio on the surface of the soil as the upper boundary, whereas the second type is the flux of water on the surface of the soil, known as flux boundary in soil physics and hydrology.

This chapter presents different types of equations for infiltration into soils, derived from fractional partial differential equations (fPDEs) mainly with the boundary condition (BC) of the first type. In addition to this form of BC, infiltration with a flux boundary condition is also discussed in this chapter. Infiltration equations derived from the boundary condition of the first type include the following situations:

1. Infiltration equations (Su 2009a, 2010, 2012, 2014) derived from fPDEs for absorption and infiltration into soils, with time-fractional PDEs, time-space-fractional PDEs for non-swelling soils, and mass-time fPDEs for swelling soils (this set of infiltration equations provides options for soils as either swelling-shrinking or non-swelling (rigid) soils, and with and without immobile pores).

2. New equations of infiltration into soils on hillslopes with different geometries, namely planar, convergent and divergent surfaces.

The relationship between the two types of BCs for infiltration (concentration BC and flux BC) is established by the flux-concentration relation, which is briefly discussed in connection with fPDEs.

Analytical solutions and their asymptotic results are presented for water exchange between large and small pores with two types of initial conditions (ICs), and the asymptotic solutions for small and large times reveal that the different sizes of pores play specific roles at different time scales.

2. Background and Connections between Different Equations of Infiltration

When infiltrating water moves into a soil vertically as a sharp front behind which the soil is saturated, the relationship between key variables and parameters during infiltration is described by the following equation known as the Green-Ampt infiltration equation (Green and Ampt 1911, Prevedello et al. 2009):

$$\frac{K_s t}{(\theta_s - \theta_i)} = z + h_z \ln\left(1 - \frac{z}{h_z}\right) \tag{6.1}$$

where θ_i and θ_s represent the initial and the saturated moisture contents, and K_s is the saturated hydraulic conductivity, respectively, h_z denotes matrix potential associated with depth z of the soil and is negative in value, t is time, and $\ln\left(1 - \frac{z}{h_z}\right)$ is the logarithmic function.

When the term $\ln\left(1 - \frac{z}{h_z}\right)$ is approximated by a series, Eq. (6.1) can be rewritten as following (Prevedello et al. 2009):

$$z(h,t) = \left(\frac{2K_s h_z}{\theta_s - \theta_i} t\right)^{1/2} + \frac{2K_s}{3(\theta_s - \theta_i)} t + \ldots + \tag{6.2}$$

which is precisely the solution for infiltration given by Philip (1954a, 1957a, 1969a). When only two terms in Eq. (6.2) are retained and used in the following integral as cumulative infiltration $I(t)$:

$$I(t) = At + \int_{\theta_i}^{\theta_s} z d\theta \qquad (6.3)$$

the well-known Philip's two-parameter equation of cumulative infiltration results:

$$I(t) = St^{1/2} + At \qquad (6.4)$$

where A represents the final infiltration rate and S is the sorptivity.

Equation (6.4) was derived with the boundary condition of the first type, and has also been shown to be related to the Green-Ampt model by Philip (1990), who detailed the connections between the Green-Ampt model in Eq. (6.1) and Eq. (6.4) through the ratio A/K_s, which lies in the range $0 \leq A/K_s \leq 2/3$, with the Green-Ampt model corresponding to $A/K_s = 2/3$. The Green-Ampt model treating infiltration as a sharp front is thus called the Delta function model implying that the diffusivity of water in the Green-Ampt model is of the form (Philip 1969a):

$$D = \frac{1}{2} S^2 (\theta_s - \theta_i) \delta(\theta_s - \theta_i) \qquad (6.5)$$

where $\delta(\theta_s - \theta_i)$ is the Dirac delta function.

In addition to the physically-based Green-Ampt and Philip equations of infiltration, several empirical models for infiltration have also been widely used. Among these are the Kostiakov equation (which is a power function) and the Horton equation. The Horton (1939) equation for the infiltration rate $i(t)$ is often written as:

$$i(t) = i_c + (i_o - i_c)e^{-bt} \qquad (6.6)$$

where i_o and i_c are the initial and the final infiltration rates, respectively, b is a constant and e is the base of natural logarithm.

As Philip (1954b) and Eagleson (1970) pointed out, even though Eq. (6.6) has been named as the Horton equation, it was proposed much earlier by Gardner and Widtsoe (1921). The cumulative infiltration is given by integrating Eq. (6.6) with time (Horton 1939):

$$I(t) = i_c t + \frac{1}{e}(i_o - i_c)(1 - e^{-bt}) \qquad (6.7)$$

These equations of infiltration have been derived or formulated in the context of integer calculus. In the rest of this chapter, new equations of infiltration, derived from fPDEs, are presented.

3. Equations of Infiltration Derived from Fractional Calculus with the Concentration Boundary Condition

Equations of infiltration for swelling-shrinking soils and non-swelling soils have been presented in this section. The two forms of models are very similar except for the term $(\gamma_n \alpha - 1)$ which appears in models for swelling-shrinking soils only and the differences in the definitions of their parameters.

3.1 Infiltration into Swelling Soils

This set of equations of infiltration into swelling soils has been discussed in detail earlier (Su 2010, 2012, 2014). As a counterpart of Eq. (6.3) for non-swelling soils, cumulative infiltration into a swelling soil is written as (Philip 1969b, Smiles 1974):

$$I(t) = At + \int_{\theta_i}^{\theta_s} m \, d\theta \tag{6.8}$$

where A represents the final infiltration rate for a large time, θ_i and θ_s are the initial and the final moisture contents or ratios, respectively, and m is the material coordinate (defined as an integral in Chapter 5).

The majority of equations of infiltration and associated solutions of fPDEs (for swelling soils) in terms of m have been presented earlier (Su 2010, 2012, 2014). Here, we summarize the different forms for equations of infiltration, extend them to infiltration on hillslopes, and discuss their implications.

3.1.1 Equation of Infiltration into Swelling Soils with Mobile and Immobile Zones Using a Distributed-order Time-fractional PDE

In this section, the equations of infiltration derived from fPDEs (Su 2012) are based on the following time-fractional PDE and initial and boundary conditions for swelling soils:

$$b_1 \frac{\partial^{\beta_1} \theta}{\partial t^{\beta_1}} + b_2 \frac{\partial^{\beta_2} \theta}{\partial t^{\beta_2}} = D_m \frac{\partial^2 \theta}{\partial m^2} - (\gamma_n \alpha - 1) \frac{dK(\theta)}{d\theta} \frac{\partial \theta}{\partial m} \tag{6.9}$$

$$\left. \begin{array}{lll} \theta = \theta_i, & t = 0, & m > 0 \\ \theta = \theta_s, & t > 0, & m = 0 \\ \theta = \theta_i, & t > 0, & m \to \infty \end{array} \right\} \tag{6.10}$$

where b_1 and b_2 are the relative porosities in immobile and mobile zones respectively, that is, $b_1 = \dfrac{\phi_{im}}{\phi}$ and $b_2 = \dfrac{\phi_m}{\phi}$, with ϕ_{im}, ϕ_m and ϕ being the porosities in the immobile and the mobile zones, and total porosity respectively (see Chapter 5). With constant D_m, and $K(\theta) = K_1 + K_0\theta$ (where K_1 and K_0 are constant), the equation of cumulative infiltration into swelling soils is given as (Su 2012):

$$I(t) = At + St^{\frac{\beta_2}{2} - \frac{\beta_1}{4}} \tag{6.11}$$

where A is the final infiltration rate expressed as:

$$A = (\gamma_n \alpha - 1)K_0 \tag{6.12}$$

K_0 being hydraulic conductivity at the surface of the medium. For saturated media, $\alpha = 1$ (Smiles and Raats 2005), and the particle specific gravity of minerals is $\gamma_n \approx 2.65 \ g/cm^3$. The sorptivity S is given as:

$$S = \frac{2\Gamma(1 + [(\beta_1/2) - \beta_2]/2)D_m^{3/2}(\theta_0 - \theta_i)\sqrt{b_1}}{\left[2D_m + K_0(\gamma_n \alpha - 1)\right]b_2} \tag{6.13}$$

which has the dimension $\left[LT^{\frac{\beta_1}{4}-\frac{\beta_2}{2}} \right]$ and ensures that cumulative infiltration has the dimension of length $[L]$.

The infiltration rate is given by differentiating Eq. (6.11) with respect to time as:

$$i(t) = A + \left(\frac{\beta_2}{2} - \frac{\beta_1}{4} \right) St^{\frac{\beta_2}{2}-\frac{\beta_1}{4}-1} \tag{6.14}$$

3.1.2 Equation of Infiltration into Swelling Soils without Immobile Zones with a Time-fractional PDE

For infiltration into swelling soils without distinguishing between mobile and immobile zones, the equation of cumulative infiltration, based on time-fractional PDEs, is given as (Su 2010):

$$I(t) = At + St^{\beta/2} \tag{6.15}$$

where the final infiltration rate A as in Eq. (6.12), and the sorptivity are expressed as:

$$S = \frac{(\theta_0 - \theta_i)\Gamma(1 - \beta/2)D_m^{1-\beta/2}}{2[K_0(\gamma_n\alpha - 1)]^{1-\beta}} \tag{6.16}$$

The infiltration rate is given by differentiating Eq. (6.15):

$$i(t) = A + \frac{\beta S}{2} t^{(\beta/2)-1} \tag{6.17}$$

In Fig. 6.1, Eq. (6.15) is plotted against the data from Talsma and van der Lelij (1976) and compared to Philip's (1957a, 1969a) two-parameter cumulative equation of infiltration. In Fig. 6.1a, the data in the cumulative infiltration equation in Eq. (6.15) is fitted to derive the order of the fractional derivative ($\beta = 0.2385$), the anomalous sorptivity ($S = 48.58$ *mm/d*$^{\beta/2}$), and the final infiltration rate ($A = 1.29$ *mm/d*).

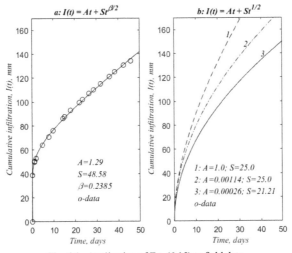

Fig. 6.1. Application of Eq. (6.15) to field data.

Philip's two-term infiltration equation, which is equivalent to $\beta = 1$ in Eq. (6.15), is also plotted in Fig. 6.1b. Herein, three different curve-fitting methods have been applied:

1. Curve 1 is based on the measured average sorptivity by Talsma and van der Lelij (1976) ($S = 25$ *mm/d*$^{1/2}$) and the theoretical final infiltration ($A/K_0 = 1/2$ for a linear model (Philip 1990) based on a geometric mean of $K_0 = 2.0$ *mm/d* of the surface soil) (Talsma and van der Lelij 1976);

2. Curve 2 is generated by fitting Philip's two-term equation to the data of Talsma and van der Lelij (1976), using the measured average sorptivity ($S = 25$ *mm/d*$^{1/2}$), which yields $A = 0.00114$ *mm/d*; and

3. Curve 3 is generated by fitting Philip's two-term equation to the same data of Talsma and van der Lelij (1976) to yield $A = 0.00026$ *mm/d* and $S = 21.21$ *mm/d*$^{1/2}$.

This comparison in Fig. 6.1 shows that the two final infiltration rates derived for Philip's two-term infiltration equation are far too small for this particular soil, given that the measured final infiltration rate is "close to 2.0 *mm/d*" (Talsma and van der Lelij 1976). With the measured sorptivity $S = 25$ *mm/d*$^{1/2}$ and the derived final infiltration rate $A = 1.0$ *mm/d* by these two authors, based on the linear model as discussed by Philip (1990), the computed cumulative infiltration is not satisfactory compared to the field data.

The near perfection of the infiltration equation in Eq. (6.15) to the field data of Talsma and van der Lelij (1976) shown in Fig. 6.1a is due to two key mechanisms which Philip's two-term infiltration does not address:

1. The fFPE accounts for anomalous diffusion due to stochastic variations in irregular pores in the soil;

2. The material coordinate which explains the realistic physical mechanics of swelling during infiltration, which affects the values of the parameters in the infiltration equation.

The order of fractional derivative, β, as the extra parameter implies that infiltration into this specific soil studied by Talsma and van der Lelij belongs to the category of slow diffusion. The derived final infiltration rate A in Eq. (6.15) is very close to the measured value and, and from Eq. (6.16), the surface saturated hydraulic conductivity of this clayey soil is determined to be $K_0 = 0.78$ *mm/d*, here for the saturated surface, $\alpha = 1$ for the studied soils (Smiles and Raats 2005) and $\gamma_n = 2.65$.

3.1.3 Equations of Infiltration in Swelling Soils with Mobile and Immobile Zones with a Distributed-order Mass-time Fractional PDE

In this case, the fPDE is of the form (Su 2014):

$$b_1 \frac{\partial^{\beta_1} \theta}{\partial t^{\beta_1}} + b_2 \frac{\partial^{\beta_2} \theta}{\partial t^{\beta_2}} = D_m \frac{\partial^{\lambda} \theta}{\partial m^{\lambda}} - (\gamma_n \alpha - 1) \frac{dK(\theta)}{d\theta} \frac{\partial \theta}{\partial m} \tag{6.18}$$

where λ is the order of mass fractional derivatives which is considered to be $\lambda = 2$ in sections 3.1.1 and 3.1.2. The values of these orders of fractional derivatives here vary in the ranges of $0 < \beta_1 \leq 2$, $0 < \beta_2 \leq 2$, $b_1 + b_2 = 1$ and $0 < \lambda \leq 2$.

With constant D_m and $K(\theta) = K_0\theta$ in Eq. (6.18), and IC and BCs in Eq. (6.10), the equation of cumulative infiltration into swelling soils is given as (Su 2014):

$$I(t) = At + S\left(\frac{t^{\beta_2+\beta_1}}{S_1 t^{\beta_2} + S_2 t^{\beta_1}}\right)^{1/(2\lambda-1)} \tag{6.19}$$

where sorptivity is given as:

$$S = \frac{(2\lambda-1)(\theta_s-\theta_i)}{2\lambda}\left[\frac{D_0^2\Gamma(1-\beta_1)\Gamma(1-\beta_2)\Gamma(2\lambda)}{(\gamma_n\alpha-1)K_0}\right]^{1/(2\lambda-1)} \tag{6.20}$$

$$S_1 = b_1\Gamma[1-\beta_2] \tag{6.21}$$

$$S_2 = b_2\Gamma[1-\beta_1] \tag{6.22}$$

with $\Gamma[1-\beta_2]$ and $\Gamma[1-\beta_1]$ being gamma functions. The final infiltration rate A in Eq. (6.19) remains as in Eq. (6.12). Equation (6.19) can be written as:

$$I(t) = At + S\left(\frac{S_1}{t^{\beta_1}} + \frac{S_2}{t^{\beta_2}}\right)^{1/(1-2\lambda)} \tag{6.23}$$

implying that the dimension of S is $L/[T^{-\beta_1} + T^{-\beta_2}]^{\frac{1}{1-2\lambda}}$, hence ensuring that the dimension of cumulative infiltration is $[L]$.

The rate of infiltration is given by differentiating Eq. (6.23) with respect to time as:

$$i(t) = A + St^{a-1}\left[a\left(S_1 t^{\beta_2} + S_2 t^{\beta_1}\right)^{1/(2\lambda-1)} + \frac{1}{(2\lambda-1)}\left(S_1\beta_2 t^{\beta_2} + S_2\beta_1 t^{\beta_1}\right)\right] \tag{6.24}$$

where

$$a = \frac{\beta_2+\beta_1}{2\lambda-1} < 1.0 \tag{6.25}$$

The final infiltration rate A and the hydraulic conductivity at the surface, K_0, relate in a similar fashion for integer-PDE-based models (Philip 1990) which were briefly reiterated in section 2. As $0 < \beta_2 \le 2$, $0 < \beta_1 \le 2$ and $0 < \lambda \le 2$, Eq. (6.23) implies that $i(t) = A$ as $t \to \infty$.

In Eq. (6.20) for sorptivity, there are measurable parameters as well as unknown parameters: θ_s, θ_i and γ_n can be measured, whereas β, λ, A and S can be derived by fitting Eq. (6.19) or (6.23) to the data. Rearranging the terms of Eq. (6.12) yields $K_0 = A/(\gamma_n\alpha - 1)$, and for the case with a saturated surface, $\alpha = 1.0$. Hence, the only unknown parameter for sorptivity in Eq. (6.20) is the diffusion coefficient D_0 which can be derived by rearranging Eq. (6.20) as:

$$D_0 = \left[\left(\frac{2\lambda S}{(2\lambda-1)(\theta_s-\theta_i)}\right)^{2\lambda-1}\frac{(\gamma_n\alpha-1)K_0}{\Gamma[1-\beta_1]\Gamma[1-\beta_2]\Gamma[2\lambda]}\right]^{1/2} \tag{6.26}$$

which offers an alternative method for determining the diffusion parameter.

3.1.4 Equation of Infiltration into Swelling Soils without Immobile Zones, with a Mass-time Fractional PDE

When soils are treated as being uniformly porous, it implies that $b_1 = 0$, $b_2 = 1$ as $b_1 + b_2 = 1$ and $\beta_1 = 0$. By writing $\beta = \beta_2$ and with constant D_m, $K(\theta) = K_0\theta$ in Eq. (6.18), and IC and BCs in Eq. (6.10), the equation of cumulative infiltration is written as (Su 2014):

$$I(t) = At + St^{\beta/(2\lambda-1)} \tag{6.27}$$

where sorptivity S is given as:

$$S = \frac{(2\lambda-1)(\theta_s - \theta_i)}{2\lambda}\left[\frac{D_0^2\Gamma(1-\beta)\Gamma(2\lambda)}{(\gamma_n\alpha-1)K_0}\right]^{1/(2\lambda-1)} \tag{6.28}$$

with the final infiltration rate A given by Eq. (6.12). Here, the dimension of S is $[L/T^{\beta/(2\lambda-1)}]$.

The rate of infiltration into soil is given by differentiating Eq. (6.27) with respect to time as:

$$i(t) = A + \frac{S\beta}{(2\lambda-1)}t^{[\beta/(2\lambda-1)]-1} \tag{6.29}$$

For $\beta = 1$ and $\lambda = 2$, Eq. (6.27) can be rewritten as:

$$I(t) = At + St^{1/3} \tag{6.30}$$

which is analogous to Philip's two-parameter equation of cumulative infiltration, with the only difference being in the power of the second term as 1/3 instead of 1/2. It should be pointed out that the power of 1/2 in Philip's series solution of the Richards equation (Philip 1957a) is the result of a numerical method. In fact, the power of 1/3 is from the exact solution of the fPDE based on fractional calculus.

The same data from field experiments published by Talsma and van der Lelij (1976), as in Fig. 6.1, is fitted to Eq. (6.27) and demonstrated in Fig. 6.2. The fitting process results in $A = 1.30$ *mm/d*, $S = 48.64$ *mm/d*$^{\beta/(2\lambda-1)}$, $\beta = 0.3445$, and $\lambda = 1.9523$, which ultimately yield a transport exponent of $\mu = 0.353$.

Similar to the procedures leading to Eq. (6.26), the constant in the expression for diffusivity of soils without immobile zones can be derived by rearranging Eq. (6.28) as:

$$D_0 = \left(\frac{2\lambda S}{(2\lambda-1)(\theta_s - \theta_i)}\right)^{\lambda-1/2}\left[\frac{(\gamma_n\alpha-1)K_0}{\Gamma(1-\beta)\Gamma(2\lambda)}\right]^{1/2} \tag{6.31}$$

3.2 Infiltration into Non-swelling Soils

Similar to section 3.1 for swelling soils, a constant diffusion coefficient D_0 and the linear hydraulic conductivity function $K(\theta) = K_0\theta$ are used for non-swelling soils, where K_0 is constant.

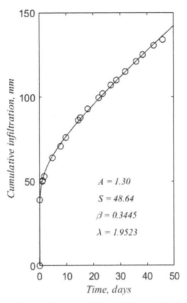

Fig. 6.2. Application of Eq. (6.27) to field data.

3.2.1 *Equation of Infiltration into Non-swelling Soils with Mobile and Immobile Zones, with a Distributed-order Time-fractional PDE*

Analogous to Eq. (6.9) for swelling soils, the fPDE with $K(\theta) = K_0\theta$ for water movement in non-swelling soils is written as:

$$b_1\frac{\partial^{\beta_1}\theta}{\partial t^{\beta_1}} + b_2\frac{\partial^{\beta_2}\theta}{\partial t^{\beta_2}} = D_0\frac{\partial^2\theta}{\partial z^2} - K_0\frac{\partial\theta}{\partial z} \tag{6.32}$$

A solution equation of Eq. (6.32) can be derived with the following IC and BCs:

$$\left.\begin{array}{lll} \theta = \theta_i, & t = 0, & z > 0 \\ \theta = \theta_s, & t > 0, & z = 0 \\ \theta = \theta_i, & t > 0, & z \to \infty \end{array}\right\} \tag{6.33}$$

Similar to swelling soils, except for the absence of the term $(\gamma_n\alpha - 1)$ in the formulation, cumulative infiltration for non-swelling soils is given as (Su 2012):

$$I(t) = At + St^{\beta_2/2 - \beta_1/4} \tag{6.34}$$

where A represents the final infiltration rate, and sorptivity S is given as:

$$S = \frac{2\Gamma(1 + [(\beta_1/2) - \beta_2]/2)D_m^{3/2}\left(\theta_0 - \theta_i\right)\sqrt{b_1}}{[2D_0 + K_0]b_2} \tag{6.35}$$

which has the dimension $\lfloor L/T^{\beta_2/2 - \beta_1/4}\rfloor$, hence ensuring that the dimension of cumulative infiltration is $[L]$.

Infiltration rate is given by differentiating Eq. (6.34) with respect to time as:

$$i(t) = A + \left(\frac{\beta_2}{2} - \frac{\beta_1}{4} \right) St^{(\beta_2/2 - \beta_1/4)-1} \tag{6.36}$$

3.2.2 Equations of Infiltration into Non-swelling Soils without Immobile Zones, with a Distributed-order Time-fractional PDE

For infiltration into non-swelling soils with no distinction between mobile and immobile zones, the equation of cumulative infiltration is of the following form (Su 2010):

$$I(t) = At + St^{\beta/2} \tag{6.37}$$

where A represents the final infiltration rate. Here, sorptivity is given as:

$$S = \frac{(\theta_0 - \theta_i)\Gamma(1 - \beta/2)D_m^{1-\beta/2}}{2K_0^{1-\beta}} \tag{6.38}$$

and infiltration rate is obtained by differentiating Eq. (6.37) as:

$$i(t) = A + \frac{\beta S}{2} t^{(\beta/2)-1} \tag{6.39}$$

For $\beta = 1$, the fPDE becomes the classic PDE and Eq. (6.37) becomes Philip's equation of cumulative infiltration in Eq. (6.4).

3.2.3 Equations of Infiltration into Non-swelling Soils with Mobile and Immobile Zones, with a Space-time Fractional PDE

For infiltration into non-swelling soils, the following fPDE, IC and BCs apply:

$$b_1 \frac{\partial^{\beta_1} \theta}{\partial t^{\beta_1}} + b_2 \frac{\partial^{\beta_2} \theta}{\partial t^{\beta_2}} = D_0 \frac{\partial^\lambda \theta}{\partial z^\lambda} - K_0 \frac{\partial \theta}{\partial z} \tag{6.40}$$

$$\left. \begin{array}{lll} \theta = \theta_i, & t = 0, & z > 0 \\ \theta = \theta_s, & t > 0, & z = 0 \\ \theta = \theta_i, & t > 0, & z \to \infty \end{array} \right\} \tag{6.41}$$

and the equation of cumulative infiltration is determined by Eq. (6.3). The cumulative infiltration equation so derived is identical to Eq. (6.19) for swelling soils, except for sorptivity, wherein the term $(\gamma_n \alpha - 1)$ is absent (Su 2014), that is:

$$I(t) = At + S \left(\frac{t^{\beta_2 + \beta_1}}{S_1 t^{\beta_2} + S_2 t^{\beta_1}} \right)^{1/(2\lambda - 1)} \tag{6.42}$$

with sorptivity written as:

$$S = \frac{(2\lambda - 1)(\theta_s - \theta_i)}{2\lambda} \left[\frac{D_0^2 \Gamma(1 - \beta_1)\Gamma(1 - \beta_2)\Gamma(2\lambda)}{K_0} \right]^{1/(2\lambda - 1)} \tag{6.43}$$

$$S_1 = b_1 \Gamma[1 - \beta_2] \tag{6.44}$$

$$S_2 = b_2 \Gamma[1 - \beta_1] \tag{6.45}$$

The rate of infiltration into soils is derived by differentiating Eq. (6.42) with respect to time as:

$$i(t) = A + St^{a-1} \left[\frac{a\left(S_1 t^{\beta_2} + S_2 t^{\beta_1}\right)^{1/(2\lambda-1)}}{+ \frac{1}{(2\lambda-1)}\left(S_1 \beta_2 t^{\beta_2} + S_2 \beta_1 t^{\beta_1}\right)} \right] \tag{6.46}$$

where

$$a = \frac{\beta_2 + \beta_1}{2\lambda - 1} < 1.0 \tag{6.47}$$

Equation (6.46) implies that for large times $t \to \infty$, the final infiltration rate is A.

3.2.4 Equations of Infiltration into Non-swelling Soils without Immobile Zones, with a Time-Space Fractional PDE

With a uniform porosity, $b_1 = 0$, $b_2 = 1$, $\beta_1 = 0$, and with $\beta = \beta_2$, the equation of cumulative infiltration into non-swelling soils without immobile zones was presented earlier (Su 2014) and found similar to that in section 3.1.4:

$$I(t) = At + St^{\beta/(2\lambda-1)} \tag{6.48}$$

In this case, for non-swelling soils, sorptivity is expressed as:

$$S = \frac{(2\lambda-1)(\theta_s - \theta_i)}{2\lambda} \left[\frac{D_0^2 \Gamma(1-\beta)\Gamma(2\lambda)}{K_0} \right]^{1/(2\lambda-1)} \tag{6.49}$$

and the equation for infiltration rate derived from Eq. (6.48) is:

$$i(t) = A + \frac{S\beta}{(2\lambda-1)} t^{[\beta/(2\lambda-1)]-1} \tag{6.50}$$

Rearranging the sorptivity equation (Eq. (6.49)) also yields the constant in the diffusivity function, D_0, given the measured and fitted values of the other parameters.

3.3 Variability of Parameters for Infiltration in a Field with Different Surface Conditions

Mbagwu's (1995) data on infiltration rates under different land covers is used here to examine the variability of parameters in Eq. (6.15) for infiltration in a field. Mbagwu (1995) compared Kostiakov's and Philip's infiltration equations and their modified forms. The conversion of the parameters in Kostiakov's and Philip's infiltration equations to the equivalent parameters in Eq. (6.15) yields the order of fractional derivatives in Eq. (6.15) as:

a. $\beta = 1.158$ for a tilled-mulched surface;
b. $\beta = 0.952$ for an untilled-unmulched surface;
c. $\beta = 0.440$ for an untilled-mulched surface;
d. $\beta = 0.386$ for a tilled-unmulched surface; and
e. $\beta = 0.058$ for continuous pasture.

Overall, Mbagwu's data gives $0.058 < \beta < 1.158$ for a single porosity fPDE to cover all these field surface conditions, which is information additional to Philip's two-term equation of infiltration (equivalent to $\beta = 1$ here).

4. Infiltration into Soils on Hillslopes

It was seen in Chapter 5 that water movement in soils on flat surfaces and that on slopes are different. Therefore, equations of infiltration into soils on hillslopes need to be derived from the fPDEs for water movement in soils on a slope. In this section, equations of infiltration are presented for planar hillslopes, with convergent and divergent hillslopes briefly discussed.

4.1 Infiltration into Non-swelling Soils on Planar Hillslopes

Philip (1991a) analyzed infiltration on planar slopes and showed that there are four components to flow in soils on slopes, that is, components in:

1. horizontal inslope, u;
2. vertical inslope, v;
3. downslope, u_d; and
4. normal to the slope, v_n.

The difficulty lies in the evaluation of the proportions of these four components.

Logically, the vertical component of the flow, v, tends to dominate the flow due to gravity as the moisture content increases, which decreases the capillary forces; on a slope, the infiltrating front is inclined due to the gradient of the slope. This was confirmed by Jackson (1992), who found that during rainfall, infiltration is nearly vertical for most (simulated) soils except those with high constant anisotropy. For this reason and for the aid of analysis in Chapter 7 on overland flow, we only discuss the vertical component in succeeding sections.

4.1.1 Equations of Vertical Infiltration into Non-swelling Soils with Mobile and Immobile Zone, on Planar Slopes, with a Distributed-order Time-space Fractional PDE

In this case, the fPDE with constant diffusivity $D = D_0$ and a linear function for hydraulic conductivity $K(\theta) = K_0\theta$, Eq. (6.32), is modified below following Philip's (1991a) approach:

$$b_1 \frac{\partial^{\beta_1}\theta}{\partial t^{\beta_1}} + b_2 \frac{\partial^{\beta_2}\theta}{\partial t^{\beta_2}} = D_0 \frac{\partial^{\lambda}\theta}{\partial z_*^{\lambda}} - K_0 \cos\lambda_s \frac{\partial\theta}{\partial z_*} \tag{6.51}$$

where λ_s is the angle of the slope. Equation (6.51) is solved with the following IC and BCs:

$$\left.\begin{array}{lll} \theta = \theta_i, & t = 0, & z_* > 0 \\ \theta = \theta_s, & t > 0, & z_* = 0 \\ \theta = \theta_i, & t > 0, & z_* \to \infty \end{array}\right\} \tag{6.52}$$

where z_* is the transformed coordinate (see Chapter 5).

With the inclusion of the slope gradient, the equations of infiltration as derived in section 3.2.3 should also be modified. Subsequently, the equations of cumulative infiltration, infiltration rate and their parameters in section 3.2.3 can be changed for vertical cumulative infiltration as (Philip 1991a):

$$I(t) = A_s t + S \left(\frac{t^{\beta_2 + \beta_1}}{S_1 t^{\beta_2} + S_2 t^{\beta_1}} \right)^{1/(2\lambda - 1)} \tag{6.53}$$

where the final infiltration rate on a slope, A_s, is then given as:

$$A_s = A \cos^2 \lambda_s + K_s \sin^2 \lambda_s \tag{6.54}$$

with K_s as the hydraulic conductivity on the surface when infiltration takes place (corresponding to $\theta = \theta_s$), and A as the final infiltration rate on a flat surface. Note that K_0 appears in the linear function $K(\theta) = K_0 \theta$, which is different from K_s in Eq. (6.54).
Here, sorptivity is expressed as:

$$S = \frac{(2\lambda - 1)(\theta_s - \theta_i)}{2\lambda} \left[\frac{D_0^2 \Gamma[1 - \beta_1] \Gamma[1 - \beta_2] \Gamma[2\lambda]}{K_0 \cos \lambda_s} \right]^{1/(2\lambda - 1)} \tag{6.55}$$

with

$$S_1 = b_1 \Gamma[1 - \beta_2] \tag{6.56}$$

$$S_2 = b_2 \Gamma[1 - \beta_1] \tag{6.57}$$

The rate of infiltration is given by differentiating Eq. (6.53) with respect to time as:

$$i(t) = A_s + S t^{a-1} \left[a \left(S_1 t^{\beta_2} + S_2 t^{\beta_1} \right)^{1/(2\lambda-1)} + \frac{1}{(2\lambda - 1)} \left(S_1 \beta_2 t^{\beta_2} + S_2 \beta_1 t^{\beta_1} \right) \right] \tag{6.58}$$

where

$$a = \frac{\beta_2 + \beta_1}{2\lambda - 1} < 1.0 \tag{6.59}$$

These equations of infiltration now incorporate the slope correction factor, $\cos \lambda_s$. By replacing the term $K_0 \cos \lambda_s$ with $(\gamma_n \alpha - 1) K_0 \cos \lambda_s$ in Eqs. (6.53) and (6.58), the equations will be applicable to infiltration into swelling soils on planar hillslopes.

4.1.2 Equations of Vertical Infiltration into Non-swelling Soils without Immobile Zones on Planar Slopes, with a Time-space Fractional PDE

In this case, cumulative infiltration is given by:

$$I(t) = A_s t + S t^{\beta/(2\lambda - 1)} \tag{6.60}$$

where A_s is obtained from Eq. (54) and sorptivity S is expressed as:

$$S = \frac{(2\lambda - 1)(\theta_s - \theta_i)}{2\lambda} \left[\frac{D_0^2 \Gamma(1 - \beta) \Gamma(2\lambda)}{K_0 \cos \lambda_s} \right]^{1/(2\lambda - 1)} \tag{6.61}$$

Here, the final infiltration rate is the same as before ($A = K_0 \cos \lambda_s$) and the rate of infiltration into soil is given by differentiating Eq. (6.60) with respect to time as:

$$i(t) = A_s + \frac{S\beta}{(2\lambda - 1)} t^{[\beta/(2\lambda-1)]-1} \tag{6.62}$$

4.1.3 Equations of Vertical Infiltration into Non-swelling Soils with Mobile and Immobile Zones on Planar Slopes with a Time-fractional PDE

With the slope correction to Eq. (6.32), the equation of cumulative infiltration on a slope is a modified form of Eq. (6.34) as follows:

$$I(t) = A_s t + St^{\beta_2/2 - \beta_1/4} \tag{6.63}$$

where A_s is given by Eq. (6.54) and sorptivity is given as:

$$S = \frac{2\Gamma(1 + [(\beta_1/2) - \beta_2]/2)D_m^{3/2}(\theta_0 - \theta_i)\sqrt{b_1}}{[2D_0 + K_0 \cos \lambda_s]b_2} \tag{6.64}$$

Again, infiltration rate is given by differentiating Eq. (6.63) with respect to time, similar to Eq. (6.36), as:

$$i(t) = A_s + \left(\frac{\beta_2}{2} - \frac{\beta_1}{4}\right)St^{(\beta_2/2 - \beta_1/4)-1} \tag{6.65}$$

4.2 Infiltration into Swelling Soils on Planar Hillslopes

Similar to the earlier discussions, equations of infiltration into swelling soils can be derived from the materials presented by replacing $K_0 \cos \lambda_s$ with $K_0 (\gamma_n \alpha - 1)\cos \lambda_s$; A_s is given by Eq. (6.54), with A determined from Eq. (6.12).

4.3 Infiltration into Soils on Convergent and Divergent Hillslopes

Flow on convergent and divergent surfaces is more complicated than flow on planar slopes. Yet, very limited investigations which are simple enough for practical applications (Philip 1991b, c, 1996) can be found in literature.

As an example of application for water movement on convergent and divergent slopes, with constant diffusivity $D = D_0$ and a linear function for hydraulic conductivity $K(\theta) = K_0\theta$, Eq. (6.40) for non-swelling soils becomes (Philip 1991b):

$$b_1 \frac{\partial^{\beta_1}\theta}{\partial t^{\beta_1}} + b_2 \frac{\partial^{\beta_2}\theta}{\partial t^{\beta_2}} = D\frac{\partial^\lambda\theta}{\partial z_*^\lambda} - \left(\frac{D}{r_0}\sin\lambda_s + \frac{dK}{d\theta}\cos\lambda_s\right)\frac{\partial\theta}{\partial z_*} \tag{6.66}$$

where the transformed coordinates z_* and r_* are given by Eqs. (5.102) and (5.103); r_0 is the value of r at the surface point $(r_*, z_*) = (r_0 \sec \lambda_s, z_*)$; and $\lambda_s < 0$ for convergent slopes and $\lambda_s > 0$ for divergent slopes.

For infiltration into swelling soils with dual porosity, Eq. (5.104) becomes:

$$b_1 \frac{\partial^{\beta_1}\theta}{\partial t^{\beta_1}} + b_2 \frac{\partial^{\beta_2}\theta}{\partial t^{\beta_2}} = D\frac{\partial^\lambda\theta}{\partial m_*^\lambda} - (\gamma_n\alpha - 1)\left(\frac{D}{r_0}\sin\lambda_s + \frac{dK}{d\theta}\cos\lambda_s\right)\frac{\partial\theta}{\partial m_*} \tag{6.67}$$

Equations (6.66) and (6.67) can be solved using the methods in the previous sections. The key difference between these formulations is the convective term $\left(\dfrac{D}{r_0}\sin \lambda_s + \dfrac{dK}{d\theta}\cos \lambda_s \right)$ for convergent and divergent surfaces. The equations of infiltration presented in section 3 can be modified by incorporating the term $\left(\dfrac{D}{r_0}\sin \lambda_s + \dfrac{dK}{d\theta}\cos \lambda_s \right)$ for infiltration on convergent and divergent slopes.

5. Infiltration Equations Derived from an fPDE with a Given Flux on the Soil Surface

5.1 Introduction

In this section, infiltration with a constant rate on the soil surface is analyzed and interpreted using fPDEs. The boundary condition associated with this form of infiltration problems is called boundary condition of the third type or the flux boundary condition (or flux BC).

To the author's knowledge, solution of fPDEs with a constant flux BC and the exact analysis of water exchange between different pores and their asymptotic results have not been reported in literature. In addition to the concentration boundary condition which was used for deriving equations of infiltration in previous sections, the flux BC is useful for analyzing infiltration and water movement in soils when the supply rate on the surface is of interest.

Examples of the mathematical analysis of PDEs with a BC of the third type include the encyclopedic treatise of Carslaw and Jaeger (Carslaw and Jaeger 1947, 1959), who solved the heat equation or the diffusion equation (with constant diffusivity) by converting the BC of the third type into a BC of the first type, through a transformation. This method has also been reported for solving the transport equation in porous media (Kreft and Zuber 1978, 1986, Parker and van Genuchten 1984, Parlange et al. 1992) and the equation for water movement in soils with a flux BC (Menziani et al. 2007).

Extensive literature exists on water movement and solute transport in soils with flux boundary conditions and flux-concentration relations (Philip 1969a, 1973, Philip and Knight 1974, Smith and Parlange 1978, van Genuchten and Alves 1982, Knight 1983, Smith 1983, Broadbridge and White 1987, 1988a, b, Sander et al. 1988, Broadbridge 1990, Broadbridge and Rogers 1990, 1993, Barry and Sander 1991, Parkin et al. 1992, Smith 2002, Yuan and Lu 2005, Hayek 2016). In particular, van Genuchten and Alves (1982) documented a number of problems associated with BCs of the third type, and Knight (1983) detailed the analysis of infiltration using Richards and Burgers equations with a BC of the third type for three stages of infiltration, namely, pre-ponding, incipient ponding and post-ponding. The monograph by Smith (2002) comprehensively documents infiltration with different approaches, including the flux BC.

Solutions of fPDEs subject to a flux BC can be determined by extending the method for solutions of PDEs documented by Carslaw and Jaeger (1947, 1959), through transformation of the BCs. The solution of these fPDEs provide improved knowledge of the stochastics of water flow in soils, water exchange between large and small pores and the mathematical and geometrical relationships between model parameters (such as the orders of fractional derivatives), scaling parameters as well as the parameters characterizing the scale-effect.

5.2 Flux-Concentration Relations in the Framework of fPDEs

In addition to the analysis of water movement in soils using fPDEs with boundary condition of the first type (Su 2010, 2012, 2014), one can further examine the processes of infiltration into and movement within soil in terms of rainfall intensity or other artificial supply rates on the surface, such as irrigation, using boundary condition of the third type.

These two types of infiltration rates can be united by using the flux-concentration relation (Philip 1973) F, which is the flux of water at a given depth $v(z, t)$ relative to the flux with which water enters the soil, or surface flux (Philip 1973, Smith 2002):

$$F = \frac{v(z,t) - K_i}{v_0 - K_i} \tag{6.68}$$

where v_0 is the flux on the surface and K_i is the initial background (gravity) flux. The notation v_0 as the surface flux is equivalent, in terms of quantity, to the infiltration rate $i(t)$, which was demonstrated by Smith (2002) in the following relationships:

$$I(t) = K_i t + \int_0^t i(t)dt = K_i t + \int_0^L (\theta - \theta_i)dz \tag{6.69}$$

with θ_i being the initial moisture ratio and L being the depth of infiltration.

By re-arranging Darcy's law and using it in the above equation, the following expression results (Smith 2002):

$$I(t) = K_i t + \int_{\theta_i}^{\theta_s} \frac{(\theta - \theta_i)D(\theta)}{v(z,t) - K(\theta)}d\theta \tag{6.70}$$

With the aid of the flux-concentration relation, Eq. (6.70) can be written as (Smith 2002):

$$I(t) = K_i t + \int_{\theta_i}^{\theta_s} \frac{(\theta - \theta_i)D(\theta)}{v(0,t)F - K(\theta)}d\theta$$

$$= K_i t + \int_{\theta_i}^{\theta_s} \frac{(\theta - \theta_i)D(\theta)}{i(t)F - K(\theta)}d\theta \tag{6.71}$$

where $v(0, t) = v_0$ is the flux on the surface.

5.3 Flux on the Surface of Non-swelling Soils

The following IC and BCs apply to flux infiltration in this case:

$$\theta(z, t) = \theta_i, \qquad\qquad t = 0, \qquad\qquad z \geq 0 \qquad\qquad (6.72)$$

$$q = -D_f \frac{\partial^\eta \theta}{\partial z^\eta} + K_0 \theta = r(t), \quad 0 < t \leq t_p, \qquad z = 0 \qquad\qquad (6.73)$$

where $r(t)$ is the effective rainfall intensity on the surface and t_p is the time for ponding; after t_p, infiltration takes place on a saturated surface.

The BC in Eq. (6.73) is the fractional flux resulting from fractional Darcy's law (He 1998), which includes gravity and implies that soil surface is unsaturated from the onset of the effective application of water until ponding time t_p, with a flux $r(t)$. When $r(t)$ is equal to or larger than the hydraulic conductivity of the surface soil, infiltration takes place on a saturated surface. For $\eta = 1$, the BC in Eq. (6.73) reduces to the standard flux BC.

In this section, we do not follow the procedures with which the flux-concentration relation is developed (Philip 1973, Smith 2002). Instead, we only demonstrate the transformation used by Carslaw and Jaeger (1947, 1959) to derive the flux-concentration relations and extend it to the fractional BC.

With the reduced moisture ratio Θ as:

$$\Theta = \frac{\theta - \theta_i}{\theta_s - \theta_i} \qquad\qquad (6.74)$$

Equation (6.73) with the reduced moisture ratio can be written as a new BC:

$$\Theta = 0, \qquad\qquad z = 0 \qquad\qquad (6.75)$$

$$\frac{\partial^\eta \Theta}{\partial z^\eta} - \frac{K_0}{D_f} \Theta = -\frac{r(z,t)}{D_f}, \qquad 0 < t \leq t_p, \qquad z = 0 \qquad\qquad (6.76)$$

With a constant flux on the surface, $r(0, t) = R$, and constant diffusion coefficient and hydraulic conductivity, Eq. (6.40) and its BCs in Eqs. (6.72) and (6.73) are transformed into:

$$b_i \frac{\partial^{\beta_1} r}{\partial t^{\beta_1}} + b_m \frac{\partial^{\beta_2} r}{\partial t^{\beta_2}} = D_f \frac{\partial^\lambda r}{\partial z^\lambda} - K_0 \frac{\partial r}{\partial z} \qquad\qquad (6.77)$$

$$r(z, t) = 0, \qquad\qquad t = 0, \qquad\qquad z > 0 \qquad\qquad (6.78)$$

$$r(z, t) = R, \qquad\qquad t > 0, \qquad\qquad z = 0 \qquad\qquad (6.79)$$

with $\lambda = \eta + 1$.

The solution of Eq. (6.77), in terms of $r(z, t)$, subject to the conditions in Eqs. (6.78) and (6.79), can now be derived and then written in terms of the original variable through Eqs. (6.75) and (6.76) without major technical difficulties.

5.4 Other Forms of Flux Boundary Conditions Relevant to Infiltration and Water Movement

In addition to BCs of the first type in Eq. (6.10) and BC of the third type in section 5.3, there are other forms of physically relevant flux boundary conditions that can be used to derive solutions of a PDE or fPDE. Examples of these boundary conditions have been discussed below.

5.4.1 Fluxes at Both Boundaries

This form of mathematical statements can be expressed as:

$$\left. \begin{array}{lll} \theta = \theta_i, & t = 0, & z > 0 \\[2ex] -D\dfrac{\partial \theta}{\partial z} + v\theta = v\theta_s, & t > 0, & z = 0 \\[2ex] \dfrac{\partial \theta}{\partial z} \to 0, & t > 0, & z \to \infty \end{array} \right\} \tag{6.80}$$

or

$$\left. \begin{array}{lll} \theta = \theta, & t = 0, & z > 0 \\[2ex] -D\dfrac{\partial}{\partial} + v\theta = v\theta, & t > 0, & z = 0 \\[2ex] \dfrac{\partial}{\partial} = g(t), & t > 0, & z = L \end{array} \right\} \tag{6.81}$$

where v is the flow velocity, $g(t)$ is a function of time and L is the depth of the soil. The function for the lower BC relates to water exchange between the soil and the aquifers such as the drainage rate.

5.4.2 Flux at the Lower Boundary

This form of the IC and BCs takes the form of

$$\left. \begin{array}{lll} \theta = \theta_i, & t = 0, & z > 0 \\[1ex] \theta = \theta_s, & t > 0, & z = 0 \\[2ex] \dfrac{\partial \theta}{\partial z} \to 0, & t > 0, & z \to \infty \end{array} \right\} \tag{6.82}$$

5.4.3 Arbitrary Flux at the Surface

The following BC is more generic:

$$\left. \begin{array}{lll} \theta = \theta_i, & t = 0, & z > 0 \\[1ex] \theta = f(t), & t > 0, & z = 0 \\[1ex] \theta \to \theta_i, & t > 0, & z \to \infty \end{array} \right\} \tag{6.83}$$

where $f(t)$ is an arbitrary function of time.

Solutions of the fPDEs with the above ICs and BCs have been developed (Su, unpublished manuscript), but they are relatively more complicated.

6. Water Exchange between Large and Small Pores

In this section, we investigate water exchange between mobile and immobile zones, which takes place during infiltration into, evaporation out of, and flow inside the soil profile.

The analysis of this kind of flow process can be performed for two types of initial conditions at the interface (Su 2012). Equation (5.141) is solved as an initial value problem with constant order fractional derivatives and two types of initial conditions (Su 2012):

$$\frac{\partial^{\beta_3} \theta_{im}}{\partial t^{\beta_3}} = \rho \theta_{im} + h \tag{6.84}$$

where

$$\rho = -\frac{\omega}{b_1} \tag{6.85}$$

$$h = \frac{\omega}{b_1} \theta_m \tag{6.86}$$

with ω being the mass exchange rate between mobile and immobile zones.

IC 1:

$$\frac{d^{\beta_3} \theta_{im}}{dt^{\beta_3}} = v_k, \qquad\qquad t = 0, \qquad\qquad k = 1, 2, \dots, n \tag{6.87}$$

IC 2:

$$\theta_{im} = \theta_{im0}, \qquad\qquad t = 0, \tag{6.88}$$

where

v_k are the fractional rates of change in water content in the immobile zone and θ_{im0} is the initial value of θ_{im}.

6.1 Water Exchange at the Interface when the Initial Rate of Change in Moisture is Known

6.1.1 The Solution

In this case, Eq. (6.84) is to be solved subject to Eq. (6.87). In the case of $\theta_m = \theta_{m0}$ with θ_{m0} being the constant value of θ_m on the surface of mobile zones, which is equivalent to a constant h, $\dfrac{\partial^{\beta_3} \theta_{im}}{\partial t^{\beta_3}}$ can be replaced with $\dfrac{d^{\beta_3} \theta_{im}}{dt^{\beta_3}}$. The solution so derived is of the following form (Su 2012):

$$\theta_{im} = \sum_{k=1}^{n} v_k t^{\beta_3 - k} E_{\beta_3, \beta_3 - k + 1}(\rho t^{\beta_3}) + \int_0^t (t - \tau)^{\beta_3 - 1} E_{\beta_3, \beta_3}[\rho(t - \tau)^{\beta_3}] h d\tau \tag{6.89}$$

where $E_{\beta_3,\beta_3-k+1}(\rho t^{\beta_3})$ and $E_{\beta_3,\beta_3}[\rho(t-\tau)^{\beta_3}]$ are the Wiman functions or two-parameter Mittag-Leffler functions (MLF).

In the case of constant θ_m, i.e., $\theta_m = \theta_0$, Eq. (6.89) can be integrated (Kilbas et al. 2004) to yield the following closed-form solution (Su 2012):

$$\theta_{im} = \sum_{k=1}^{n} v_k t^{\beta_3-k} E_{\beta_3,\beta_3-k+1}\left(-\frac{\omega}{b_1}t^{\beta_3}\right) + \frac{\omega\theta_0}{b_1}t^{\beta_3} E_{\beta_3,\beta_3+1}\left(-\frac{\omega}{b_1}t^{\beta_3}\right) \qquad (6.90)$$

where $E_{\beta_3,\beta_3-k+1}\left(-\frac{\omega}{b_1}t^{\beta_3}\right)$ and $E_{\beta_3,\beta_3+1}\left(-\frac{\omega}{b_1}t^{\beta_3}\right)$ are MLFs and can be expressed as a series:

$$E_{\beta_3,\beta_3-k+1}\left(-\frac{\omega}{b_1}t^{\beta_3}\right) = \sum_{n=0}^{\infty}\frac{1}{\Gamma(\beta_3 n + \beta_3 - k + 1)}\left(-\frac{\omega}{b_1}t^{\beta_3}\right)^{n} \qquad (6.91)$$

$E_{\beta_3,\beta_3+1}\left(-\frac{\omega}{b_1}t^{\beta_3}\right)$ can also be expanded as a similar series with a different gamma function $\Gamma(\beta_3 n + \beta_3 + 1)$.

An approximate expression of Eq. (6.90) can be derived by retaining the first term only in the first part of the right hand side of the equation, that is, $k = 1$ in Eq. (6.90), which results in the following approximation:

$$\theta_{im} = v_1 t^{\beta_3-1} E_{\beta_3,\beta_3}\left(-\frac{\omega}{b_1}t^{\beta_3}\right) + \frac{\omega\theta_0}{b_1}t^{\beta_3} E_{\beta_3,\beta_3+1}\left(-\frac{\omega}{b_1}t^{\beta_3}\right) \qquad (6.92)$$

As the movement of water across the mobile-immobile boundary is more likely to be sub-diffusion, it is more likely that $\beta_3 < 1$.

6.1.2 *Asymptotic Solutions when the Initial Rate of Change in the Moisture Content is Known*

To carry out asymptotic analysis, we write the temporal part of the function in Eq. (6.90) as the generalized Prabhakar function (GPF) (Prabhakar 1971, Friedrich 1991, Tomovski et al. 2014, Mainardi and Garrappa 2015):

$$e_{\alpha,\beta}^{\gamma}(t;\rho) = t^{\beta-1}E_{\alpha,\beta}^{\gamma}(-\rho t^{\alpha}), \qquad t \geq 0 \qquad (6.93)$$

which has two forms of asymptotic solutions under different conditions as stated below.

For asymptotic analysis, the parameters in Eq. (6.90) corresponding to those in Eq. (6.93) are listed in Table 6.1.

6.1.2.1 Asymptotic Solution for $t \to \infty$

For the first MLF or Wiman function in Eq. (6.90), the condition (Mainardi and Garrappa 2015) is:

$$0 < \beta_3 < \beta_3 - k + 1 \leq 1.0 \qquad (6.94)$$

where k is the counter in Eq. (6.90).

Table 6.1. Parameters in Eq. (6.93) for asymptotic analysis.

	First GPF	Second GPF
	$t^{\beta_3-k}E_{\beta_3,\beta_3}\left(-\dfrac{\omega}{b_1}t^{\beta_3}\right)$	$t^{\beta_3}E_{\beta_3,\beta_3+1}\left(-\dfrac{\omega}{b_1}t^{\beta_3}\right)$
γ	1	1
α	β_3	β_3
β	β_3	β_3+1

Physically, the only case which applies to this condition is $k = 1$ in the first MLF in Eq. (6.90), which justifies the approximate solution in Eq. (6.92) as the only physically relevant solution. Subsequently, the condition for Eq. (6.93) is simplified as:

$$0 < \alpha = \beta \leq 1.0 \tag{6.95}$$

which is the second condition given by Mainardi and Garrappa (2015). It is also a very important theoretical justification of the limits of the parameter β_3.

For the second MLF in Eq. (6.92), we only need to extend the upper limit of the first condition by Mainardi and Garrappa (2015) as:

$$0 < \alpha < \beta \leq 2.0 \tag{6.96}$$

Hence, the asymptotic solution of Eq. (6.92) or Eq. (6.90) with $k = 1$ as $t \to \infty$ can be derived by using the second condition by Mainardi and Garrappa (2015) for the first MLF and their first condition for the second MLF in Eq. (6.90), resulting in the following expression:

$$\theta_{im} \sim \left(\frac{\beta_3 v_1}{\Gamma(1-\beta_3)t^{\beta_3+1}} + \theta_0\right)\frac{\omega}{b_1}, \qquad t \to \infty \tag{6.97}$$

6.1.2.2 Asymptotic Solution of the GPF in Eq. (6.92) as $t \to 0$

The asymptotic solution under this condition is given by (Friedrich 1991):

$$\theta_{im} \sim \left[\frac{v_1}{\Gamma(\beta_3)}t^{-1} + \frac{\theta_0}{\Gamma(\beta_3+1)}\right]\frac{\omega}{b_1}t^{\beta_3}, \qquad t \to 0 \tag{6.98}$$

which implies that as the time approaches the onset, the moisture ratio increases within the soil and decreases on the surface of the soil due to $\beta_3 \leq 1$.

6.2 Water Exchange at the Interface when the Initial Water Content is Known

6.2.1 The Solution

In this case, Eq. (6.84) is to be solved with Eq. (6.88) and the solution is given as (Kilbas et al. 2006, Su 2012):

$$\theta_{im} = \theta_{im0}E_{\beta_3}(\lambda t^{\beta_3}) + \int_0^t (t-\tau)^{\beta_3-1}E_{\beta_3,\beta_3}[\lambda(t-\tau)^{\beta_3}]h d\tau \tag{6.99}$$

where $E_{\beta_3}(\lambda t^{\beta_3})$ is the MLF and $E_{\beta_3,\beta_3}[\lambda(t - \tau)^{\beta_3}]$ is the Wiman function or two-parameter MLF.

With $\theta_m = \theta_0$ on the saturated surface during infiltration, Eq. (6.99) is integrated to yield:

$$\theta_{im} = \theta_{im0}E_{\beta_3}\left(-\frac{\omega}{b_1}t^{\beta_3}\right) + \frac{\omega\theta_0}{b_1}t^{\beta_3}E_{\beta_3,\beta_3+1}\left(-\frac{\omega}{b_1}t^{\beta_3}\right) \tag{6.100}$$

which can be written as

$$\theta_{im} = \theta_{im0}E_{\beta_3}\left(-\frac{\omega}{b_1}t^{\beta_3}\right) + \frac{\omega\theta_0}{b_1}e_{\beta_3,\beta_3+1}\left(t;\frac{\omega}{b_1}\right) \tag{6.101}$$

where

$$e_{\beta_3,\beta_3+1}\left(t;\frac{\omega}{b_1}\right) = t^{\beta_3}E_{\beta_3,\beta_3+1}\left(-\frac{\omega}{b_1}t^{\beta_3}\right) \tag{6.102}$$

is the GPF, and

$$E_{\beta_3}\left(-\frac{\omega}{b_1}t^{\beta_3}\right) = \sum_{k=0}^{\infty}\frac{1}{\Gamma(\beta_3k+1)}\left(-\frac{\omega}{b_1}t^{\beta_3}\right)^k \tag{6.103}$$

is the MLF.

6.2.2 Asymptotic Solution when the Initial Water Content on the Surface is Known

The solution of Eq. (6.100) has two cases of asymptotes as discussed in the following sections.

6.2.2.1 Asymptotic Solutions for $t \to \infty$

For the MLF, the condition is for $t \to \infty$ (Mainardi and Garrappa 2015):

$$0 < \beta_3 \leq 1.0 \tag{6.104}$$

For the GPF, the condition by Mainardi and Garrappa (2015) is reset to be:

$$0 < \beta_3 < \beta_3 + 1 \leq 2.0 \tag{6.105}$$

With the above conditions, the asymptotic solution of Eq. (6.100) for $t \to \infty$ is then given by the summation of the two terms of the asymptotes in Eq. (6.100):

$$\theta_{im} \sim \left(\frac{\theta_{im0}\beta_3 t^{-\beta_3-1}}{\Gamma(1-\beta_3)} + \theta_0\right)\frac{\omega}{b_1}, \qquad t \to \infty \tag{6.106}$$

6.2.2.2 Asymptotic Solution of the GPF as $t \to 0$

The asymptotic solution under this condition is given by (Friedrich 1991):

$$\theta_{im} \sim \left(\theta_{im0} + \theta_0 t^{\beta_3}\right)\frac{\omega}{b_1}, \qquad t \to 0 \tag{6.107}$$

7. Example of Solutions for Water Movement in a Soil of Finite Depth

The solution of Eq. (5.90) with $b = 0$ and $k = 1$ was presented earlier (Su 2017b) and the IC and the BCs used for the solution are specified as:

$$\theta(z, t) = \theta(z, 0), \qquad\qquad t = 0 \qquad\qquad 0 < z < L \qquad\qquad (6.108)$$

$$\theta(z, t) = \theta(0, t), \qquad\qquad t = 0 \qquad\qquad z = 0 \qquad\qquad (6.109)$$

$$\theta(z, t) = \theta(L, t), \qquad\qquad t > 0 \qquad\qquad z = L \qquad\qquad (6.110)$$

where L is the depth of the soil profile. The moisture ratios on the surface and at depth $z = L$ are $\theta(0, t)$ and $\theta(L, t)$ respectively. By retaining only the first term in the series solution, the solution is of the form (Su 2017b):

$$\theta(z,t) = \theta(0,t) + \frac{\left[\theta(L,t) - \theta(0,t)\right]z}{L} +$$
$$\frac{1}{\pi}\left\{\frac{a_2 L}{\pi}\left[1 - \cos\left(\frac{2\pi z}{L}\right)\right] - (a_1 + a_2 z)\sin\left(\frac{2\pi z}{L}\right)\right\}F(t) \qquad (6.111)$$

where

$$a_1 = \theta(z, 0) - \theta(0, 0) \qquad\qquad (6.112)$$

$$a_2 = -\frac{\left[\theta(L,0) - \theta(0,0)\right]}{L} \qquad\qquad (6.113)$$

$$F(t) = 1 - K_1 t^{\beta_2} E_{(\beta_1,\beta_2),1+\beta_2}\left(-\frac{b_1}{b_2}t^{\beta_1}, -K_1 t^{\beta_2}\right) \qquad (6.114)$$

$$E_{(\beta_1,\beta_2),1+\beta_2}\left(-\frac{b_1}{b_2}t^{\beta_1}, -K_1 t^{\beta_2}\right)$$
$$= \sum_{j=0}^{\infty}\sum_{i=0}^{j}\binom{j}{i}\left(-\frac{b_1}{b_2}t^{\beta_1}\right)^i \frac{(-K_1 t^{\beta_2})^{j-i}}{\Gamma[(1+\beta_2) + \beta_1 i + \beta_2(j-i)]} \qquad (6.115)$$

such that:

$$K_1 = \frac{1}{b_2}\left[D_0\left(\frac{\pi}{L}\right)^\lambda - K_0\frac{\pi}{L}\right] \qquad\qquad (6.116)$$

The full Eq. (6.115) is complex and, for practical applications, an approximation can be made by retaining limited terms. Here, only two leading terms in Eq. (6.115) are retained, which corresponds to $j = 0,1$ and $i = 0,1$, hence resulting in the following expression:

$$E_{(\beta_2-\beta_1,\beta_2),1+\beta_2}\left(-\frac{b_1}{b_2}t^{\beta_2-\beta_1}, -K_1 t^{\beta_2}\right)$$
$$\approx \frac{1}{\Gamma_{[1+\beta_2]}} + \frac{b_1 t^{-\beta_1}}{K_1 b_2 \Gamma_{[1+\beta_2-\beta_1]}} - \frac{K_1 t^{\beta_2}}{\Gamma_{[1+2\beta_2]}} - \frac{b_1 t^{\beta_2-\beta_1}}{b_2 \Gamma_{[1+2\beta_2-\beta_1]}} \qquad (6.117)$$

With hypothetical data, Eq. (6.111) is plotted in Fig. 6.3 to demonstrate the effects of the order of fractional derivatives on changes in moisture ratio.

The temporal part of the solution in Eq. (6.111) is Eq. (6.114) for $F(t)$, which is depicted in Fig. 6.4 to examine the effects of the order of fractional derivatives on the moisture ratio.

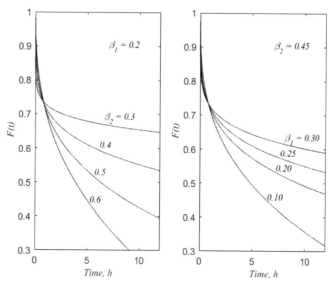

Fig. 6.3. Moisture distribution affected by the order of fractional derivatives in Eq. (6.111); left: fixed β_1, with variable β_2, indicating the facilitating role of β_2 in large pores; right: fixed β_2, with variable β_1, indicating the trapping role of β_1 in small pores; in both cases, the joint parameter is $K_1 = 0.01$.

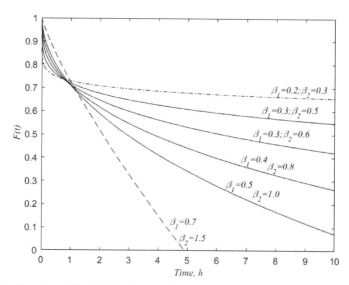

Fig. 6.4. The illustration of Eq. (6.111) with pairs of β_2 and β_1 (joint parameter: $K_1 = 0.01$); the facilitating role of β_2 overtaking β_1 in increasing the flow rate when both orders increase.

This section demonstrates the application of the fPDE for water movement in soils as the second example in addition to the application of the fPDEs to infiltration. A more detailed discussion of the model and solution are given can be found in Su (2017b).

8. Summary

This chapter applies some of the fPDE-based models presented in Chapter 5 for water movement in soils. The applications of fPDEs have been demonstrated on three issues: [1] infiltration into soils; [2] water exchange between large and small pores in soils, and [3] water movement in soils for a finite depth. There are more issues in practice which can be investigated using the fPDEs: some of the approaches developed for analyzing the Richards equation during the past eight decades, and approaches using the two-phase flow and multiphase models can be used to analyze the coupled fPDEs for water flow associated with heat transfer and solute transport.

Chapter 7

Fractional Partial Differential Equations for Solute Transport in Soils

1. Introduction

The objective of this chapter is to develop a set of fPDE-based models for solute transport in soils to be consistent with the models presented for water flow in soils discussed in Chapters 5 and 6. Because solute transport in soils is driven by water movement and modulated by differentiated pore structures, fPDEs are presented here for concurrent water movement and solute transport in soils. The soils discussed here include swelling and non-swelling soils with and without immobile pores.

With a very brief discussion of conventional solute transport models in non-swelling soils, the developments in the study of solute transport in swelling soils in different material coordinates by Smiles (2000a, b) are reconciled in the same material coordinates consistent with the fPDE-based models for water movement in swelling soils. Subsequently, a series of fPDEs are presented as options for modelling solute transport under various scenarios, which include distributed-order time fPDEs and space-time fPDEs for non-swelling soils, and distributed-order and mass-time fPDEs for swelling soils. This chapter also deals with multi-term fPDEs and functional-order fPDEs.

2. Solute Transport in Non-swelling Soils

The movement of solutes in unsaturated soils involves many processes with water movement as the major driving force, in addition to molecular diffusion, adsorption due to physico-chemical forces, and transformation from one form to other forms, etc. A mathematical model in one dimension for describing these processes is written as (Nielsen et al. 1986):

$$\frac{\partial}{\partial t}(\gamma s) + \frac{\partial}{\partial t}(\theta c) = \frac{\partial}{\partial z}\left(\theta D \frac{\partial c}{\partial z} - qc\right) - \sum_{i=1}^{n} \phi_i(c, s, \ldots) \tag{7.1}$$

where c and s are solute concentrations in solution phase and solid phase of the soil respectively, θ is the volumetric water content, γ is the bulk density of soil, D

is the dispersion coefficient, q is the flux of soil water, and ϕ_i are the source-sink terms which account for the rates of solute addition and/or removal which are not accounted for in the adsorbed phase s.

The dispersion coefficient D is a combination of molecular diffusion and mechanical dispersion due to water flow:

$$D = D_0\tau + \alpha_L v^n \tag{7.2}$$

where D_0 is the molecular diffusion coefficient, τ is the tortuosity factor, v is the velocity of soil water, and α_L and n are treated as constants which, in fact, may vary.

For relatively homogeneous saturated media, α_L is called dispersivity, which has been shown to be approximately unity ($\alpha_L = 1$); $\alpha_L \approx 0.05$ or less for disturbed soils in laboratory conditions and $\alpha_L \approx 0.1$ or more for field-scale experiments (Nielsen et al. 1986). The value of α_L depends on the extent at which it is measured and this phenomena is often referred to as scale-dependence in literature. In saturated media, α_L is a function of the observation scale x and can be approximated by $\alpha_L \propto 0.1x^{2d-1}$ as one of the scale-dependent models, where d is the fractal dimension of the medium (Wheatcraft and Tyler 1988, Su 1995).

Another component of processes during solute transport is adsorption, which can be described by either equilibrium adsorption or non-equilibrium adsorption.

2.1 Equilibrium Solute Transport

Equilibrium transport is often modelled as instantaneous adsorption and, with this approach, the simplest model for relating the adsorption concentration on a porous medium, s, with the solute concentration in soil solution, c, is:

$$s = kc \tag{7.3}$$

which is known as equilibrium isotherm, with k as a constant. With first-order and/or zero-order source-sink terms, Eq. (7.1) is rewritten as (Nielsen et al. 1986):

$$R\frac{\partial c}{\partial t} = D\frac{\partial^2 c}{\partial z^2} - v\frac{\partial c}{\partial z} - \mu c + \sigma \tag{7.4}$$

where R is the retardation factor given as:

$$R = 1 + \gamma k/\theta \tag{7.5}$$

with γ as the soil bulk density. For non-reactive solutes, $R = 1$.

The decay term is expressed as:

$$\mu = \mu_l + \mu_s\gamma/\theta \tag{7.6}$$

which accounts for solute removal in liquid phase, μ_l, and that in solid phase, μ_s. The solute production term is written as:

$$\sigma = \sigma_l + \sigma_s\gamma/\theta \tag{7.7}$$

where σ_l and σ_s are zero-order production terms for solutes in liquid and solid phases respectively.

For conservative solutes, with no source and sink, Eq. (7.4) is reduced to:

$$\frac{\partial c}{\partial t} = D\frac{\partial^2 c}{\partial z^2} - v\frac{\partial c}{\partial z} \tag{7.8}$$

Models with equilibrium reactions assume that absorption and production of solutes take place in an instantaneous manner, with no delay and transit between particles of porous media.

2.2 Non-equilibrium Solute Transport in Non-swelling Soils with Immobile Pores

2.2.1 The Mobile-Immobile Model

Soil aggregates form sophisticated connected networks and barricades at micro-scales. Instantaneous or equilibrium transport of solutes in such complex porous media is less likely to be true. Instead, stochastic movements of solutes with differentiated flow velocity distributions and diffusion gradients exist. One model for explaining such mechanisms is the dual-porosity model which dates back to Rubinshtein (1948).

Since its inception, an extensive literature exists on its applications to a range of unsaturated and saturated media in laboratory and field conditions. Some of the analyses on this topic can be found in the works of Barenblatt et al. (1960), Nielsen et al. (1986), Bond and Wierenga (1990), Gerke and van Genuchten (1993), Haggerty and Gorelick (1995), Griffioen et al. (1998), Gamerdinger and Kaplan (2000), Culligan et al. (2001), Haggerty et al. (2004), Culkin et al. (2008), Gao et al. (2009, 2010) and Khuzhayorov et al. (2010).

Mobile-immobile models have been applied to investigate both single-rate transfer and multiple-rate mass transfer (Haggerty and Gorelick 1995, Haggerty et al. 2004) between mobile and immobile zones. The concept has been applied to investigate both unsaturated (Nielsen et al. 1986) and saturated media (Culkin et al. 2008, Khuzhayorov et al. 2010), and tested using data from outflow methods or non-destructive methods such as nuclear magnetic resonance (Culligan et al. 2001). It has been successfully applied for describing solute transport in different forms of porous media and under different flow conditions such as density-driven flow (Starr and Parlange 1976), flow in macropores (Seyfried and Rao 1987), flow in randomly heterogeneous media (Barry and Sposito 1989), stratified porous media (Li et al. 1994), and scale-dependent transport of reactive solutes in porous media with mobile-immobile zones (Gao et al. 2010).

When complicated pore structures are conceptualized and mobile-immobile represented, one such model is of the following form (Nielsen et al. 1986):

$$\theta_m R_m \frac{\partial c_m}{\partial t} + \theta_{im} R_{im} \frac{\partial c_{im}}{\partial t} = \theta_m D_m \frac{\partial^2 c_m}{\partial z^2} - \theta_m v_m \frac{\partial c_m}{\partial z} \tag{7.9}$$

with

$$\theta_{im} R_{im} \frac{\partial c_{im}}{\partial t} = \varepsilon(c_m - c_{im}) \tag{7.10}$$

where the subscripts m and im represent the quantities for mobile and immobile pores respectively and ε is the mass transfer rate between mobile and immobile pores.

The models in Eqs. (7.4), (7.9), and (7.10) have been widely used for modelling solute transport in soils for past several decades. In section 7, exact solutions and related asymptotic solutions of an fPDE as a generalized Eq. (7.10) have been presented.

2.2.2 The Multi-rate Model

In addition to Eq. (7.10) as the model for mass transfer between mobile and immobile zones, a model for multi-rate mass transfer was reported by Haggerty and Gorelick (1995) to account for mass transfer between one mobile zone and multiple immobile zones. This model is written as:

$$\frac{\partial c_m}{\partial t} + \sum_{j=1}^{n} b_j \varepsilon_j \left(c_m - c_{im,j} \right) = L(c_m) \tag{7.11}$$

where $L(c_m)$ is the linear operator representing the terms on the right-hand side of Eq. (7.9) (with sources/sinks, if required), namely, $L(c_m) = \theta_m D_m \dfrac{\partial^2 c_m}{\partial z^2} - \theta_m v_m \dfrac{\partial c_m}{\partial z} - \mu c_m + \sigma; \varepsilon$ is the mass transfer rate with j being the *jth* immobile zone; c_m is the solute concentration in the mobile zone; $c_{im,j}$ is the solute concentration in the *jth* immobile zone; and b_j is the capacity ratio given as:

$$b = \sum_{j=1}^{n} b_j = \frac{(R_{im})_j (\theta_{im})_j}{R_m \theta_m} \tag{7.12}$$

with R_{im} and θ_{im} being the retardation factor and the water content in the immobile zone, and R_m and θ_m being the retardation factor and the water content in the mobile zone.

2.3 Relationship between Flux and Residential Solute Concentrations

The concentration of solutes can be measured by two different quantities (Kreft and Zuber 1978). The first is the measure of solutes in soils in residence, which is the mass of solutes per unit volume of fluid contained in an elementary volume of the porous medium at a given time interval; this quantity is termed as *residential concentration* and denoted by c_r. The second is the measure of solutes as mass of solutes per unit volume of fluid passing through a given cross section at a given time interval, which is termed as *flux concentration* and denoted by c_f.

The two quantities above are related by the following relationships (Kreft and Zuber 1978):

$$Q \int_0^t c_f(x,\tau)d\tau = \phi A \int_x^\infty c_r(\xi,t)d\xi \tag{7.13}$$

where Q is the volumetric flow rate of the fluid $[L^3 T^{-1}]$, A is the cross-sectional area, and ϕ is the effective porosity. When the volumetric flow rate per unit area, $q = Q/A$, is used, Eq. (7.13) is written as:

$$q \int_0^t c_f(x,\tau)d\tau = \phi \int_x^\infty c_r(\xi,t)d\xi \qquad (7.14)$$

With the two quantities of solutes measured by different references, the fluxes of the two quantities are related through the following expression (Kreft and Zuber 1978, Parker and van Genuchten 1984):

$$c_f = c_r - \frac{D}{v} \frac{\partial c_r}{\partial x} \qquad (7.15)$$

and the flux density of the flux-averaged concentration is related to the volumetric flow rate as (Parker and van Genuchten 1984):

$$J = qc_f \qquad (7.16)$$

With these definitions and identities, the following relationships result (Parker and van Genuchten 1984):

$$\frac{\partial c_r}{\partial t} = -v \frac{\partial c_f}{\partial x} \qquad (7.17)$$

Differentiating Eq. (7.15) with respect to time yields the following:

$$\frac{\partial c_f}{\partial t} = \frac{\partial c_r}{\partial t} - \frac{D}{v} \frac{\partial^2 c_r}{\partial x \partial t} \qquad (7.18)$$

which, in combination with Eqs. (7.17) and (7.18), gives:

$$\frac{\partial c_f}{\partial t} = D \frac{\partial^2 c_f}{\partial x^2} - v \frac{\partial c_f}{\partial x} \qquad (7.19)$$

Equation (9.19) is mathematically identical to (7.8), but with a different definition of solute concentration. In fact, the physical implications differ significantly.

Equation (7.15) implies that as $\dfrac{D}{v} \to 0$, $c_f \to c_r$. Physically, this implies that as variations in the microscopic pore water velocities approach zero (for small D), flux concentration and residential concentration tend to be equal. As variations in the microscopic pore water velocities increase, the spatial distributions of c_r and c_f gradually diverge (Parker and van Genuchten 1984). Clearly, the two solute concentration definitions are only meaningful when convection is concerned and, for solute transport due to diffusion only, there is no need to define flux concentration.

The above formulations in terms of flux and residential solute concentrations can be applied to analyze total solute concentrations in all levels of pores in a porous media and/or that in large and small pores if the quantity in immobile pores is appreciable and its measurement possible. As we will see in the later sections of this chapter, these concepts can be extended by employing fPDEs to account for anomalous solute transport in porous media.

The transform in Eq. (7.15) used by Kreft and Zuber (1978) for particle transport is, in fact, a standard transformation in solving the heat equation widely used in applied mathematics (Carslaw and Jaeger 1947, 1959). This transformation changes the basic measurement quantity into its flux. In a report on the application of fPDEs to flux and residential solute concentrations, Zhang et al. (2006a) outlined the relationship between the two types of solute concentrations in three forms: one time-space-fractional ADE and two special cases.

3. Concurrent Water Flow and Solute Transport in Swelling Soils

Smiles (2000a, b), using material coordinates, presented partial differential equations of water flow and solute movement in swelling soils. Two types of material coordinates were used by Smiles: one material coordinate, m, was based on the solid distribution, and the second type, p, was based on the water distribution in the medium. The two types of material coordinates are related by the following relationship (Smiles 2000a, b):

$$dp = \left(\frac{\partial p}{\partial m}\right) dm + \left(\frac{\partial p}{\partial t}\right) dt = \theta \, dm - v \, dt \tag{7.20}$$

where v is the Darcy flux (or velocity) relative to the solid ($[LT^{-1}]$) and θ is the moisture ratio. Based on these definitions, Smiles (2000b) presented diffusion equations for water flow and solute movement in different material coordinates, that is, for water flow in swelling soils, the diffusion equation is of the form:

$$\frac{\partial \theta}{\partial t} = \frac{\partial}{\partial m}\left(D_m(\theta)\frac{\partial \theta}{\partial m}\right) \tag{7.21}$$

and the formulation for solute movement in swelling soils is:

$$\frac{\partial c}{\partial t} = \frac{\partial}{\partial p}\left(D(c)\theta_i^2 \frac{\partial c}{\partial p}\right) \tag{7.22}$$

where c denotes solute concentration and $D(c)$ is solute diffusivity (or dispersion coefficient).

For the vertical movement of water in swelling soils, which is the case of flow subject to gravity, the governing equation is (Smiles and Raats 2005):

$$\frac{\partial \theta}{\partial t} = \frac{\partial}{\partial m}\left[D_m(\theta)\frac{\partial \theta}{\partial m}\right] - (\gamma_n \alpha - 1)\frac{\partial K(\theta)}{\partial m} \tag{7.23}$$

where $K(\theta)$ and $D_m(\theta)$ are hydraulic conductivity and diffusivity for swelling soils respectively.

For the same swelling soil, Eqs. (7.21) and (7.23) for water movement and Eq. (7.22) for solute movement have been formulated in different material coordinates. This inconsistency is inconvenient for analyzing the concurrent flow of water and solute in the same swelling soil. To overcome this problem, let us reconcile the two material coordinates by re-writing Eq. (7.20) as:

$$\frac{dp}{dm} = \theta - \frac{v}{\frac{dm}{dt}} \tag{7.24}$$

where the term $\dfrac{dm}{dt}$ accounts for the rate of change of soil mass during swelling or shrinking. It is intuitive by observation to reason that during the wetting process, the rate of change in the swelling soil, $\dfrac{dm}{dt}$, is much larger than the flow velocity, v, due to absorption of water into pores being faster than the flow between pores, particularly at small times when advection is negligibly small (Philip 1992a). In the latter case, the relation $\dfrac{dm}{dt} \gg v$ holds. This relationship enables Eq. (7.20) to be approximated, by neglecting the second term on the right-hand side, as:

$$dp = \left(\frac{\partial p}{\partial m}\right) dm + \left(\frac{\partial p}{\partial t}\right) dt = \theta \, dm \tag{7.25}$$

which enables Eq. (7.22) to be written as:

$$\theta \frac{\partial c}{\partial t} = \frac{\partial}{\partial m}\left(D_m(c)\theta_l \theta_s \frac{\partial c}{\partial m} \right) \tag{7.26}$$

with the identity $\theta = \theta_l/\theta_s$.

Similar to Eq. (7.23) for water flow subject to gravity or advection, Eq. (7.26) for solute movement can be extended to account for convection as:

$$\theta \frac{\partial c}{\partial t} = \frac{\partial}{\partial m}\left(D_m(c)\theta_l \theta_s \frac{\partial c}{\partial m} \right) - \frac{\partial(vc)}{\partial m} \tag{7.27}$$

which will be extended later in this chapter to formulate fPDEs for solute transport in swelling media.

Water velocity in swelling soils, v, is given as (Smiles and Raats 2005):

$$v = -k(\theta_l)\theta_s \frac{\partial \theta}{\partial m}\frac{d\psi}{d\theta} - k(\theta_l)\left[(1 - \gamma\alpha) + \frac{1}{\theta_s}\frac{d\alpha}{d\theta} \int_m^0 \gamma \, dm \right] \tag{7.28}$$

where $k(\theta_l)$ denotes water-based hydraulic conductivity and ψ represents water potential.

As in many practical situations where the value of α is nearly constant for ranges of water content (Smiles and Raats 2005), the second term in the square brackets in Eq. (7.28) can be neglected for cases where the soil surface has a constant moisture content or is saturated, which leads to the following approximation:

$$v = -k(\theta_l)\left[\theta_s \frac{\partial \theta}{\partial m}\frac{d\psi}{d\theta} + (1 - \gamma\alpha) \right] \tag{7.29}$$

It has also been shown (Philip 1992b) that there exists a practical relationship $\dfrac{d\psi}{d\theta} \sim \theta^{-2}$ which can be used to simplify the procedure for determining water velocity in practice.

There are several mathematical models which relate the dispersion coefficient $D_m(c)$ in Eqs. (7.26) and (7.27) to velocity v for unsaturated soils, in addition to other variables such as space and time for non-swelling soils (Su et al. 2005). If we are concerned with the effects of velocity and concentration only, we have a similar expression for swelling soils:

$$D_m = av^{b0}c^b \tag{7.30}$$

where a, b_0 and b are constants. Consequently, with Eq. (7.30), Eq. (7.27) reads as:

$$\theta \frac{\partial c}{\partial t} = \frac{\partial}{\partial m}\left(D_{m0}\frac{\partial c^{b+1}}{\partial m}\right) - \frac{\partial(vc)}{\partial m} \tag{7.31}$$

where

$$D_{m0} = \frac{av^{b0}\theta_l\theta_s}{b+1} \tag{7.32}$$

with v given by Eq. (7.28) or approximated by Eq. (7.29).

Equation (7.31) is particularly interesting as it is a form of non-linear porous medium equation characterized by the term $\dfrac{\partial\theta^{b+1}}{\partial m}$; a similar form of fPDE has also been studied in terms of water flow in soils (Su 2014).

4. Fractional Partial Differential Equations for Anomalous Solute Transport in Soils

4.1 fPDEs for Conservative Solute Transport in Non-swelling Soils

4.1.1 fPDEs for Diffusion Only

The governing equations for anomalous solute transport in non-swelling porous media can be derived from the CTRW concept outlined in Chapter 4. Hence, Eq. (4.38) for solute transport in soils can be written as:

$$\frac{\partial^\beta c(\mathbf{x},t)}{\partial t^\beta} = D\frac{\partial^\lambda c(\mathbf{x},t)}{\partial|\mathbf{x}|^\lambda}, \qquad c(\mathbf{x},0) = \delta(\mathbf{x}) \tag{7.33}$$

Similar to the anomalous diffusivity of water in soils as discussed in Chapter 5, the anomalous diffusivity of solutes in soils in Eq. (7.33), according to Uchaikin and Saenko (2003), is:

$$D = \frac{\Gamma(n/2)\Gamma(1-\lambda/2)A}{2^\lambda\Gamma[(\lambda+n)/2]\Gamma(1-\beta)B} \tag{7.34}$$

for n dimensions. In one dimension, Eq. (7.34) is simplified as:

$$D = \frac{\pi^{1/2}\Gamma(1-\lambda/2)A}{2^\lambda\Gamma[(\lambda+1)/2]\Gamma(1-\beta)B} \tag{7.35}$$

where A and B are coefficients in the transitional probability density functions of CTRW theory.

Equation (7.33) in one dimension can be written as fractional solute gradient of the form (Podlubny 1999):

$$\frac{\partial^{\beta} c}{\partial t^{\beta}} = D \frac{\partial}{\partial x} \left(\frac{\partial^{n} c}{\partial x^{n}} \right) \tag{7.36}$$

or as fractional derivative of the solute gradient as:

$$\frac{\partial^{\beta} c}{\partial t^{\beta}} = D \frac{\partial^{n}}{\partial x^{n}} \left(\frac{\partial c}{\partial x} \right) \tag{7.37}$$

with $\eta = \lambda - 1$ when Riemman-Liouville (RL) fractional derivatives are used (Podlubny 1999).

The orders of fPDEs are related both to the exponents of the two probability functions in the CTRW model and the fractal geometry of porous media. The latter relationship was given by Zaslavsky (2002) as the space-time fractional diffusion-wave equation (fDWE), which can be written for solute transport as:

$$\frac{\partial^{\beta} c(x,t)}{\partial t^{\beta}} = D \left(\frac{L_s^{\lambda}}{L_t^{\beta}} \right)^{n} \frac{\partial^{\lambda} c(x,t)}{\partial x^{\lambda}} \tag{7.38}$$

where L_s and L_t are space and time scaling parameters respectively, n is the number of renormalization transformations and D is diffusivity.

If the condition $\lim_{n \to \infty} (L_s/L_t) = 1$ is met and when the space-time fPDE in Eq. (7.38) 'survives', the following relationship holds for the renormalization group of kinetics (RGK):

$$\frac{\beta}{\lambda} = \frac{\ln L_s}{\ln L_t} = \frac{\mu}{2} \tag{7.39}$$

which relates the orders of fractional derivatives to the spatio-temporal-scaling parameters, L_s and L_t. Subsequently, Eq. (7.39) can be written as:

$$\mu = \frac{2 \ln L_s}{\ln L_t} = \frac{2\beta}{\lambda} \tag{7.40}$$

which is the transport exponent defining the patterns of diffusion, that is, $\mu < 1$ for sub-diffusion or slow diffusion, $\mu > 1$ for super-diffusion or fast diffusion, and $\mu = 1$ for normal diffusion or classical diffusion. The transport exponent for the fDWE for swelling soils takes an identical form as for non-swelling soils, but with different definitions and values of parameters.

When symmetrical backward and forward fractional derivatives are considered for non-swelling soils, the moment equation for mean fractional spatial displacement can be written as (Zaslavsky 2002):

$$\left\langle |x|^{\lambda} \right\rangle = \frac{\Gamma(1+\lambda)}{\Gamma(1+\beta)} D t^{\beta} \tag{7.41}$$

The fPDEs presented in this book do not always explicitly imply backward motion. However, as we have explained in section 6.4 in Chapter 2 that we adopt identical signs for Riesz symmetric fractional derivatives (SFDs) $\dfrac{\partial^{\lambda} \theta}{\partial |x|^{\lambda}}$ and $D_0^{\lambda} \theta$, hence $D_0^{\lambda} \theta = \dfrac{\partial^{\lambda} \theta}{\partial x^{\lambda}}$ is used throughout the text as a synonym of SFDs.

4.1.2 Dimensions of the Parameters in fPDEs

For the fractional diffusion coefficient in an fDE, Kilbas et al. (2006) suggested a new parameter to be added to the classical diffusion coefficient so that the dimension of the conventional parameter is retained, while ensuring a correct dimension in the time-fractional diffusion equation of order β. Their approach for the new fractional diffusion coefficient in the fDE is given as:

$$D = D_c \tau^{1-\beta} \tag{7.42}$$

where D_c is the classic diffusion coefficient with dimension $[L^2 T^{-1}]$ and τ is the new time constant parameter to accommodate the new dimensions.

Similar to the dimensions for water movement in soils discussed in Chapter 4, the approach by Compte (1997) can be used where the parameters $A = \dfrac{\tau_a}{\tau^\beta}$ and $D = \dfrac{\sigma^2}{\tau^\beta}$ are incorporated in the formulation of fPDEs.

4.2 fPDEs for Solute Transport with Diffusion and Convection in Non-swelling Soils

The fDWE in Eq. (7.33) in one dimension can be extended to account for convection as:

$$\frac{\partial^\beta c}{\partial t^\beta} = D \frac{\partial^\lambda c}{\partial x^\lambda} - \frac{\partial}{\partial x}(vc) \tag{7.43}$$

which is a result of the shift jump size distribution $(f^*(x - Vt)) = f(x)$; for water movement in higher dimensions, Eq. (7.43) can be extended with $(f^*(\mathbf{x} - Vt)) = f(\mathbf{x})$ in multiple dimensions (Zhang et al. 2009) as:

$$\frac{\partial^\beta c}{\partial t^\beta} = D \frac{\partial}{\partial x_i}\left(\frac{\partial^\eta c}{\partial x_i^\eta}\right) - \frac{\partial}{\partial z}(vc) \tag{7.44}$$

Equation (7.43) in the vertical direction can be written as fractional solute flux of the form (Podlubny 1999):

$$\frac{\partial^\beta c}{\partial t^\beta} = D \frac{\partial}{\partial z}\left(\frac{\partial^\eta c}{\partial z^\eta}\right) - \frac{\partial}{\partial z}(vc) \tag{7.45}$$

or as fractional derivative of gradient in the following form:

$$\frac{\partial^\beta c}{\partial t^\beta} = D \frac{\partial^\eta}{\partial z^\eta}\left(\frac{\partial c}{\partial z}\right) - \frac{\partial}{\partial z}(vc) \tag{7.46}$$

4.3 fPDEs for Conservative Solute Transport in Swelling Soils

The fPDEs for anomalous solute transport in swelling soils can also be derived with the aid of the CTRW concept, similar to fPDEs for water flow in swelling soils (Su 2014).

4.3.1 fPDEs for Diffusion in Swelling Soils

With the integral transform in Eq. (3.17), which changes the physical coordinate to the material coordinate, the fDWE for solute transport in swelling-shrinking media is written as:

$$\frac{\partial^\beta c}{\partial t^\beta} = D_{cm} \frac{\partial^\lambda c^{b+1}}{\partial m^\lambda} \tag{7.47}$$

where

$$D_{cm} = \frac{av^{b_0} \theta_l \theta_s}{\theta(b+1)} \tag{7.48}$$

is identical to Eq. (7.32). Eq. (7.47) can also be expressed as fractional solute flux in the form:

$$\frac{\partial^\beta c}{\partial t^\beta} = D_{cm} \frac{\partial}{\partial m} \left(\frac{\partial^\eta c^{b+1}}{\partial m^\eta} \right) \tag{7.49}$$

or as fractional derivative of gradient in the following form:

$$\frac{\partial^\beta c}{\partial t^\beta} = D_{cm} \frac{\partial^\eta}{\partial m^\eta} \left(\frac{\partial c^{b+1}}{\partial m} \right) \tag{7.50}$$

with $\lambda = \eta + 1$.

For swelling soils, the moment equation is written as:

$$\langle |m|^\lambda \rangle = \frac{\Gamma(1+\lambda)}{\Gamma(1+\beta)} D t^\beta \tag{7.51}$$

and the parameters are related through the following relationship:

$$\frac{\beta}{\lambda} = \frac{\ln L_m}{\ln L_t} = \frac{\mu}{2} \tag{7.52}$$

which connects the orders of fractional derivatives to the mass-temporal-scaling parameters, L_m and L_t. Equation (7.52) can also be written as:

$$\mu = \frac{2 \ln L_m}{\ln L_t} = \frac{2\beta}{\lambda} \tag{7.53}$$

which connects the orders of fractional derivatives in time and mass with time- and mass-fractal measures of the medium.

4.3.2 fPDEs for Solute Transport with Diffusion and Convection in Swelling Soils

When solute transport takes place in a convective field or in flowing water in a medium, Eq. (7.49) is extended to the following form:

$$\frac{\partial^\beta c}{\partial t^\beta} = D_{cm} \frac{\partial^\lambda c^{b+1}}{\partial m^\lambda} - \frac{\partial}{\partial m}(vc) \tag{7.54}$$

with velocity v given by Eq. (7.28) and its approximation in Eq. (7.29). The diffusion term in Eq. (7.54) can be written either as a derivative of fractional flux or as fractional derivative of flux (see Eqs. (7.49) and (7.50)).

4.4 fPDEs for Anomalous Reactive Solute Transport in Non-swelling and Swelling Soils

4.4.1 Anomalous Equilibrium Solute Transport

For *anomalous equilibrium* solute transport, Eq. (7.4) without source terms reads as:

$$R\frac{\partial^\beta c}{\partial t^\beta} = D\frac{\partial^\lambda c}{\partial z^\lambda} - \frac{\partial}{\partial z}(vc) \tag{7.55}$$

where R is the retardation factor given by Eq. (7.5). For solute transport in swelling soils, the equivalent formulation is:

$$R\frac{\partial^\beta c}{\partial t^\beta} = D_{cm}\frac{\partial^\lambda c^{b+1}}{\partial m^\lambda} - \frac{\partial}{\partial m}(vc) \tag{7.56}$$

4.4.2 Non-equilibrium Anomalous Solute Transport in Soils with Mobile and Immobile Zones

For non-equilibrium solute transport, different approaches can be taken with the mobile-immobile model as the most popular one (Nielsen et al. 1986). The corresponding fPDEs have been outlined in Chapters 4, 5, and 6 for water flow in soils. Here, these concepts are applied to anomalous solute transport in soils.

To describe solute transport in soils with a mobile-immobile structure, most of the quantitative reports published since Rubinshtein's work in 1948 till 2003 are based on integer PDEs. Schumer et al. (2003b) presented a distributed-order fPDE (dofPDE) for solute transport in aquifers. Recently, the mobile-immobile concept has also been incorporated in a set of fPDE-based models for water movement in non-swelling and swelling soils and aquifers (Su 2012, 2014, 2017a). The orders of dofPDEs, which represent mobile and immobile zones of porous media, could change subject to the levels of swelling/shrinking measured by saturation or moisture ratio in each zone of swelling aggregates, such as in the works of Gerasimov et al. (2010). In other words, the two corresponding orders of the dofPDE, β_1 and β_2, which are defined below representing the different flow patterns in the two regions, differ in the two zones.

With the nomenclature used here, we have $0 < \beta_1 < 2$ for flow in immobile zones, and $0 < \beta_2 < 2$ for flow in mobile zones. With these ranges, it is accepted that the flow is subject to:

1. slow diffusion if $0 < \beta_1 < 1$ and/or $0 < \beta_2 < 1$;
2. classical diffusion if $\beta_1 = 1$ and/or $\beta_2 = 1$; and
3. super diffusion if $1 < \beta_1 < 2$ and/or $1 < \beta_2 < 2$.

Further mathematical background on distributed-order fractional differential equations (dofDEs), dofPDEs and their applications can be found in literature

(Caputo 1995, 2001, Lorenzo and Hartley 2000, Chechkin et al. 2002, 2003, Sokolov et al. 2004, Gorenflo and Mainardi 2005, Umarov and Gorenflo 2005a, Zhang et al. 2006a, 2008, 2009, 2014, Mainardi et al. 2007, 2008, Atanackovic et al. 2009a, 2009b, Sun et al. 2019).

By introducing distributed orders in fPDEs to account for solute transport in two forms of pores in soils, consistency can be achieved with fPDE-based models for water flow in unsaturated soils (Su 2012, 2014) and saturated groundwater (Su et al. 2015, Su 2017a). Hence, the equation for solute transport in non-swelling soils with retardation, solute production and decay included (Nielsen et al. 1986) can be modified as a distributed-order fractional form of Eq. (7.4):

$$b_{im}R_{im}\frac{\partial^{\beta_1}c}{\partial t^{\beta_1}}+b_m R_m\frac{\partial^{\beta_2}c}{\partial t^{\beta_2}}=D\frac{\partial^{\lambda}c}{\partial z^{\lambda}}-\frac{\partial^{\eta}}{\partial z^{\eta}}(vc)-\mu c+\sigma \qquad (7.57)$$

where R is retardation factor, based on linear adsorption, given as:

$$R=1+\rho k/\theta \qquad (7.58)$$

$$\mu=\mu_l+\mu_s\rho/\theta \qquad (7.59)$$

$$\gamma=\gamma_l+\gamma_s\rho/\theta \qquad (7.60)$$

λ and η being orders of space fractional derivatives. The relative water fractions (or porosities) in the immobile and mobile zones are expressed as:

$$b_{im}=\theta_{im}\tau^{\beta_1-1} \qquad (7.61)$$

$$b_m=\theta_m\tau^{\beta_2-1} \qquad (7.62)$$

with τ as a parameter, which follows Kilbas et al. (2006), by modifying the diffusion coefficient for the fractional diffusion equation.

As for Eqs. (7.36) and (7.37) for fDWEs for non-swelling soils, when RL fractional derivatives are used, an alternative formulation for Eq. (7.57) based on fractional flux is:

$$b_{im}R_{im}\frac{\partial^{\beta_1}c}{\partial t^{\beta_1}}+b_m R_m\frac{\partial^{\beta_2}c}{\partial t^{\beta_2}}=D\frac{\partial}{\partial z}\left(\frac{\partial^{\eta}c}{\partial z^{\eta}}\right)-\frac{\partial^{\eta}}{\partial z^{\eta}}(vc)-\mu c+\sigma \qquad (7.63)$$

Equation (7.57) can also be formulated as fractional derivative of solute gradient as:

$$b_{im}R_{im}\frac{\partial^{\beta_1}c}{\partial t^{\beta_1}}+b_m R_m\frac{\partial^{\beta_2}c}{\partial t^{\beta_2}}=D\frac{\partial^{\eta}}{\partial z^{\eta}}\left(\frac{\partial c}{\partial z}\right)-\frac{\partial^{\eta}}{\partial z^{\eta}}(vc)-\mu c+\sigma \qquad (7.64)$$

or in the following mixed form:

$$b_{im}R_{im}\frac{\partial^{\beta_1}c}{\partial t^{\beta_1}}+b_m R_m\frac{\partial^{\beta_2}c}{\partial t^{\beta_2}}=D\frac{\partial^{\lambda_1}}{\partial z^{\lambda_1}}\left(\frac{\partial^{\lambda_2}c}{\partial z^{\lambda_2}}\right)-\frac{\partial^{\eta}}{\partial z^{\eta}}(vc)-\mu c+\sigma \qquad (7.65)$$

with $\lambda=\lambda_1+\lambda_2$.

Similar to Eq. (7.57) for non-swelling soils, the fractional equivalent for swelling soils is a modified form of Eq. (7.31):

$$b_{im}R_{im}\frac{\partial^{\beta_1}c}{\partial t^{\beta_1}}+b_mR_m\frac{\partial^{\beta_2}c}{\partial t^{\beta_2}}=D_{cm}\frac{\partial^{\lambda}c^{b+1}}{\partial m^{\lambda}}-\frac{\partial^{\eta}}{\partial m^{\eta}}(vc)-\mu c+\sigma \tag{7.66}$$

where the material coordinate m is used (see also Su 2014, 2017a). It should be noted that for swelling soils, the material coordinate m is used instead of the conventional physical coordinate z. The different fractional combinations of the diffusion term in Eq. (7.66) for swelling soils can be formulated in a similar manner as Eqs. (7.63), (7.64) and (7.65) for non-swelling soils.

The parameters in the above dofPDEs vary in the following ranges:

$$\left.\begin{array}{ll} b_{im}>0, b_m>0; & b_{im}+b_{im}=1 \\ 0<\beta_1\le 1; & 0<\beta_2\le 1 \\ 0<\lambda\le 2; & 0<\eta<1 \end{array}\right\} \tag{7.67}$$

4.5 Tempered Anomalous Solute Transport

The two-term fPDEs presented above incorporate different retardation factors and different orders of fractional derivatives for mobile and immobile zones. Meerschaert et al. (2008) presented a modified two-term distributed-order time-fractional PDE to model solute transport in heterogeneous porous media with tempered diffusion. The tempered anomalous transport is featured with an exponential memory function for delayed power-law waiting times in the immobile zone. With this approach, with Eq. (7.57) as an example without source/sink terms, the fPDE for *total solute* is expressed as:

$$\frac{\partial c}{\partial t}+b_{im}e^{-\varepsilon t}\frac{\partial^{\beta_1}}{\partial t^{\beta_1}}(e^{\varepsilon t}c)=D\frac{\partial^2 c}{\partial z^2}-\frac{\partial}{\partial z}(vc)+b_{im}\varepsilon^{\beta_1}c+m_0b_{im}g(t)\delta(x) \tag{7.68}$$

where ε is the tempering parameter, m_0 is the total mass initially at the boundary of the mobile-immobile interface, e is the exponential function, and $g(t)$ is the memory function written as:

$$g(t)=\int_t^{\infty}e^{-\varepsilon\tau}\frac{\beta_1}{\Gamma(1-\beta_1)}\tau^{-\beta_1-1}d\tau \tag{7.69}$$

Equation (7.68) is formulated for total solute concentration in both mobile and immobile zones (Meerschaert et al. 2008). The tempered solute transport in the *mobile zone* is of the form (Meerschaert et al. 2008):

$$\frac{\partial c_m}{\partial t}+b_{im}e^{-\varepsilon t}\frac{\partial^{\beta_1}}{\partial t^{\beta_1}}(e^{\varepsilon t}c_m)=D\frac{\partial^2 c_m}{\partial z^2}-\frac{\partial}{\partial z}(vc_m)+b_{im}\varepsilon^{\beta_1}c_m \tag{7.70}$$

The formulation by Meerschaert et al. (2008) does not incorporate spatial fractional derivatives; once spatial fractional derivatives are considered, the diffusion term can be modified as $D\dfrac{\partial^{\lambda}c}{\partial z^{\lambda}}$ which can be split into different forms such as in Eqs. (7.63) to Eq. (7.65). The material coordinate can also be introduced in the tempered fPDEs to account for the swelling-shrinking properties of soils.

4.6 Multi-term fPDEs for Solute Transport in Multifractal Media

Pores within soil particles and aggregates essentially have multiple levels of structures which can be interpreted by the concept of multifractals (Martin et al. 2009). Multifractals are geometrical patterns generated by continuously distributed variables (such as pore size, porosity, etc.), a property exhibiting a multitude of power-law scaling relations (Perfect et al. 2009). For anomalous solute transport in multifractal media, multi-term fPDEs can be used.

In addition to the above fPDEs which were specified with two terms for mobile and immobile zones, a more generic multi-term fPDE was given by Jiang et al. (2012a):

$$P(D_t)c(x,t) = R\sum_{j=1}^{s} b_i D_t^{\beta_i} \tag{7.71}$$

where $0 \leq \beta_s < ...\beta_1 < \beta \leq 1$ are the orders of the time-fractional diffusion equation, or $0 \leq \beta_s < ...\beta_1 < \beta \leq 2$ are the orders of the time-fractional diffusion-wave equation with $j = 1,2,...s$; D_t^β and $D_t^{\beta_1}$ are Caputo fractional derivatives; and the spatial part of the fPDE with the source and sink takes the following form:

$$\frac{\partial^\lambda (k_\lambda c)}{\partial |x|^\lambda} - \frac{\partial^\eta (k_\eta c)}{\partial |z|^\eta} - \mu c + \sigma \tag{7.72}$$

where $\dfrac{\partial^\lambda (k_\lambda c)}{\partial |x|^\lambda}$ and $\dfrac{\partial^\eta (k_\eta c)}{\partial |z|^\eta}$ are Riesz space-fractional derivatives (Saichev and Zaslavsky 1997, Jiang et al. 2012a) for diffusion and convection respectively.

With symmetrical fractional derivatives, the time- and space-fractional PDE for solute movement in soils, with a fractional convection term included, the multi-term fPDE can be written as follows:

$$b_1 R_1 \frac{\partial^{\beta_1} c}{\partial t^{\beta_1}} + b_2 R_2 \frac{\partial^{\beta_2} c}{\partial t^{\beta_2}} + ...b_i R_i \frac{\partial^{\beta_i} c}{\partial t^{\beta_i}} = \frac{\partial^\lambda (Dc)}{\partial |z|^\lambda} - \frac{\partial^\eta (vc)}{\partial |z|^\eta} - \mu c + \sigma, \quad i = 1,2,...n, \tag{7.73}$$

where the multiple temporal terms on the left-hand side of Eq. (7.73) represent solute movement at different levels of pores, consistent with a quantitative representation of the multifractal concept for water movement and solute transport in porous media.

The above formulations are in one dimension. Umarov and Gorenflo (2005b) showed that the CTRW concept can be applied to multi-dimensional problems, which results in an n-dimensional fPDE. For solute transport in mobile and immobile zones, Eq. (7.73) can be extended to different dimensions ($i = 1,2,3$) by expressing it in a vector form:

$$\sum_{i=1}^{n} b_i R_i \frac{\partial^{\beta_i} c}{\partial t^{\beta_i}} = \frac{\partial^\lambda (\mathbf{D}c)}{\partial |\mathbf{x}|^\lambda} - \frac{\partial^\eta (vc)}{\partial |z|^\eta} - \mu c + \sigma \tag{7.74}$$

Note that when the swelling property of soil is included, the original fPDE is formulated for the vertical direction so that the material coordinate m applies to the one-dimensional case only, that is, the multi-term fPDE for solute transport in swelling soils takes the following form:

$$b_1 R_1 \frac{\partial^{\beta_1} c}{\partial t^{\beta_1}} + b_2 R_2 \frac{\partial^{\beta_2} c}{\partial t^{\beta_2}} + ... b_i R_i \frac{\partial^{\beta_i} c}{\partial t^{\beta_i}} = \frac{\partial^{\lambda}(D_{cm} c^{b+1})}{\partial |m|^{\lambda}} - \frac{\partial^{\eta}(vc)}{\partial |m|^{\eta}} - \mu c + \sigma \qquad (7.75)$$

While multi-term time-fractional derivatives are mathematically ideal for describing fluid mechanics in multifractal porous media, the difficulty in parameter estimation increases as the number of terms increases. The two-term temporal fPDEs are equivalent to:

$$P(D_t)c(x, t) = (\theta_{im}D_t^{\beta_1} + \theta_m D_t^{\beta_2}) \qquad (7.76)$$

For further information on backward and forward fractional derivatives, the reader is referred to Saichev and Zaslavsky (1997) and Jiang et al. (2012a). Justification for the two-term dofPDE for solute movement has been provided in the works of Chechkin et al. (2002), Schumer et al. (2003b) and Su (2012, 2014). In the following section, we discuss dimensions and present solutions of the above dofPDEs for solute movement in soils.

5. Dimensions of the Parameters in Multi-term fPDEs

In section 4.1.2, the dimension of the diffusion coefficient is modified with the approach taken by Kilbas et al. (2006). Here, we modify the other parameters in two-term fPDEs for solute transport in soils with mobile-immobile zones as follows:

$$b_{im} = \frac{\theta_{im}}{\theta} \tau^{\beta_1 - 1} \qquad (7.77)$$

$$b_m = \frac{\theta_m}{\theta} \tau^{\beta_2 - 1} \qquad (7.78)$$

and the new fractional diffusion coefficient in the fPDE is given by:

$$D_f = D_0 \tau_d^{\lambda - 2} \qquad (7.79)$$

Hence, velocity is expressed as:

$$v_f = v \tau_v^{\eta - 1} \qquad (7.80)$$

where τ, τ_d and τ_v are parameters for dimension amendments, and b_1, b_2, θ_{im}, θ_m and θ have been defined earlier.

When multi-term fPDEs are used, the amended dimensions for multi-term fPDEs are as follows:

$$\left. \begin{aligned} b_1 &= \frac{\theta_1}{\theta} \tau^{\beta_1 - 1} \\ b_2 &= \frac{\theta_2}{\theta} \tau^{\beta_2 - 1} \\ &\ldots\ldots \\ b_n &= \frac{\theta_n}{\theta} \tau^{\beta_n - 1} \end{aligned} \right\} \qquad (7.81)$$

where θ_1, θ_2, ...,θ_n are the moisture ratios in the 1,2, ...,n levels of pores.

Using two-term fPDEs for reactive solute transport, the dimensions in Eq. (7.73) are amended as:

$$b_{im} R_{im} \frac{\partial^{\beta_1} c}{\partial t^{\beta_1}} + b_m R_m \frac{\partial^{\beta_2} c}{\partial t^{\beta_2}} = D_f \frac{\partial^{\lambda} c}{\partial z^{\lambda}} - \frac{\partial^{\eta}}{\partial z^{\eta}}(v_f c) - \mu c + \sigma \tag{7.82}$$

where the diffusion term is written as a fractional derivative which can be formulated as follows:

$$b_{im} R_{im} \frac{\partial^{\beta_1} c}{\partial t^{\beta_1}} + b_m R_m \frac{\partial^{\beta_2} c}{\partial t^{\beta_2}} = D_f \frac{\partial}{\partial z}\left(\frac{\partial^{\eta} c}{\partial z^{\eta}}\right) - \frac{\partial^{\eta}}{\partial z^{\eta}}(vc) - \mu c + \sigma \tag{7.83}$$

For reactive transport in swelling soils, the equation takes the following forms:

$$b_{im} R_{im} \frac{\partial^{\beta_1} c}{\partial t^{\beta_1}} + b_m R_m \frac{\partial^{\beta_2} c}{\partial t^{\beta_2}} = D_f \frac{\partial^{\lambda} c^{b+1}}{\partial m^{\lambda}} - \frac{\partial^{\eta}}{\partial m^{\eta}}(v_f c) - \mu c + \sigma \tag{7.84}$$

$$b_{im} R_{im} \frac{\partial^{\beta_1} c}{\partial t^{\beta_1}} + b_m R_m \frac{\partial^{\beta_2} c}{\partial t^{\beta_2}} = D_f \frac{\partial}{\partial m}\left(\frac{\partial^{\eta} c^{b+1}}{\partial m^{\eta}}\right) - \frac{\partial^{\eta}}{\partial m^{\eta}}(v_f c) - \mu c + \sigma \tag{7.85}$$

The above equations can also be written in terms of fractional derivatives of gradient. With the above formulations, this set of fPDEs for solute transport in both swelling and non-swelling soils is comparable to the fPDEs for water flow in similar soils (Su 2014).

6. Functional-order fPDEs

6.1 Introduction

As briefly outlined in section 6 of Chapter 4, the orders of fractional derivatives can be a number or a function. This approach was pioneered by Unterberger and Bokobza (1965a) and Višik and Èskin (1967) and followed by Samko and Ross (1993), Jacob and Leopold (1993) and many others. When symmetrical notations are used, functional-order or variable-order fPDEs with two temporal terms can be written as:

$$b_{im} R_{im} \frac{\partial^{\beta_1(z,t)} c}{\partial t^{\beta_1(z,t)}} + b_m R_m \frac{\partial^{\beta_2(z,t)} c}{\partial t^{\beta_2(z,t)}} = \frac{\partial^{\lambda(z,t)}}{\partial |z|^{\lambda(z,t)}}(D_f c) - \frac{\partial^{\eta(z,t)}(vc)}{\partial |z|^{\eta(z,t)}} - \mu c + \sigma \tag{7.86}$$

for non-swelling soils, and as the following equation for swelling soils:

$$b_{im} R_{im} \frac{\partial^{\beta_1(m,t)} c}{\partial t^{\beta_1(m,t)}} + b_m R_m \frac{\partial^{\beta_2(m,t)} c}{\partial t^{\beta_2(m,t)}} = \frac{\partial^{\lambda(m,t)}}{\partial |m|^{\lambda(m,t)}}(D_f c^{b+1}) - \frac{\partial^{\eta(m,t)}(vc)}{\partial |m|^{\eta(m,t)}} - \mu c + \sigma \tag{7.87}$$

Furthermore, the order of fractional derivatives can be any function.

6.2 Functional-order fPDEs for Solute Transport in Multifractal Media

In section 4.6, multi-term fPDEs of fractional orders were outlined. When the orders of fractional derivatives are functions, the mathematical representation of solute transport in multifractal porous media is modified as:

$$\sum_{j=1}^{n} a_j D_t^{\beta_j(c(x_i,t))} c(x_i,t) = \nabla^{\lambda(c(x_i,t))}(\mathbf{D}c) - \frac{\partial}{\partial z}(vc) \tag{7.88}$$

with $i = 1,2,3$ and $j = 1,2, \dots, n$, which can also incorporate retardation, solute decay and production terms when needed.

With a large number of fractional terms in the formulation, the difficulty of evaluating the parameters β_j with real data increases and some compromises are needed to balance the accuracy of the model with the available data for parameter estimation.

The one-dimensional form of Eq. (7.88) for non-swelling soils can be written in three different forms with the following forms as examples:

1. fractional solute gradient:

$$\sum_{j=1}^{n} a_j D_t^{\beta_j(c(x,t))} c(x,t) = \frac{\partial}{\partial x}\left(D(x,c) \frac{\partial^{\eta(c(x,t))}}{\partial x^{\eta(c(x,t))}} c(x,t) \right) - \frac{\partial}{\partial z}(vc(x,t)) \tag{7.89}$$

2. fractional derivative of solute gradient:

$$\sum_{j=1}^{n} a_j D_t^{\beta_j(c(x,t))} c(x,t) = \frac{\partial^{\eta(c(x,t))}}{\partial x^{\eta(c(x,t))}}\left(D(z,c) \frac{\partial c(x,t)}{\partial z} \right) - \frac{\partial}{\partial z}(vc(x,t)) \tag{7.90}$$

3. combination of fractional flux and fractional derivative of fractional flux:

$$\sum_{j=1}^{n} a_j D_t^{\beta_j(c(x,t))} c(x,t) = \frac{\partial^{\eta_1(c(x,t))}}{\partial x^{\eta_1(c(x,t))}}\left(D(x,c) \frac{\partial^{\eta_2(c(x,t))} c(x,t)}{\partial x^{\eta_2(c(x,t))}} \right) - \frac{\partial}{\partial x}(vc(x,t)) \tag{7.91}$$

Following the discussions on solute transport in swelling media, material coordinates can be used for an easy modification of Eqs. (7.89), (7.90) and (7.91) by changing the fixed coordinate x to the material coordinate m and using v for flow velocity in swelling soils, as discussed earlier. One form of fPDEs equivalent to Eq. (7.90) for swelling soils is:

$$\sum_{j=1}^{n} a_j D_t^{\beta_j(c(m,t))} c(x,t) = \frac{\partial^{\eta(c(m,t))}}{\partial m^{\eta(c(m,t))}}\left(D(m,c) \frac{\partial c(m,t)}{\partial m} \right) - \frac{\partial}{\partial m}(vc(m,t)) \tag{7.92}$$

As discussed in Chapter 4, the orders of fractional derivatives can be functions of time, space, solute concentration, temperature, or functions of any one or more of these variables, such that the functional order is in the form of $\beta(c(x, t))$, $\beta(x, t)$, $\beta(x)$ or $\beta(t)$. These combinations apply to λ and η as well.

In addition to the fPDEs in different forms presented above, some special forms of fPDEs can be found in the works of Sandev et al. (2015) and Jiang et al. (2012a, 2012b).

7. The fPDE and Its Solution for Solute Exchange between Mobile and Immobile Zones

7.1 The fPDE for Solute Exchange between Mobile and Immobile Zones

In soils with mobile and immobile zones, water flow and solute movement take place in the two zones at different rates and with different flow patterns (captured by the parameters for each zone). As solutes are entrained in flowing water, the analysis of solute movement in soils with mobile and immobile zones is similar to that for water movement in such soils (Su 2012, 2014).

The fractional models for analyzing the exchange of solute between the two zones are nearly identical to the conventional mobile-immobile models widely used in literature (Nielsen et al. 1986, Sardin et al. 1991), except for the order of fractional derivatives. Similar to water exchange between large and small pores (Su 2012), the equivalent fractional model for solute exchange can be written as:

$$b_{im}R_{im}\frac{\partial^{\beta_3}c_{im}}{\partial t^{\beta_3}} = \varepsilon(c - c_{im}), \qquad 0 < \beta_3 \le 1 \tag{7.93}$$

or

$$\frac{\partial^{\beta_3}c_{im}}{\partial t^{\beta_3}} = \rho c_{im} + h \tag{7.94}$$

with

$$\rho = -\frac{\varepsilon}{b_{im}R_{im}} \tag{7.95}$$

and

$$h = \frac{\varepsilon c}{b_{im}R_{im}} \tag{7.96}$$

where c and c_{im} are the concentrations of solutes in mobile and immobile zones respectively; β_3 is the order of fractional derivatives (it is reasonable to assume that $\beta_3 = \beta_1$, but β_3 can be variable or identical for moisture and solute exchanges between mobile and immobile zones); R_{im} denotes retardation factor (for conservative solutes, $R_{im} = 1$); and ε is the mass transfer rate between mobile and immobile zones.

In the present fractional formulation, the dimension of ε is different from its conventional counterpart. If the conventional dimension is to be retained, another dimension correction has to be added, such as $b_{im} = \frac{\theta_{im}}{\theta}\tau_3^{\beta_3-1}$.

We examine solute exchanges between mobile and immobile zones with two types of ICs at the interface of these zones:

IC 1:

$$\frac{\partial^{\beta_3}c_{im}}{\partial t^{\beta_3}} = v_k, \qquad k = 1, 2, \ldots, n, \qquad t = 0, \tag{7.97}$$

where v_k is the fractional rate of change in solute contents in the immobile zone.

IC 2:

$$c_{im} = c_{im0}, \qquad\qquad t = 0, \qquad\qquad (7.98)$$

where c_{im0} is the initial value of c_{im}.

7.2 Exact Solution for Solute Exchange at the Interface when the Initial Rate of Change in Solute Concentration is Known

In this case, Eq. (7.94) is solved with the condition in Eq. (7.97), which is similar to the method presented for distributed-order movement of water in mobile and immobile pores (Su 2012). In the case of $c = c_0$, with c_0 being the constant value of c on the surface of the mobile zone (equivalent to constant h), $\dfrac{\partial^{\beta_3} c_{im}}{\partial t^{\beta_3}}$ is replaced by $\dfrac{d^{\beta_3} c_{im}}{dt^{\beta_3}}$. Hence, the derived solution is of the following form:

$$c_{im} = \sum_{k=1}^{n} v_k t^{\beta_3-k} E_{\beta_3,\beta_3-k+1}(\rho t^{\beta_3}) + \int_0^t (t-\tau)^{\beta_3-1} E_{\beta_3,\beta_3}[\rho(t-\tau)^{\beta_3}]h\,d\tau \qquad (7.99)$$

where $E_{\beta_3,\beta_3-k+1}(\rho t^{\beta_3})$ and $E_{\beta_3,\beta_3}[\rho(t-\tau)^{\beta_3}]$ are the Wiman functions or two-parameter Mittag-Leffler functions (MLFs).

With constant h, Eq. (7.99) can be integrated (Kilbas et al. 2004) to give:

$$c_{im} = \sum_{k=1}^{n} v_k t^{\beta_3-k} E_{\beta_3,\beta_3-k+1}\left(-\frac{\varepsilon}{b_{im}R_{im}}t^{\beta_3}\right) + \frac{\varepsilon c_0}{b_{im}R_{im}}t^{\beta_3} E_{\beta_3,\beta_3+1}\left(-\frac{\varepsilon}{b_{im}R_{im}}t^{\beta_3}\right) \qquad (7.100)$$

where $E_{\beta_3,\beta_3+1}(\rho t^{\beta_3})$ is the WF.

By retaining the first term in the summation, Eq. (7.100) yields:

$$c_{im} = v_1 t^{\beta_3-1} E_{\beta_3,\beta_3}\left(-\frac{\varepsilon}{b_{im}R_{im}}t^{\beta_3}\right) + \frac{\varepsilon c_0}{b_{im}R_{im}}t^{\beta_3} E_{\beta_3,\beta_3+1}\left(-\frac{\varepsilon}{b_{im}R_{im}}t^{\beta_3}\right) \qquad (7.101)$$

As it has been seen in Chapter 6, $k = 1$ in Eq. (7.100) is, in fact, the only physically relevant solution. Hence, the solution in Eq. (7.101) is the physically exact solution to be analyzed further in section 7.4.

7.3 Exact Solution for Solute Exchange at the Interface when the Initial Solute Concentration is Known

With a constant solute concentration ($c = c_0$) on the mobile-immobile zone interface of the porous medium, Eq. (7.94) is to be solved with Eq. (7.98). Subsequently, the solution is given as (Kilbas et al. 2006):

$$c_{im} = c_{im0} E_{\beta_3}(\rho t^{\beta_3}) + \int_0^t (t-\tau)^{\beta_3-1} E_{\beta_3,\beta_3}[\rho(t-\tau)^{\beta_3}]h\,d\tau \qquad (7.102)$$

where $E_{\beta_3}(\rho t^{\beta_3})$ is the MLF and $E_{\beta_3,\beta_3}[\rho(t-\tau)^{\beta_3}]$ is the Wiman function.

With $c = c_0$, which is equivalent to constant h, Eq. (7.102) is integrated (Kilbas et al. 2004) to yield:

$$c_{im} = c_{im0} E_{\beta_3}\left(-\frac{\varepsilon}{b_{im}R_{im}}t^{\beta_3}\right) + \frac{\varepsilon c_0}{b_{im}R_{im}}t^{\beta_3} E_{\beta_3,\beta_3+1}\left(-\frac{\varepsilon}{b_{im}R_{im}}t^{\beta_3}\right) \tag{7.103}$$

where $E_{\beta_3}\left(-\dfrac{\varepsilon}{b_{im}R_{im}}t^{\beta_3}\right)$ and $E_{\beta_3,\beta_3+1}\left(-\dfrac{\varepsilon}{b_{im}R_{im}}t^{\beta_3}\right)$ are MLFs.

As the movement of water and solute in the immobile zone has a greater tendency to be subject to sub-diffusion, the case of $0 < \beta_3 < 1$ is more likely.

7.4 Asymptotic Solutions

7.4.1 Asymptotic Solutions when the Initial Rate of Change of Solute Concentration is Known

To carry out the asymptotic analysis, we write the temporal part of the function in Eq. (7.101) as the generalized Prabhakar function (GPF) (Mainardi and Garrappa 2015):

$$e_{\alpha,\beta}(t;\rho) = t^{\beta-1}E_{\alpha,\beta}(-\rho t^{\alpha}), \qquad t \geq 0 \tag{7.104}$$

which has two forms of asymptotic solutions (under the two conditions: $t \to +\infty$ and $t \to 0$) as discussed in asymptotic solutions relevant to water flow in soils in Chapter 6:

1. $t \to +\infty$: In this case, the asymptotic solution of Eq. (7.101) can be derived by following the procedures presented by Mainardi and Garrappa (2015) (see also Chapter 6):

$$c_{im} \sim \left[\frac{\beta_3 v_1}{\Gamma(1-\beta_3)t^{\beta_3+1}} + c_0\right]\frac{\varepsilon}{b_{im}R_{im}}, \qquad t \to +\infty \tag{7.105}$$

2. $t \to 0$: Under this condition, the asymptotic solution is given by (Friedrich 1991):

$$c_{im} \sim \left[\frac{v_1}{t\Gamma(\beta_3)} + \frac{c_0}{\Gamma(\beta_3+1)}\right]\frac{\varepsilon}{b_{im}R_{im}}t^{\beta_3}, \qquad t \to 0 \tag{7.106}$$

which implies that as time approaches the onset, the solute ratio decreases on the mobile-immobile interface, whereas it increases on the surface of the soil as a result of $\beta_3 \leq 1$.

The asymptotic analyses by Mainardi and Garrappa (2015) (also in Chapter 6) establish the fact that $k = 1$, indicating that the one-term approximate solution in Eq. (7.101) is essentially the only physically relevant solution and is exact.

7.4.2 Asymptotic Solutions when the Initial Solute Concentration at the Interface of Mobile and Immobile Zones is Known

In this case, Eq. (7.94) is solved with the condition in Eq. (7.98) to yield a solution in Eq. (7.103). The following two cases of asymptotes are considered here:

1. $t \to +\infty$: The asymptotic solution in this case is:

$$c_{im} \sim \left[\frac{c_{im0}\beta_3 t^{-\beta_3-1}}{\Gamma(1-\beta_3)} + c_0 \right] \frac{\varepsilon}{b_{im} R_{im}}, \qquad t \to +\infty \qquad (7.107)$$

2. $t \to 0$: Under this condition, the asymptotic solution is given as (Friedrich 1991):

$$c_{im} \sim (c_{im0} + c_0 t^{\beta_3}) \frac{\varepsilon}{b_{im} R_{im}}, \qquad t \to 0 \qquad (7.108)$$

7.5 A Note on the Solutions of Eq. (7.93)

Regarding the solutions Eq. (7.93), two issues are worth a brief mention:

1. The above analyses have been demonstrated for reactive solutes with $R_{im} \neq 1$. The above results also apply for conservative solutes, with $R_{im} = 1$.

2. The solutions discussed in the previous sections have been derived for the case of $c = c_0$ (equivalent to constant h), resulting in $\dfrac{\partial^{\beta_3} c_{im}}{\partial t^{\beta_3}}$ being replaced by $\dfrac{d^{\beta_3} c_{im}}{dt^{\beta_3}}$ so that the fDE is solved. When $c \neq c_0$, other methods can be used to derive the solution, such as the Laplace transform demonstrated in the Appendix of this chapter.

8. Fractional Flux-Residential Solute Concentration Relationships during Anomalous Transport

In section 2.3, the relationships between flux and residential solute concentrations were outlined on the basis of integer PDEs for transport of solutes in porous media. When anomalous transport is considered, the original formulation (Carslaw and Jaeger 1947, 1959, Kreft and Zuber 1978) should be modified.

Depending on the units used for quantifying solute concentration, the latter can be measured either by flux concentration $c_f(x, t)$ defined as the mass of solutes per unit of fluid discharge or as residential concentration $c_{r,t}(x, t)$ defined as the total mass of solutes per unit of total fluid volume in both mobile and immobile zones. Of course, these terminologies can be defined for mobile or immobile zones only and the two components for residential solutes are simply related as follows:

$$c_{r,t}(x, t) = c_{r,m}(x, t) + c_{r,im}(x, t) \qquad (7.109)$$

with $c_{r,m}(x, t)$ and $c_{r,im}(x, t)$ being mobile and immobile residential solute components respectively. In a detailed report, Zhang et al. (2006a) outlined the relationship between the two types of concentrations, which were categorized into three cases: time-space-fractional ADE and two special cases.

With the distinction of total porosity of soil, ϕ, from porosity of mobile zones, ϕ_m, Zhang et al. (2006a) rewrote the relationship between flux concentration and residential solute concentration in Eq. (7.14) by differentiating it with respect to time:

$$c_f(x,t) = \frac{\phi}{v\phi_m} \frac{\partial}{\partial t} \int_x^\infty c_{r,t}(\xi,t) d\xi \qquad (7.110)$$

where c_f and $c_{r,t}$ refer to total flux concentration and residential solute concentration, with no distinction between mobile and immobile components. With Eq. (7.110), three cases have been discussed in the following sections.

8.1 Space-fractional PDE for Flux-Residential Solute Relationships

Based on the above relationship between flux concentration and residential solute concentration, the space fPDE presented by Zhang et al. (2006a) takes the following form:

$$\frac{\partial c_{r,t}}{\partial t} = D\frac{\partial^{\lambda} c_{r,t}}{\partial x^{\lambda}} - v\frac{\partial c_{r,t}}{\partial x} \tag{7.111}$$

The following relationship between flux and residential concentrations for anomalous solute transport also holds (Zhang et al. 2006a):

$$c_f(x,t) = c_{r,t}(x,t)\left(1 + \frac{x - vt}{\lambda vt}\right) \tag{7.112}$$

The above relationship applies to non-swelling soils; for swelling soils, the following relationship holds:

$$c_f(m,t) = c_{r,t}(m,t)\left(1 + \frac{m - vt}{\lambda vt}\right) \tag{7.113}$$

where velocity v is given by Eq. (7.28) or its approximation in Eq. (7.29).

8.2 Distributed-order Time-fractional Advection for Flux-Residential Solute Relationships

A special case of the distributed-order fPDE for solute transport in mobile and immobile zones (Schumer et al. 2003b) under strong convection or fast flow is the space-fractional advection equation (sfAE) without diffusion (Zhang et al. 2006a):

$$b_{mim}\frac{\partial^{\beta} c_{r,t}}{\partial t^{\beta}} + \frac{\partial c_{r,t}}{\partial t} = -v\frac{\partial c_{r,t}}{\partial x} \tag{7.114}$$

where the capacity coefficient is written as (Schumer et al. 2003b):

$$b_{mim} = \frac{\phi_{mim}}{\phi_m} \tag{7.115}$$

This special form of Eq. (7.114), when written in terms of flux concentration, is expressed as (Zhang et al. 2006a):

$$b_{mim}\frac{\partial^{\beta} c_f}{\partial t^{\beta}} + \frac{\partial c_f}{\partial t} = -v\frac{\partial c_f(x,t)}{\partial x} + b_{mim}\frac{t^{-\beta}\delta(x)}{\Gamma(1 - \beta)} \tag{7.116}$$

where $\delta(x)$ is the Dirac delta function. Under advection-dominated solute transport conditions, Schumer et al. (2003b) and Zhang et al. (2006a) showed that Eq. (7.116) also applies to the mobile phase residential concentration $c_{r,m}(x, t)$, that is, $c_f(x, t) = c_{r,m}(x, t)$ (Zhang et al. 2006a).

The relationship between flux and residential concentrations under advection is given as (Zhang et al. 2006a):

$$c_f(x,t) = \frac{\beta x}{b_2 \left[vt - x(1-\beta) \right]} c_{r,t}(x,t) \tag{7.117}$$

where

$$b_2 = \frac{\phi_m}{\phi_{total}} \tag{7.118}$$

is the ratio of porosity in the mobile zone, ϕ_m, to the total porosity, ϕ_{total}.

8.3 Time-space fPDE for Flux-Residential Solute Relationships

A more generic fPDE is the distributed-order fPDE for residential solute concentrations (Zhang et al. 2006a):

$$b_{mim} \frac{\partial^\beta c_{r,t}}{\partial t^\beta} + \frac{\partial c_{r,t}}{\partial t} = D \frac{\partial^\lambda c_{r,t}}{\partial x^\lambda} - v \frac{\partial c_{r,t}}{\partial x} \tag{7.119}$$

where b_{mim} is the capacity ratio defined in Eq. (7.115) for non-reactive solutes.

It has been shown (Zhang et al. 2006a) that the relationship between flux concentration c_f and mobile and immobile residential concentrations of solutes in non-swelling soils is described by the following expression:

$$c_f(x,t) = \left(\frac{x-vt}{\lambda vt} + 1 \right) c_{r,m}(x,t) + \frac{\beta b_{mim} x}{\lambda vt} c_{r,im}(x,t) \tag{7.120}$$

where $c_{r,m}(x, t)$ and $c_{r,im}(x, t)$ are the residential concentrations of solutes in mobile and immobile zones respectively.

For swelling soils, similar procedures, with the material coordinate transform, yield:

$$c_f(m,t) = \left(\frac{m-vt}{\lambda vt} + 1 \right) c_{r,m}(x,t) + \frac{\beta b_{mim} m}{\lambda vt} c_{r,im}(m,t) \tag{7.121}$$

where velocity v is given by Eq. (7.28) or its approximation in Eq. (7.29).

With a detailed analysis of solute transport in mobile and immobile zones, Zhang et al. (2014) discovered that a dofPDE in terms of temporally tempered Lévy motion is precisely the multi-rate mass transfer model for late-time solute behavior.

9. Applications of fPDEs for Coupled Solute Transport in Swelling and Non-swelling Soils

9.1 A Brief Description of the Methods for Solutions of the fPDEs

As for the solutions of integer PDEs, solutions of fPDEs for coupled water movement and solute movement in swelling and non-swelling soils can also be developed using analytical and/or numerical methods. In literature, fPDEs have been investigated using various methods, which include the following categories:

A. Decomposition iteration methods: This category includes the Adomian decomposition methods (Adomian 1983) and the variation iteration method (He 1998), etc. (Baleanu et al. 2014).

B. Similarity transformations: This method is similar to those used for integer PDEs (Buckwar and Luchko 1998, Djordjevic and Atanackovic 2008, Lizama 2012, Costa et al. 2015).

C. Separation of variables methods: This traditional method can also be applied to derive solutions of fPDEs (Chen et al. 2012, Jiang et al. 2012a, 2012b, Ding and Jiang 2013).

D. The Green function: This approach forms the solution of the desired problem by the superposition of the Green function and the input function (Podlubny 1999, Debnath 2003b). The method requires both the Green function and the desired solution to satisfy the zero boundary condition (Kevorkian 1990).

E. Numerical methods: Although numerical methods for fPDEs were pioneered only recently (Liu et al. 2002, 2003), their development over the past few decades has been rapid. According to Meerschaert and Tadjeran (2004, 2006), the first numerical methods for fPDEs were developed by Liu et al. (2002, 2003). Li and Chen (2018) provided an extensive review of numerical methods for fPDEs and noted that, compared to numerical methods for integer PDEs, efficient numerical methods for fPDEs are relatively sparse.

The IC and BCs for solving fPDEs complement those conditions for integer PDEs which include the following four types of conditions (Myint-U and Debnath 1987):

1. The first boundary-value problem or the Dirichlet problem: This BC requires a given value or function on the boundary B. Examples of this type include those for solving the fractional Basset equation (Debnath 2003a) in fluid mechanics and for water movement in soils (Su 2014).

2. The second boundary-value problem or the Neumann problem: This BC satisfies
$$\frac{\partial(vc)}{\partial x} - \frac{\partial^\lambda (Dc)}{\partial x^\lambda} = f(t)$$ on a boundary B. Examples in this category include the developments in the works of Ding and Jiang (2013).

3. The third boundary-value problem: Here, the rate of a quantity is given at the specified boundary, such as $\frac{\partial^\lambda c}{\partial x^\lambda} = f(t)$. This type of solution was given by Atangana and Kilicman (2013) for the lower boundary of groundwater flow.

4. The fourth boundary-value problem or the Robin problem: This BC involves different types of BCs on different portions of the boundary.

9.2 Solutions of the fPDEs for Solute Transport in Soils with Large and Small Pores

In this section, we present solutions of fPDEs for a generic case and demonstrate the procedures for explicit solutions of Eq. (7.82) for solute transport in non-swelling soils with a constant velocity v and constant D_f.

9.2.1 Solutions Subject to Instantaneous and Arbitrary Inputs: The Green Function

As an example, we analyze Eq. (7.82) with constant v and D_f, $R = 1$ and $\mu c + \sigma = 0$, subject to the following IC and BCs:

$$c(z, t) = \delta(z, t), \qquad\qquad t = 0, \qquad\qquad z = 0 \qquad\qquad (7.122)$$

$$c = 0, \qquad\qquad t > 0, \qquad\qquad z = 0 \qquad\qquad (7.123)$$

$$c = 0, \qquad\qquad t > 0, \qquad\qquad z \to \infty \qquad\qquad (7.124)$$

where $\delta(z, t)$ is the Dirac delta function.

The Green function is the solution of an fPDE or a partial differential equation (PDE) subject to a Dirac delta function input, $\delta(z, t)$, which is an instantaneous input released at the origin at $t = 0$. The Green function is of two types: the first kind deals with the problem wherein $c(0, t) = \delta(z, t)$, whereas the second kind is for $\dfrac{dc}{dz}(0,t) = \delta(z,t)$ (Kevorkian 1990). In our case, we discuss the Green function of the first kind.

Once the Green function is derived, it can then be used as a kernel in convolution to construct a solution of the fPDE or PDE subject to an arbitrary input $f(t)$. It should be reiterated here that the Green function with a zero BC (or a homogeneous BC) is equivalent to the mathematical problem which consists of the fPDE with $\delta(z, t)$ as the source term (Kevorkian 1990). The derivation of the Green function of Eq. (7.82), $G(z, t)$, subject to the IC and the BCs in Eqs. (7.122), (7.123) and (7.124) has been detailed in Appendix A, with the final solution as Eq. (A20)); the approximate asymptotic result of the solution as $t \to +\infty$ is of the form (Eq. (A24)):

$$G(z,t) \sim \frac{z^{2\lambda-1}}{\Gamma(\beta_1)D_f} \exp\left[-\frac{1}{D_f}(1+v_f z)\right]\left(\frac{b_i}{b_m}t\right)^{\beta_1-1} \qquad\qquad (7.125)$$

which is a decreasing function of time as a result of $\beta_1 < 1$. Equation (7.125) is a special form of the hyper-gamma function (Lehnig 1993) for which some special properties discussed by Lehnig are applicable.

9.2.2 The Solution Subject to an Arbitrary Input

Equation (A21) as the solution of Eq. (7.82) is a consequence of the instantaneous input, which is equivalent to a forcing source/sink term in the fPDE. With Eq. (A21), the solution of a non-homogeneous Eq. (7.82) with an arbitrary forcing term $f(t)$ can be derived by the following convolution integral:

$$c(z,t) = \int_0^t G(z,t-\tau)f(\tau)d\tau \qquad\qquad (7.126)$$

9.3 Solutions of the fPDEs for Solute Transport in Swelling Soils and Other Issues

Solutions of differential equations for solute transport in swelling soils can be derived by following similar procedures as for water movement presented in the above sections. To derive solutions and their properties for solute transport in swelling soils, for the case of $b = 0$ in Eq. (7.84), the only changes needed are: (a) replacement of the physical coordinate z with the material coordinate m, and (b) noting the definition of velocity v for swelling soils from Eq. (7.28) or its approximation in Eq. (7.29), compared to non-swelling soils defined by conventional methods.

The solutions in Appendix A are derived for solute transport in soils with mobile and immobile zones. When there is no distinction between mobile and immobile zones, we only need to replace $(\beta_2 - \beta_1)$ with a single parameter, β.

10. Summary

This chapter presents a set of fPDEs formulated for concurrent water flow and solute movement in both swelling and non-swelling soils. A link between fPDEs and the CTRW concept has also been provided and applied to develop new models for concurrent water flow and solute movement. An fPDE has been solved as an example for solute movement in soils with two zones (large and small pores). These developments complement mass-time and space-time fPDEs developed for water flow in swelling and non-swelling soils (Su 2014).

Based on the analytical solution of the fPDE in Eq. (7.82) subject to an instantaneous input, the solution of the fPDE subject to an arbitrary input has also been presented. Analytical solutions and asymptotic solutions of a new fPDE-based model were also developed for solute exchange between large and small pores, subject to two types of initial conditions, namely, the known rate of change of solutes at the interface and the initial concentration of solutes at the interface of mobile and immobile zones.

We now have consistent frameworks for investigating solute movement and water flow (Su 2014) in swelling and non-swelling soils. These new models that incorporate the orders of fractional derivatives in time and mass or space also have physical definitions in terms of the CTRW concept. Hence, we have moved a step further from the classical ADE which has been investigated for solute transport in soils since the early 20th century.

The introduction of these new models and concepts for solute transport has clearly provided new insights into our understanding of solute movement and water flow in soils. However, challenges follow these developments, which include (a) establishment of links between orders of fractional derivatives and properties of soils, and flow dynamics in soils in which stochastic processes take place, and (b) acquisition of accurate data on water flow and solute movement in soils, with information on water flow and solute transport at different scales ranging from pores to fields.

Appendix A: Solving the Two-term Distributed-order fPDE in Eq. (7.82)

With the Dirac delta function as the IC, which can be written as (Evans et al. 1999):

$$\delta(z, t) = \delta(z)\delta(t) \tag{A1}$$

Equation (7.82) for conservative solutes without source-sink terms reads as the following dofPDE:

$$\frac{\partial^\lambda}{\partial z^\lambda}\left(D_f c\right) - \frac{\partial^\eta}{\partial z^\eta}\left(v_f c\right) - b_i \frac{\partial^{\beta_1} c}{\partial t^{\beta_1}} - b_m \frac{\partial^{\beta_2} c}{\partial t^{\beta_2}} = \delta(z)\delta(t) \tag{A2}$$

which is solved subject to the following IC and BCs:

$$c(z, t) = 0, \qquad\qquad t = 0, \qquad\qquad z = 0 \tag{A3}$$

$$c(z, t) = 0, \qquad\qquad t > 0, \qquad\qquad z = 0 \tag{A4}$$

$$c(z, t) = 0, \qquad\qquad t > 0, \qquad\qquad z \to \infty \tag{A5}$$

The above equivalence in Eq. (A1) enables the inhomogeneous fPDE with a Dirac delta function IC to be converted to an inhomogeneous fPDE with a homogeneous BC which can be easily solved with the Laplace transform for solutions of the original problem.

For constant D_f and v_f in Eq. (A2), and by denoting $c = c(z, t)$, the Laplace transform of Eq. (A2), and the IC and the BCs with respect to time yield:

$$a_2 \frac{d^\lambda \tilde{c}}{dz^\lambda} + b_2 \frac{d^\eta \tilde{c}}{dz^\eta} + c_2 \tilde{c} = \delta(z) \tag{A6}$$

$$a = D_f \tag{A7}$$

$$b = -v_f \tag{A8}$$

$$c = -(b_i s^{\beta_1} + b_m s^{\beta_2}) \tag{A9}$$

with s as the Laplace variable. The second Laplace transform of Eq. (A6) with the transformed boundary conditions:

$$\tilde{c} = 0, \qquad\qquad z = 0 \tag{A10}$$

$$\tilde{c} \to 0, \qquad\qquad z \to \infty \tag{A11}$$

yields the solution of Eq. (A6) in terms of the Green function in the Laplace domain, $\tilde{G}(z, s)$, subject to the Dirac delta function input (Podlubny 1999), as:

$$\tilde{G}(z, s) = \frac{1}{D_f}\sum_{k=0}^{\infty}\frac{1}{k!}\left[-\frac{b_i s^{\beta_1} + b_m s^{\beta_2}}{D_f}\right]^k z^{\lambda(k+1)-1} E_{\lambda-\eta,\lambda+\eta k}^{(k)}\left(-\frac{v_f z^{\lambda-\eta}}{D_f}\right) \tag{A12}$$

where $E_{\lambda-\eta,\lambda+\eta k}^{(k)}\left(-\dfrac{v_f z^{\lambda-\eta}}{D_f}\right)$ is the *kth* derivative of the two-parameter Mittag-Leffler function or the Wiman function (with $k = 0,1,2,...$) and is expressed as:

$$E_{\lambda-\eta,\lambda+\eta k}^{(k)}\left(-\frac{v_f z^{\lambda-\eta}}{D_f}\right) = \frac{d^k}{dz^k} E_{\lambda-\eta,\lambda+\eta k}\left(-\frac{v_f z^{\lambda-\eta}}{D_f}\right)$$

$$= \sum_{j=0}^{\infty} \frac{(j+k)!}{j![\Gamma[(\lambda-\eta)j+\lambda+\eta k]}\left(-\frac{v_f z^{\lambda-\eta}}{D_f}\right)^j$$

(A13)

Subsequently, Eq. (A12) can be written as:

$$\tilde{G}(z,s) = \frac{1}{D_f}\sum_{k=0}^{\infty}\frac{1}{k!}\left[-\frac{b_i s^{\beta_1}+b_m s^{\beta_2}}{D_f}\right]^k z^{\lambda(k+1)-1}\sum_{j=0}^{\infty}\frac{(j+k)!}{j![\Gamma[(\lambda-\eta)j+\lambda+\eta k]}\left(-\frac{v_f z^{\lambda-\eta}}{D_f}\right)^j$$ (A14)

The inversion of Eq. (A14) can be completed with the results of Kilbas et al. (2004). The term $\left[-\dfrac{(b_i s^{\beta_1}+b_m s^{\beta_2})}{D_f}\right]^k$ in Eq. (A14) can be rewritten as:

$$\left[-\frac{(b_i s^{\beta_1}+b_m s^{\beta_2})}{D_f}\right]^k = \left(-\frac{b_m}{D_f}\right)^k s^{\beta_2}\left(1+\frac{b_i}{b_m}s^{-(\beta_2-\beta_1)}\right)^k$$

$$= \left(-\frac{b_m}{D_f}\right)^k s^{k\beta_2}\left(\frac{b_m}{b_m+b_i s^{-(\beta_2-\beta_1)}}\right)^{-k} = \left(-\frac{1}{D_f}\right)^k \frac{s^{k\beta_2}}{b_m\left(1+\dfrac{b_i}{b_m}s^{-(\beta_2-\beta_1)}\right)^k}$$

(A15)

which has the Laplace inversion of the form known as the generalized Prabhakar function (GPF):

$$e_{\beta_2-\beta_1,-k\beta_2}^k\left(t;\frac{b_i}{b_m}\right) = t^{-k\beta_2-1}E_{\beta_2-\beta_1,-k\beta_2}^k\left(-\frac{b_i}{b_m}t^{\beta_2-\beta_1}\right), \qquad k=0,1,2,...,\infty$$ (A16)

with the Prabhakar function (PF) given by:

$$E_{\alpha,\beta}^k(x) = \sum_{n=0}^{\infty}\frac{(k)_n x^n}{n![\Gamma[\alpha n+\beta]}$$ (A17)

where $\Gamma[\alpha n + \beta]$ is the gamma function of $\alpha n + \beta$. In Eq. (A16), the parameter k of the GPF can be a non-zero positive, negative or complex number (Prabhakar 1971, Prajapati et al. 2013).

With Eq. (A16) and the following identities:

$$\lambda = \eta + 1$$ (A18)

$$\sum_{k=0}^{\infty} \frac{1}{k!} \left(-\frac{1}{D_f}\right)^k = \exp\left(-\frac{1}{D_f}\right) \tag{A19}$$

Equation (A14) can be inverted (Kilbas et al. 2004) to yield:

$$G(z,t) = \frac{1}{D_f} \sum_{k=0}^{\infty} \left[\exp\left(-\frac{1}{D_f}\right) e_{\beta_2-\beta_1,-k\beta_2}^{k} \left(t; \frac{b_i}{b_m}\right) z^{\lambda(k+1)-1} \sum_{j=0}^{\infty} \frac{(j+k)!}{j! \Gamma[j+\lambda(1+k)-k]} \left(-\frac{v_f z}{D_f}\right)^j \right] \tag{A20}$$

which is the solution we seek for Eq. (7.82) subject to the Dirac delta function input or an instantaneous input. A solution derived from such an input is known as the Green function in mathematics (Kevorkian 1990). According to Podlubny (1999), the solution derived using the Laplace transform in Eq. (A20) was first presented by Doetsch (1956).

Mainardi and Garrappa (2015) showed that the counter k in Eq. (A20) takes the value as $k = 1$ for $\beta_2 \geq 0.5$ for the parameters in the MLF to be physically relevant. This implies that Eq. (A20) can be simplified by retaining two terms (with $k = 0,1$) only, resulting in the following:

$$G(z,t) = \frac{1}{D_f} \begin{bmatrix} \exp\left(-\frac{1}{D_f}\right) \frac{z^{\lambda-1}}{\Gamma(\beta)t} \sum_{j=0}^{\infty} \frac{1}{\Gamma[j+\lambda]} \left(-\frac{v_f z}{D_f}\right)^j + \\ \exp\left(-\frac{1}{D_f}\right) z^{2\lambda-1} t^{-\beta_2-1} E_{\beta_2-\beta_1,-\beta_2}\left(-\frac{b_i}{b_m} t^{\beta_2-\beta_1}\right) \sum_{j=0}^{\infty} \frac{(j+1)!}{j! \Gamma[j+2\lambda-1]} \left(-\frac{v_f z}{D_f}\right)^j \end{bmatrix} \tag{A21}$$

As $\Gamma[j + 2\lambda - 1] \to (j + 1)!$ with $\lambda < 2$, the summation in Eq. (A21) can be approximated as an exponential function as follows:

$$\sum_{j=0}^{\infty} \frac{(j+1)!}{j! \Gamma[j+2\lambda-1]} \left(-\frac{v_f z}{D_f}\right)^j = \exp\left(-\frac{v_f z}{D_f}\right) \tag{A22}$$

Furthermore, as showed in Chapter 2, the asymptotic results for the GPF as $t \to +\infty$ are:

$$t^{\beta-1} E_{\alpha,\beta}(-\rho t^{\alpha}) \sim \frac{(\rho t)^{\beta-\alpha-1}}{\Gamma(\beta-\alpha)}, \qquad 0 < \alpha < \beta \leq 2 \tag{A23}$$

which enables Eq. (A21) to be written as follows (for $t \to +\infty$):

$$G(z,t) \sim \frac{z^{2\lambda-1}}{D_f} \exp\left[-\frac{1}{D_f}(1+v_f z)\right] \frac{1}{\Gamma(\beta_1)} \left(\frac{b_i}{b_m} t\right)^{\beta_1-1}, \qquad t \to +\infty \tag{A24}$$

Chapter 8

Hydraulics of Anomalous Flow on Hillslopes, in Catchment Networks and Irrigated Fields

1. Introduction

Overland flow is a very important process during rainfall on hillslopes in natural terrains and engineered land surfaces during surface irrigation. In dry areas, where soil is not saturated, runoff forms on the surface of the soil when the rainfall intensity exceeds the infiltration rate; this mechanism of runoff generation is called *infiltration access* accredited to Horton (1933). In areas with saturated soil surfaces, runoff forms as soon as rain falls onto the surface; this form of runoff generation is called *saturation access* (Zhao and Zhuang 1963, Zhao 1992). These two major forms of runoff generation mechanisms appear in different climatic regions with the infiltration access mainly taking place in dry regions while saturation access in humid regions.

In addition to these two major forms of mechanisms of runoff generation, *spatially*, a portion of a catchment in humid regions either with a higher moisture content near stream channels or topographic depressions could contribute to runoff only. Then a concept of *partial area runoff* or *variable source area* was initiated (Hewlett 1961, Hewlett and Hibbert 1967). *Temporally*, in catchments in dry regions where the top soil has a higher soil water content due to prior rainfall either near stream channels or in depressions, both infiltration access and saturation access can develop in the same location (Su 1993). Irrespective of the form of runoff generation, the hydraulics of runoff, coupled water and solute movement, and soil erosion on hillslopes are of great environmental and engineering concerns.

The previous chapters were mainly concerned with water flow and solute transport within porous media under unsaturated and saturated conditions. In this chapter, the topics are relevant to water flow and solute movement on the surface of porous media (land surface or natural hillslopes). As such, four topics have been discussed in this chapter:

1. Fractional Kinematic Wave (fKW) models: These models have been outlined and the methods for solutions given as examples of applications of these

models. fKW models are non-linear fractional differential equations (fDEs) and solutions for one form of fKW models have been presented for specified initial and boundary conditions; asymptotic solutions on the depth of water and discharge overland have also been developed for large times.

2. A fractional model for solute transport in overland flow on hillslopes has been outlined to be consistent with the fKW model for overland flow.

3. Anomalous flow in catchments has been discussed in the context of network flow in which catchment is represented by a topographically random channel-hillslope network model and streamflow in networks is determined as the convolution of effective rainfall input and instantaneous unit hydrograph (IUH) model. Effective rainfall is calculated as gross rainfall minus interception and anomalous infiltration. It is anomalous infiltration which enters the discussion on network flows.

4. The integrated processes of water advance-infiltration-evaporation have been discussed in relation with border-check irrigation or flood irrigation with anomalous infiltration into soils as the focal point in these integrated processes.

2. Rainfall-Infiltration-Runoff Relations on a Planar Hillslope

2.1 The Kinematic Wave Model of Overland Flow

Runoff on land surface, widely known as overland flow, can be modelled by either a complete set of equations in hydraulics known as the de Saint-Venant equations (SVEs) (de Saint-Venant 1871) or the simplified kinematic wave model (KWM) proposed by Lighthill and Whitham (1955) (Eagleson 1970). While SVEs and the KWM share the same conservation equation, the former considers all the forces such as acceleration (both local and convective), pressure, gravity and friction in the momentum equation, while the latter considers the gravity and friction terms only (Chow et al. 1988).

The conservation equation for overland flow along the x plane is expressed as (Eagleson 1970):

$$\frac{\partial h}{\partial t} + \frac{\partial q}{\partial x} = r(t) - i(t) \tag{8.1}$$

where h is the depth of runoff, $r(t)$ is the effective rainfall intensity, $i(t)$ is the infiltration rate, t is time and q is the volumetric discharge of runoff which is represented by the simplified momentum equation as:

$$q = \alpha h^m \tag{8.2}$$

where α is a coefficient and m is a parameter depending on flow patterns.

The velocity of overland flow, V, is expressed as:

$$V = \frac{q}{A} \tag{8.3}$$

where A is the cross-sectional area of the flow. The velocity of overland flow can be determined by different methods, the Manning equation, written as follows, being one of the commonly used models (Chow et al. 1988):

$$V = \frac{R^{2/3} S_0^{1/2}}{n} \tag{8.4}$$

where S_0 is the friction slope for overland flow (given as $S_0 = \cos \gamma$, with γ being the angle of the slope), n is the Manning roughness coefficient and R is the hydraulic radius defined as:

$$R = \frac{A}{p} \tag{8.5}$$

where p is the *wetted perimeter* of flow, which represents that portion of the perimeter of the cross section where direct contact between the flow and the solid boundary exists (Daugherty et al. 1989).

The theoretical values of m in the integer-based KWM are different for the two types of flow: $m = 3$ for laminar overland flow and $m = 5/3$ for turbulent overland flow (Eagleson 1970). However, for natural surfaces based on experiments, the earlier observations by Horton (1938) suggested that $m \approx 2$ (Eagleson 1970).

With the discharge of overland flow on a planar surface, the cross sectional area of the flow is given as $A = wh$ and discharge is expressed as:

$$q = wVh \tag{8.6}$$

where w is the width of the flow. For a unit width in Eq. (8.6), the Manning equation for velocity V can be written as:

$$V = \frac{S_0^{1/2}}{n} h^{m-1} \tag{8.7}$$

Subsequently, the KWM for overland flow on a planar slope in Eq. (8.1) can be written as:

$$\frac{\partial h}{\partial t} + \frac{S_0^{1/2}}{n} \frac{\partial h^m}{\partial x} = r(t) - i(t) \tag{8.8}$$

Owing to its simplicity, the KWM has been widely applied for analyzing overland flow and related problems for over six decades. As Eagleson (1970) pointed out, the kinematic approximation as a simplification of full SVEs requires that the wave-like properties of flow must be accounted for in the model, through the continuity equation, and the assumption of dynamic uniformity in the kinematic model precludes solutions that exhibit changes in the surface profile, which is a characteristic of dynamic waves. With such simplifications, the initial and boundary conditions for solving the KWM apply only to the continuity equation or its joint form so that changes in the flow profile can be transmitted downstream without the backwater effect. In the meantime, the one-dimensional KWM neglects the difference between the slopes of the water surface and the bottom; hence, variations in the surface and the bottom of the flow are not considered.

2.2 The Fractional Kinematic Wave Model of Overland Flow

In a fashion similar to the fractional conservation of mass for subsurface flow in porous media (Wheatcraft and Meerschaert 2008), the fractional conservation of mass for overland flow can be introduced to Eq. (8.8) to form an fKW for runoff as:

$$\frac{\partial h}{\partial t} + \frac{S_0^{1/2}}{n} \frac{\partial^\rho}{\partial x^\rho} h^m = r(t) - i(t) \tag{8.9}$$

Equation (8.9) is a non-linear fractional partial differential equation (fPDE) and is essentially a fractional advection equation (fAE) with a negative convective coefficient in contrast to the fractional diffusion-wave equation which has a positive diffusion coefficient. The fKW with a spatial fractional derivative ρ offers more flexibility on the spatial variability of discharge (or unit discharge, discharge per unit width).

In a similar fashion, Deng et al. (2005) applied a fractional advection-dispersion equation (fADE) for solute transport in overland flow, with the overland flow modelled by the non-fractional Eq. (8.8), which, however, raises the question of consistency between the two models for the same flow. Deng et al. (2004, 2006) and Shen and Phanikumar (2009) also applied the fADE for modelling anomalous solute movement in stream flows. However, their fractional approaches were based on fractional dispersion rather than fractional advection as suggested in Eq. (8.9) here.

In the following sections, applications of the fKW model to overland flow have been demonstrated with specific initial and boundary conditions.

2.3 Rainfall-Runoff Relations with Infiltration into Soils with Mobile and Immobile Zones

A common situation to which Eq. (8.9) applies can be described by the following initial and boundary conditions on a semi-infinite slope:

$$h(x, t) = 0, \qquad\qquad t = 0, \qquad\qquad x > 0 \tag{8.10}$$

$$h(x, t) = h_0, \qquad\qquad t > 0, \qquad\qquad x = 0 \tag{8.11}$$

where the scenario of $h_0 = 0$ can be included.

In this example, we demonstrate the application of the equation of infiltration rate given by Eq. (6.65) in Chapter 6, that is, $i(t) = A_s + \left(\frac{1}{2} \beta_2 - \frac{1}{4} \beta_1 \right) S t^{(\beta_2/2 - \beta_1/4) - 1}$,

where the rate of infiltration was derived from a new fPDE for water flow in soils on a slope. Of course, other forms of infiltration equations can be used such as those derived from the mass-time fractional PDE for swelling soils (see section 4.2 in Chapter 6).

The IC describes the initial profile of water level on the surface of the soil, along the x axis. With a non-zero initial condition, it is more convenient to handle the fractional differential equation (fDE) with Caputo's fractional definition (Agarwal et al. 2010). Applying Laplace transform to Eq. (8.9), with the IC and the BCs in Eqs. (8.10) and (8.11), yields:

$$s\tilde{h} + \frac{S_0^{1/2}}{n}\frac{d^\rho}{dx^\rho}(\tilde{h}^m) = \frac{\tilde{r}(s)}{s} - \left[\frac{A_s}{s} + \left(\frac{\beta_2}{2} - \frac{\beta_1}{4}\right)\frac{S\Gamma(\beta_2/2 - \beta_1/4)}{s^{\beta_2/2 - \beta_1/4}}\right] \quad (8.12)$$

$$\tilde{h}(x,t) = \frac{h_0}{s}, \qquad\qquad x = 0 \qquad\qquad (8.13)$$

where $\tilde{r}(s)$ is the Laplace transform of $r(t)$, S denotes sorptivity, A_s is the final infiltration rate on the hillslope, and β_2 and β_1 are the orders of fractional derivatives (representing large and small pores respectively).

With the following transformation:

$$g(x, s) = \tilde{h}^m(x, s) \qquad\qquad (8.14)$$

or $\tilde{h} = g^{1/m}$, Eq. (8.12) is written as:

$$\frac{d^\rho g}{dx^\rho} = f(g(x)) \qquad\qquad (8.15)$$

where

$$f(g(x)) = \frac{n}{S_0^{1/2}}\left\langle\left\{\frac{\tilde{r}(s)}{s} - \left[\frac{A_s}{s} + \left(\frac{\beta_2}{2} - \frac{\beta_1}{4}\right)\frac{S\Gamma[(\beta_2/2) - \beta_1/4]}{s^{(\beta_2/2) - \beta_1/4}}\right]\right\} - sg^{1/m}\right\rangle \quad (8.16)$$

Here $f(g(x))$ is a function of $g(x)$, and the BC becomes:

$$g = \frac{h_0^m}{s}, \qquad\qquad x = 0 \qquad\qquad (8.17)$$

Equation (8.15) is a non-linear fDE and its solutions are known (Lubich 1986, Podlubny 1999, Kilbas and Marzan 2005, Su 2009). The solution of Eq. (8.15) subject to the condition in Eq. (8.17) is of the form:

$$g(x) = \sum_{k=0}^{n-1}\frac{1}{sk!}h_0^m x^k + \frac{1}{\Gamma(\rho)}\int_0^x (x - \xi)^{\rho-1} f(g(\xi))d\xi \qquad (8.18)$$

which was derived for x in the range $[a, x]$, with $a = 0$ as a special case. In the Laplace domain, the solution in Eq. (8.18) is written through the transform $g = \tilde{h}^m$ in Eq. (8.14) as:

$$\tilde{h}^m = \sum_{k=0}^{n-1}\frac{h_0}{sk!}(x - a)^k + \frac{1}{\Gamma(\rho)}\int_a^x \frac{f(\tilde{h}(\xi))}{(x - \xi)^{1-\rho}}d\xi \qquad (8.19)$$

where $f(\tilde{h}(\xi))$ is a function of $\tilde{h}(\xi)$.

Equation (8.19) is the non-linear Abel integral equation of the second kind (AIE2); however, when $h_0 = 0$, it is the non-linear Abel integral equation of the first kind (AIE1) (Gorenflo 1987). AIE1 is also called the Generalized Abel equation (Weiss 1972) or the Abel-Volterra integral equation (Gorenflo and Kilbas 1995,

Kilbas et al. 1995, Karapetyants et al. 1996, Kilbas and Saigo 1999a). In addition to these studies, there have been further extensive investigations on these equations in literature (Atkinson 1974, Gorenflo and Vessella 1980, Linz 1985, Gorenflo 1987, Srivastava and Buschman 1992, Polyanin and Manzhirov 1998, Karapetiants et al. 2001, Umarov and Saydamatov 2006). The works of Gorenflo and Vessella (1980) and Gorenflo (1987) are particularly extensive, with the latter providing a review of literature published in the previous 35 years.

A special form of an integral equation similar to Eq. (8.19) has been extensively investigated by Gorenflo and Kilbas (1995), Kilbas et al. (1995), and Kilbas and Saigo (1999a). In particular, Kilbas et al. (1995) presented the exact solutions for such integral equations similar to Eq. (8.19), besides asymptotic solutions.

Apart from those derived from numerical methods, two types of solutions are particularly relevant here to solve Eq. (8.19): (a) asymptotic solutions as shown by Gorenflo and Kilbas (1995), wherein many forms of asymptotic solutions are available for both small and large x, that is, solutions for $x \to 0$ and $x \to \infty$ respectively; (b) the second type of solutions can be derived by the step-by-step method (Kilbas and Marzan 2005) with which the solution of Eq. (8.19) can be derived for the range $[x_1, x_2]$. The second method is similar to the successive approximation method (Polyanin and Manzhirov 1998).

Once a solution of Eq. (8.19) is derived, its inverse Laplace transform yields the solution in h.

2.4 Anomalous Rainfall-Runoff-Infiltration Relation with a Constant Rainfall Intensity on a Dry Surface

In this instance, the conditions of $h_0 = 0$ and a constant rainfall intensity, and the IC and the BCs in Eqs. (8.10) and (8.11) are written as:

$$h(x, t) = 0, \qquad t = 0, \qquad x > 0 \qquad\qquad (8.20)$$

$$h(x, t) = 0, \qquad t > 0, \qquad x = 0 \qquad\qquad (8.21)$$

With the limit of integration in the range of the slope length $[0, x]$, the solution in Eq. (8.18) becomes:

$$g(x) = k(x) - \frac{ns}{\Gamma(\rho)S_0^{1/2}} \int_0^x (x - \xi)^{\rho-1} g(\xi)^{1/m} d\xi \qquad\qquad (8.22)$$

with

$$k(x) = \frac{n}{\Gamma(\rho)(\rho - 2)S_0^{1/2} s} \left[(r_0 - A_s) + \left(\frac{\beta_2}{2} - \frac{\beta_1}{4} \right) \frac{S\Gamma[(\beta_2/2) - \beta_1/4]}{s^{[(\beta_2/2) - \beta_1/4]-1}} \right] x^{\rho-2} \qquad (8.23)$$

The original variable can be restored in Eq. (8.22) using Eq. (8.14) to yield:

$$\tilde{h}^m(x) = f(x) - \frac{ns}{\Gamma(\rho)S_0^{1/2}} \int_0^x \frac{\tilde{h}(x)}{(x - \xi)^{1-\rho}} d\xi \qquad\qquad (8.24)$$

Kilbas et al. (1995, Eq. (6.3)) presented asymptotic solutions of Eq. (8.24) in the following form:

$$\tilde{h}(x) = cx^{1-\rho}$$
(8.25)

where the constant c appears in the following algebraic function (Kilbas et al. 1995):

$$c^m + ac - b = 0$$
(8.26)

with

$$a = \frac{\Gamma(1-\rho)ns}{\Gamma(\rho)S_0^{1/2}}$$
(8.27)

and

$$b = \frac{n}{\Gamma(\rho)(\rho-2)S_0^{1/2}s}\left[(r_0 - A_s) + \left(\frac{\beta_2}{2} - \frac{\beta_1}{4}\right)\frac{S\Gamma[(\beta_2/2)-\beta_1/4]}{s^{[(\beta_2/2)-\beta_1/4]-1}}\right]$$
(8.28)

When an explicit solution for c is derived by solving the algebraic equation, the solution in Eq. (8.24) needs to be transformed using the inverse Laplace transform. Here, we use $m = 2$ from earlier experimental findings (Horton 1938, Eagleson 1970) to solve Eq. (8.26) which has two solutions (or roots), c_1 and c_2, such that:

$$c_{1,2} = \frac{-a \pm \left[a^2 + 4b\right]^{1/2}}{2}$$
(8.29)

With the first (positive) solution in Eq. (8.29) and for $m = 2$, Eq. (8.25) is expressed as:

$$\tilde{h}(x) = \frac{1}{2}\left[\left(a^2 + 4b\right)^{1/2} - a\right]x^{1-\rho}$$
(8.30)

One of the most important situations for overland flow is at large times as $t \to \infty$ or $s \to 0$, which yields $a \to 0$; then Eq. (8.30) can be approximated as:

$$\tilde{h}(x) = b^{1/2}x^{1-\rho}$$
(8.31)

The inverse Laplace transform of Eq. (8.31), with b from Eq. (8.28), yields (Kilbas et al. 2004):

$$h(x) = \left[\frac{n(r_0 - A_s)}{\Gamma(\rho)(\rho-2)S_0^{1/2}}\right]^{1/2} e_{\beta-1,1/2}^{-1/2}\left(\beta S\Gamma(\beta)t^{\beta-1}\right)x^{1-\rho}$$
(8.32)

where

$$e_{\beta-1,1/2}^{-1/2}\left(\beta S\Gamma(\beta)t^{\beta-1}\right) = t^{-1/2}E_{\beta-1,1/2}^{-1/2}\left(\beta S\Gamma(\beta)t^{\beta-1}\right)$$
(8.33)

is the generalized Prabhakar function (GPF), and $E_{\beta-1,1/2}^{-1/2}\left(\beta S\Gamma(\beta)t^{\beta-1}\right)$ is the Prabhakar function with

$$\beta = \frac{\beta_2}{2} - \frac{\beta_1}{4}$$
(8.34)

Here, we only present its asymptotic solution to be consistent with Eq. (8.25), as studied by Kilbas et al. (1995). To carry out the asymptotic analysis, the temporal part of the GPF in Eq. (8.32) can be analyzed. Following procedures similar to those used by Mainardi and Garrappa (2015), the asymptotic solution of the GPF in Eq. (8.33) is:

$$e_{\beta-1,1/2}^{-1/2}\left(\beta S\Gamma_{(\beta)}t^{\beta-1}\right) \sim \beta St^{\beta-1}, \qquad t \to \infty \qquad (8.35)$$

Subsequently, with the asymptotic form of Eq. (8.32) for $t \to \infty$, Eq. (8.32) is written as:

$$h(x) = \left[\frac{n(r_0 - A_s)}{\Gamma(\rho)(\rho-2)S_0^{1/2}}\right]^{1/2}\left(\frac{\beta_2}{2} - \frac{\beta_1}{4}\right)\frac{Sx^{1-\rho}}{t^{1-[\beta_2/2-\beta_1/4]}}, \qquad t \to \infty \qquad (8.36)$$

As $1 - \beta_2/2 + \beta_1/4 > 0$, Eq. (8.36) explains the effect of infiltration parameters on overland flow, with the term $\left(\dfrac{\beta_2}{2} - \dfrac{\beta_1}{4}\right)\dfrac{S}{t^{1-\beta_2/2+\beta_1/4}}$ representing the recession of overland flow over time. In the meantime, the above solutions include three cases: [1] the case of $\rho < 1$ implies an increasing water level downslope; [2] $\rho > 1$ implies a decreasing water level downslope; and [3] $\rho = 1$ in fact implies the recession of water level at any given location.

With Eq. (8.36) for asymptotic depth of overland flow, derived from an approximate exponent $m = 2$, the asymptotic moment equation given by Eq. (8.2) can be approximated by a more generic case which includes both laminar flow ($m = 3$) and turbulent flow ($m = 5/3$) as follows (Eagleson 1970):

$$q \sim \frac{S_0^{(2-m)/4}}{n^{(2-m)/2}}\left[\frac{(r_0 - A_s)}{\Gamma(\rho)(\rho-2)}\right]^{m/2}\left[\left(\frac{\beta_2}{2} - \frac{\beta_1}{4}\right)\frac{Sx^{1-\rho}}{t^{1-[\beta_2/2-\beta_1/4]}}\right]^m, \qquad t \to \infty \qquad (8.37)$$

Thus, the asymptotic overland flow velocity in Eq. (8.7) can now be written as:

$$V \sim \frac{S_0^{(3-m)/4}}{n^{(3-m)/2}}\left[\frac{(r_0 - A_s)}{\Gamma(\rho)(\rho-2)}\right]^{(m-1)/2}\left[\left(\frac{\beta_2}{2} - \frac{\beta_1}{4}\right)\frac{Sx^{1-\rho}}{t^{1-[\beta_2/2-\beta_1/4]}}\right]^{m-1}, \qquad t \to \infty \qquad (8.38)$$

The length dimension of h ([L]) should be retained, which requires other traditional dimensions to be changed in order to accommodate the fractional dimensions in time and space variables. One way to achieve this is for changes in dimensions to be lumped in the Manning roughness coefficient so that other dimensions remain the same as in the classical models.

2.5 Fractional Kinematic Wave Models for Overland Flow and Solute Transport on Hillslopes

As briefly stated in section 2.1, the kinematic wave model simulates convective overland flow, subject to gravity and friction only, by neglecting other forces which are accommodated in SVEs. The introduction of spatial fractional derivatives in the kinetic wave equation in Eq. (8.8) results in the fKW or the fractional advection equation (fADE) in Eq. (8.9). With the fKW, questions can be raised here as to whether overland flow also changes its flow patterns from sub-convective or classical to super-convective flows, and whether there is a criterion in the fKW for classifying flow patterns if such variable flow patterns exist. With these questions in mind, let us further introduce time fractional derivatives into Eq. (8.9) to arrive at a time-space fractional kinetic wave equation:

$$\frac{\partial^\beta h}{\partial t^\beta} + \frac{S_0^{1/2}}{n}\frac{\partial^\rho}{\partial x^\rho}h^m = r(t) - i(t) \tag{8.39}$$

For simplicity in analysis, the source terms are omitted so that Eq. (8.39) is written as:

$$\frac{\partial^\beta h}{\partial t^\beta} = D_{of}\frac{\partial^\rho h^m}{\partial x^\rho} \tag{8.40}$$

where the convection coefficient of overland flow is expressed as:

$$D_{of} = -\frac{S_0^{1/2}}{n} \tag{8.41}$$

Equation (8.40) resembles the fractional diffusion-wave equation (fDWE) discussed in previous chapters (such as in section 3 of Chapter 4, where Zaslavsky's RGK theory was outlined). In the RGK theory, the following relationships hold:

$$\frac{\beta}{\rho} = \frac{\ln L_s}{\ln L_t} = \frac{\mu}{2} \tag{8.42}$$

where μ is called transport exponent (Zaslavsky 2002), which connects the order of fractional derivatives and the fractal scaling parameters, L_s and L_t. Equation (8.42) can also be rewritten as:

$$\mu = \frac{2\ln L_s}{\ln L_t} = \frac{\ln L_s^2}{\ln L_t} \tag{8.43}$$

and

$$\mu = \frac{2\beta}{\rho} \tag{8.44}$$

The transport exponent as defined above is a criterion for assessing flow patterns:

1. sub-diffusion when $\mu < 1$;
2. classical when $\mu = 1$; and
3. super-diffusion when $\mu > 1$.

Zaslavsky (2002) demonstrated the applications of the RGK theory in anomalous transport; the RGK method has been applied to evaluate flow patterns during infiltration into soils (Su 2010, 2012, 2014). For overland flow, there is a need to assess how the orders of time and space fractional derivatives change.

Zhang et al. (2016) presented a distributed-order fKW (dofKW) for overland flow, which resembles the tempered fPDE presented by Meerschaert et al. (2008) for anomalous transport in heterogeneous systems. For small depths of water overland, the dofKW could capture more features of overland flow due to the effects of heterogeneous pores in the top soil where the flow takes place.

3. Rainfall-Infiltration-Runoff Relation on Convergent and Divergent Hillslopes

Natural surfaces of landscapes are variable in their geometries and the continuity equation of flow on a planar surface above is only an approximation in limited cases. Convergent and divergent surfaces are more frequently intertwined to form channels and ridges. Hence, the governing equations for flow on convergent and divergent hillslopes are different from Eq. (8.8) or Eq. (8.9) for a planar slope.

The original model for flow on a convergent slope was proposed by Veal (1966) and is written as (Kibler and Woolhiser 1970):

$$\frac{\partial h}{\partial t} + \frac{\partial}{\partial x}(uh) = r(t) + \frac{uh}{L_0 - x} \tag{8.45}$$

or

$$\frac{\partial h}{\partial t} + \frac{\partial}{\partial x}(uh) + \frac{uh}{x - L_0} = r(t) \tag{8.46}$$

where r is a parameter defining the convergence and L_0 is the diameter of the convergent cone (which is different from the flow length x).

Campbell and Parlange (1984) and Campbell et al. (1984) used the continuity equation of flow in a radial coordinate, which, when infiltration is considered, takes the following form:

$$\frac{\partial h}{\partial t} + \frac{S_0^{1/2}}{nr} \frac{\partial}{\partial r}(rh^m) = r(t) - i(t) \tag{8.47}$$

Equations (8.46) and (8.47) are more complex than their counterpart for planar flow in Eq. (8.8), and analytical solutions of these equations can only be derived for highly simplified situations and numerical methods are needed to solve them for more generic cases. Note that, to be consistent with the geometry of hillslopes, the equation of infiltration rate for the convergent surface should be used, which can be derived from equations based on the planar surface by introducing a slope factor (see Chapter 6).

The geometry of slopes is of great importance as it determines how overland flows form runoff on a hillslope. With a convergent geometry, the depth of water increases towards the downslope direction due to the convergence of slope, thus increasing the flow velocity, whereas divergent slopes disperse water towards the downslope direction so that a divergent slope reduces the depth of water.

When the fractional order is introduced to Eq. (8.47), the following equation results:

$$\frac{\partial h}{\partial t} + \frac{S_0^{1/2}}{nr} \frac{\partial^p}{\partial r^p}(rh^m) = r(t) - i(t) \tag{8.48}$$

Equally, the order of fractional derivatives in time can also be introduced into Eq. (8.48). In addition to convergence and divergence of the surface, the downslope change in the slope elevation is another issue. Philip (1991c) analyzed flow equations for concave and convex slopes in different dimensions, along with convergent and divergent geometries (Philip 1991b). Agnese et al. (2007) investigated nine types of hillslopes as a matrix of convergent, divergent and planar geometries corresponding to concave, straight and convex downslope shapes. They introduced a shape factor ϕ_s to account for variations of a slope in two aspects: the planform geometry function $\phi_{pl}(x)$ to represent the horizontal view of a slope, and the profile shape function $\phi_{pr}(x)$ to represent the elevation changes along the downslope direction.

Zaslavsky and Rogowski (1969) used a different function for hillslope profiles, describing convergent, parallel and divergent slopes.

4. Solute Transport by Runoff on Hillslopes

Overland flows on loose soil surfaces on hillslopes can erode soil and generate solutes and sediments so that the water flow is often accompanied by coupled solute and sediment movement in runoff. Here, these two issues are briefly discussed in association with the kinematic wave overland flow.

Coupled overland flow and solute transport by runoff in overland flow on a planar surface can be described by the conservation of mass of solutes (Deng et al. 2005, 2006):

$$\frac{\partial}{\partial t}(ch) + \frac{\partial}{\partial x}(cuh) - \frac{\partial}{\partial x}(-Jh) = \alpha h_m(c_s - c) \tag{8.49}$$

where c is the solute concentration averaged over the cross-section of the flow and J is the fractional flux of dispersion into overland flow from the soil, given by:

$$J = -K_c \frac{\partial^{\eta-1} c}{\partial x^{\eta-1}} \tag{8.50}$$

with K_c as the longitudinal dispersion coefficient, $\eta - 1$ as the order of fractional derivatives, c_s being the solute concentration in the mixing zone, α as the mass transfer rate, and h_m as the depth of the mixing zone.

With this formulation, the equation for solute transport in kinematic waves is written as (Deng et al. 2006):

$$\frac{\partial c}{\partial t} = K \frac{\partial^{\eta} c}{\partial x^{\eta}} - u \frac{\partial c}{\partial x} + \frac{1}{h}\{\alpha h_m(c_s - c) - [r(t) - i(t)]c\} \tag{8.51}$$

Equation (8.51) takes the form of the fractional advection-dispersion equation (fADE) and the extra source/sink terms are coupled with the depth of overland flow. For solute transport in overland flow, the values of η depend on flow conditions and the properties of the porous medium on which the overland flow develops ($\eta = 2$ for isotropic media and $\eta < 2$ for porous media); the more heterogeneous the medium is, the smaller is the value of η. Moreover, for natural rivers, $1.4 < \eta < 2$ (Deng et al. 2005).

Equation (8.51) was formulated for planar slopes. Similar to water movement in soils on divergent and convergent slopes, the formulations for solute transport by runoff on divergent and convergent slopes can also be developed. Note that Eq. (8.51) is based on non-fractional conservation of mass of water for the kinematic wave model; when the fKW is used, a fractional version of overland flow can be used to derive a different form of the model comparable to Eq. (8.51). Hence, a consistent coupling is achieved by combining the fKW in Eq. (8.9) for anomalous overland flow and Eq. (8.49) for solute transport on planar slopes.

5. Related Topics

The introduction of the order of fractional derivatives in the kinematic wave equation enables the consistent coupling of the fADE for solute movement in overland flow. In addition, many related issues naturally arise when the fKW is used for overland flow.

One of the potentially interesting issues is the functional order as discussed in Chapter 4, which refers to the order of fractional derivatives, ρ, in Eq. (8.9) as well as β in Eq. (8.39) as functions of related variables. In the context of overland flow, they could be functions of depth of flow, space, and time, that is, $\rho = \rho(h, x, t)$ and $\beta = \beta(h, x, t)$. These extensions await experimental verification. The fKW is a fractional advection equation without a diffusive term and associated properties discussed by Eagleson (1970) apply.

6. Streamflow through Catchment Networks

6.1 Flow in Catchment Networks

The concept of Geomorphologic Instantaneous Unit Hydrograph (GIUH) conceived by Rodriguez-Iturbe and Valdés (1979) is regarded as one of the most important developments in hydrology since the 1930s when Sherman (1932) initiated the unit hydrograph (UH) method for rainfall-runoff analysis (Maidment 1993). In an analytical mathematical form, GIUH is the output of an instantaneous input to a catchment, similar to the Dirac delta function in physics and mathematics.

The GIUH links the hydrological response of a catchment network to rainfall, through geomorphologic modulation in terms of hydro-geomorphic parameters. Following the inception of the GIUH concept, there has been widespread interest in the application of this approach (Gupta et al. 1986). Singh et al. (2014) presented an extensive review of the development of the UH concept, which tracked the GIUH as a major direction in the UH evolution. Referring to this actively and intensely pursued topic since its inception 40 years ago, Dooge (1986) pointed out that "further work is continuing in this field which represents one possible route to the development of hydrologic laws on a catchment scale."

6.2 Mathematical Models for Flow in Topographically Random Hillslope-Channel Networks

Among the findings of the GIUH, a significant progress in the understanding of geomorphic links to network flows is the link-based model developed by Troutman and Karlinger (1984, 1985), which incorporates the concept of topographically random channel networks (TRCN) (Shreve 1967) and all pertinent hydraulic information important to flows in such networks. Of central importance to Troutman and Karlinger's model (the TK model) are three asymptotic results based on three different routing methods linking the catchment response to an input through a scaling parameter. Troutman and Karlinger (1985) highlighted the important properties of these asymptotic results and presented these findings from assessment and simulation (Karlinger and Troutman 1985, Troutman and Karlinger 1986). Their asymptotic results are regarded as being "of fundamental significance to both hydrology and geomorphology" (Gupta and Mesa 1988) and a "powerful concept" in hydrological analysis (Bras 1990).

Following simulations and comparisons, Troutman and Karlinger (1986) concluded that all linear routing schemes (linear, diffusion, and translation) lead to the same asymptotic expected IUH, which suggests that translation routing may be entirely adequate for routing through networks. The asymptotic result of translation routing for a large number of first-order streams is the Weibull distribution and is written as:

$$\bar{u}(t) = \frac{\sigma^2 t}{2n} \exp\left[-\left(\frac{\sigma t}{2\sqrt{n}}\right)^2\right] \tag{8.52}$$

where $\bar{u}(t)$ is the topologically expected IUH as a function of time, n is the number of first-order streams, and

$$\sigma = \frac{\beta_c}{\alpha} \tag{8.53}$$

with α as the mean internal link length of the channel segment, and β_c being the hydraulic parameter ($\beta_c = 1.0$ for a width function if one is interested in only the width function, which depends only on the distances of points in the network to the outlet; $\beta_c = \frac{3}{2}V_0$ for celerity, V_0 being the equilibrium velocity of channel flow).

The time to peak, t_p, and peak discharge, Q_p, of the GIUH are expressed as (Troutman and Karlinger 1985):

$$t_p = \frac{\alpha\sqrt{2n}}{\beta_c} \tag{8.54}$$

and

$$Q_p = \frac{\beta_c}{\alpha\sqrt{2ne}} \tag{8.55}$$

where $e = 2.718$. Eliminating β_c in Eqs. (8.54) and (8.55) yields:

$$t_p = \frac{1}{\sqrt{e}Q_p} \approx \frac{1}{1.65Q_p} \tag{8.56}$$

which indicates that the larger the flood peak, the shorter is the time to peak.

Other interesting parameters in the TK GIUH model (Troutman and Karlinger 1985) include the mean holding time per link, t_h, which is the inverse of Eq. (8.53):

$$t_h = \frac{1}{\sigma} = \frac{\alpha}{\beta_c} \tag{8.57}$$

and the scaling parameter expressed as:

$$\xi = \frac{\alpha\sqrt{n}}{\beta_c} \tag{8.58}$$

which is very close to the GIUH time to peak in Eq. (8.54).

Troutman and Karlinger (1984, 1985, 1986) outlined the structures of the model and stated that the key idea is that the two upstream links joining the outlet link at the junction may themselves be taken to be the outlets of two subnetworks. The model offers a simple and hydrologically and topographically meaningful tool for comparative studies of catchment responses to rainfall since this model with the scaling parameter ξ makes inter-catchment and inter-regional comparisons possible. Since the asymptotic solution of the TK model is an analytical expression for IUH, it can be used as a kernel in the system hydrological model, that is, the output is a convolution integral of the kernel and effective input. This approach is important for analyzing geomorphic parameters which control hydrological response in catchment networks.

Equation (8.57) for mean holding time can be determined from the topographical data of a catchment and the flow properties of the catchment. As a very approximate estimate, the holding time or residence time of water on hillslopes and in channels was given by Kirby (1988) as follows:

surface detention: $0.1 \sim 1.0$ hour;
downslope flow: $1.0 \sim 12$ hours;
infiltration: $1.0 \sim 20$ hours;
percolation: $1.0 \sim 50$ hours;
channel flow: 0.5 hour for an area of 1 km², 7 hours for an area of 100 km², and 100 hours for an area of 1×10^4 km².

With $\beta_c = \frac{3}{2}V_0$, and the above estimates of holding times, the mean internal link length α can be estimated using Eq. (8.57).

6.3 The Topographical Model and Its Applications in Streamflow and Flood Routing

Since the asymptotic solution of the TK model is an expression for IUH, the outflow from catchment networks as a system can be derived as a convolution integral of the kernel and the input to the system (Chow et al. 1988, Bras 1990) as:

$$Q(t) = \int_0^t f(\tau)\overline{u}(t-\tau)d\tau \tag{8.59}$$

where $Q(t)$ denotes outflow, $f(t)$ is the effective (or net) input function and $\bar{u}$ is the kernel, the IUH in Eq. (8.52). The effective input $f(t)$ is the difference between the effective rainfall intensity $r(t)$ and infiltration rate $i(t)$:

$$f(t) = r(t) - i(t) \tag{8.60}$$

which has also been used in Eq. (8.9) and other equations for flow on hillslopes.

At present, Eq. (8.52) was derived by Troutman and Karlinger without introducing any anomalous parameter in the derivations, and the mechanism of anomalous flow only enters the streamflow in catchment networks through Eq. (8.60) for anomalous infiltration. It is expected that when fPDEs are used for deriving an IUH similar to Eq. (8.52) by Troutman and Karlinger (1984, 1985, 1986), a new IUH can be found, which can incorporate the anomalous flow in channel networks and anomalous infiltration.

With the change of variables, $\zeta = t - \tau$, Eq. (8.59) incorporating Eq. (8.52) can be written as:

$$Q(t) = \int_0^t f(t - \zeta) \frac{\sigma^2 \zeta}{2n} \exp\left[-\left(\frac{\sigma \zeta}{2\sqrt{n}} \right)^2 \right] d\zeta \tag{8.61}$$

With digitized data for a small catchment, Su et al. (1997) also demonstrated the applications of the TK model.

7. Anomalous Flow during Irrigation

Irrigation is one of the topics where infiltration plays an important role in water management. This section demonstrates the application of the anomalous equations of infiltration by incorporating them into the Lewis-Milne equation for border-check irrigation and the Rasmussen equation for radial flow.

7.1 Water Advance-Infiltration-Evaporation during Border-check Irrigation

The widely-used method for analyzing water flow during border-check irrigation (or flood irrigation), known as the Lewis-Milne equation (LME), is accredited to Lewis and Milner (1938) and is written as:

$$qt = hx + \int_0^t I(t - \tau) \frac{\partial x}{\partial \tau} d\tau \tag{8.62}$$

where q is the volumetric water flow rate per unit width $[L^2 T^{-1}]$, h is the mean depth of water on soil surface, $I(t)$ denotes cumulative infiltration, and x is the distance of advance at time t.

Equation (8.62) can also be written as follows (Rasmussen 1994):

$$Qt = hwx + w \int_0^t I(t - \tau) v(t) d\tau \tag{8.63}$$

where Q is the volumetric rate of water supply (having the dimension $[L^3T^{-1}]$), $v(t) = \dfrac{\partial x}{\partial t}$ represents advance velocity and w is the width of the flow surface.

Equation (8.62) has been extensively investigated for the past eight decades, where cumulative infiltration is represented by different forms of models such as a power function and a two-term equation (Philip and Farrell 1964), etc.

Irrigated regions are generally characterized by dry climates with high evaporation potentials as the reason for irrigation. The Lewis-Milne equation for surface irrigation considers only water advance and infiltration without evaporation. The neglect of evaporation in this approach raises the issue of accuracy, particularly for large areas which require longer times to deliver water to cover the whole area, which increases evaporation as water flows on its way. Without taking evaporation and transpiration in dry weather conditions in large areas into account, such models for analyzing irrigation carry certain inaccuracies. To remediate this problem, an evaporation term was introduced in the Rasmussen equation (Su 2007) for radial water advance-infiltration-evaporation. Similarly, a cumulative evaporation term, $E_T(t)$, can be introduced to the Lewis-Mile equation to yield:

$$qt = hx + \int_0^t [I(t-\tau) - E(t-\tau)] \frac{\partial x}{\partial \tau} d\tau \tag{8.64}$$

which can also be written as:

$$qt = hx + \int_0^t I(t-\tau) \frac{\partial x}{\partial \tau} d\tau + \int_0^t E(t-\tau) \frac{\partial x}{\partial \tau} d\tau \tag{8.65}$$

When the equations of cumulative infiltration and cumulative evaporation take the same form, solutions of Eq. (8.64) are simple (Su 2007). Otherwise, Eq. (8.64) can be treated as the Hammerstein equation of the second kind, with polynomial kernels (Polyanin and Manzhirov 1998).

7.2 Solutions of Water Advance-Anomalous Infiltration during Border-check Irrigation

In this situation, Eq. (8.64) is solved and analyzed. To aid in the analysis, Eq. (8.64) is written in the following form by applying the Laplace transform (Philip and Farrell 1964):

$$x = q \mathscr{L}^{-1} \left\{ \frac{1}{s^3 \mathscr{L}[I(t) - E(t)] + hs^2} \right\} \tag{8.66}$$

where $\mathscr{L}$ and $\mathscr{L}^{-1}$ denote the Laplace transform and the inverse Laplace transform (with the variable s) respectively.

Depending on the form of the equation for cumulative infiltration, Eq. (8.66) yields different results. In the following sections, a new equation for anomalous cumulative infiltration from Eq. (6.11) in Chapter 6 has been used in Eq. (8.66) as an example. Even though evaporation can be included, such as in the previous analysis (Su 2007), it has not been considered here for the sake of simplicity.

7.2.1 Solution of Water Advance-Infiltration on Swelling Soils with Mobile and Immobile Zones

7.2.1.1 Full Solutions

As an example, substitution of the cumulative infiltration equation (without evaporation) given by Eq. (6.11), that is, $I(t) = At + St^{\beta_2/2 - \beta_1/4}$, in Eq. (8.66) yields:

$$x = \frac{q}{h} \mathcal{L}^{-1} \left\{ s^{-1} \left[s + \frac{S\Gamma[1+(\beta_2/2)+\beta_1/4]}{h} s^{1-(\beta_2/2)+\beta_1/4} + \frac{A}{h} \right]^{-1} \right\} \tag{8.67}$$

which has the following inverse Laplace transform (Haubold et al. 2011):

$$x = \frac{q}{h} \sum_{k=0}^{\infty} \left(-\frac{S\Gamma[1+(\beta_2/2)+\beta_1/4]}{h} \right)^k t^{1+(\beta_2/2-\beta_1/4)k} E_{1,2+(\beta_2/2-\beta_1/4)k}^{k+1} \left(-\frac{A}{h} t \right) \tag{8.68}$$

The identity for summation in Eq. (8.68) can be expressed as (Gradshhteyn and Ryzhik 1996):

$$\sum_{k=0}^{\infty} \left(-\frac{S\Gamma[1+(\beta_2/2)-\beta_1/4]}{h} \right)^k = \frac{h}{h+S\Gamma[1+(\beta_2/2)-\beta_1/4]} \tag{8.69}$$

and the term $t^{(\beta_2/2-\beta_1/4)k+1} E_{1,(\beta_2/2-\beta_1/4)k+2}^{k+1} \left(-\frac{A}{h} t \right)$ is a special form of the GPF (see previous section and Tomosvki et al. 2014):

$$e_{1,\beta}^{\gamma} \left(t; \frac{A}{h} \right) = t^{\beta+1} E_{1,\beta+2}^{\gamma} \left(-\frac{A}{h} t \right), \qquad k = 0,1,2,... \tag{8.70}$$

where

$$\beta = \left(\frac{\beta_2}{2} - \frac{\beta_1}{4} \right) k + 2 \tag{8.71}$$

$$\gamma = k + 1 \tag{8.72}$$

With the GPF, the solution in Eq. (8.68) can be written in a compact form as:

$$x = \frac{q}{h + S\Gamma[1+(\beta_2/2) - \beta_1/4]} e_{1,\beta}^{\gamma} \left(t; \frac{A}{h} \right) \tag{8.73}$$

7.2.1.2 Asymptotic Solution

The GPF in Eq. (8.70) has interesting asymptotic solutions for $t \to +\infty$, which have been investigated by Mainardi and Garrappa (2015). By extending the upper limit in their analysis to 2, the asymptotic solution of Eq. (8.70) is physically meaningful only when the first term is used, that is, $k = 0$ in Eqs. (8.68)–(8.72). With $k = 0$ or $\gamma = 1$, the asymptotic solutions of Eq. (8.70) take the following form:

$$e_{1,\beta}^{\gamma}\left(t;\frac{A}{h}\right) \approx 1, \qquad \text{for} \quad 0 < 1 < \beta \le 2 \tag{8.74}$$

Hence, Eq. (8.73) yields:

$$x = \frac{q}{h + S\Gamma[1 + (\beta_2/2) - \beta_1/4]}, \qquad t \to +\infty \tag{8.75}$$

which is the steady-state part (after a large time) of the solution.

For uniform soils (without considering immobile zones), $\beta_1 = 0$ and let $\beta_2 = \beta$. Subsequently, Eq. (8.75) can be used to estimate the value of β, given the measured values of other variables, by rearranging it:

$$\Gamma[1 + \beta/2] = \frac{1}{S}\left(\frac{q}{x} - h\right) \tag{8.76}$$

The above equation can be inverted to find the value of β using tables of the gamma function (Abramowitz and Stegun 1965) or computer programs.

7.2.1.3 Approximate Solutions

The asymptotic analysis above implies that $k = 0$ is the only case for a physically meaningful solution. Hence, by retaining only the first term in Eq. (8.73), the following solution is arrived at:

$$x = \frac{q}{h + S\Gamma[1 + (\beta_2/2) - \beta_1/4]}\left(\frac{h}{A}\right)\left[1 - \exp\left(-\frac{A}{h}t\right)\right] \tag{8.77}$$

For large times, the exponential function in Eq. (8.77) diminishes so that a steady-state develops from Eq. (8.77) as:

$$x = \frac{q}{h + S\Gamma[1 + (\beta_2/2) - \beta_1/4]}\left(\frac{h}{A}\right) \tag{8.78}$$

7.2.2 Water Advance-Infiltration in Swelling Soils without Immobile Zones

In this case, the equation of cumulative infiltration in the form of $I(t) = At + St^{\beta/(2\lambda-1)}$ is used and the rest of the analysis is similar to the case in section 7.2.1 as follows:

$$e_{1,\rho}^{\gamma}\left(t;\frac{A}{h}\right) = t^{1+k\rho}E_{1,2+k\rho}^{k+1}\left(-\frac{A}{h}t\right), \qquad k = 0,1,2,\dots \tag{8.79}$$

where $E_{1,2+k\beta/(2\lambda-1)}^{k+1}\left(-\frac{A}{h}t\right)$ is the Prabhakar function with

$$\rho = \beta/(2\lambda - 1) \tag{8.80}$$

and the solution in Eq. (8.68) is written in the following compact form:

$$x = \frac{q}{h + S\Gamma[1 + \beta/(2\lambda - 1)]} e_{1,\rho}^{\gamma}\left(t; \frac{A}{h}\right) \tag{8.81}$$

The simplified solution for uniform pores, analogous to Eq. (8.77), is:

$$x = \frac{q}{h + S\Gamma[1 + \beta/(2\lambda - 1)]} \left(\frac{h}{A}\right)\left[1 - \exp\left(-\frac{A}{h}t\right)\right] \tag{8.82}$$

and the steady-state solution from Eq. (8.82), for large times, is written as:

$$x = \frac{q}{h + S\Gamma[1 + \beta/(2\lambda - 1)]} \left(\frac{h}{A}\right) \tag{8.83}$$

7.3 Anomalous Radial Flow

Based on the principles of water balance, for deriving the classical Lewis-Milne equation, Rasmussen (1994) presented the following mathematical model for analyzing the dynamics of radial water movement from a point source:

$$qt = \frac{\phi_s hr^2}{2} + \phi_s \int_0^t I(t - \tau) \frac{\partial}{\partial \tau}\left[\frac{r^2(\tau)}{2}\right] d\tau \tag{8.84}$$

where q is the supply rate, h is the mean depth of water on the surface, r is the radial distance of advance reached at time t, $I(t)$ denotes cumulative infiltration, and ϕ_s is the angle of water spreading. The value of ϕ_s defines the geometry of water spreading and varies between 0 to 2π, with $\phi_s = 2\pi$ for spreading in a full circle and $\phi_s = \pi$ for a half circle.

When evaporation is considered, the Rasmussen equation takes the following form (Su 2007):

$$qt = \frac{\phi_s hr^2}{2} + \phi_s \int_0^t [I(t - \tau) + E(t - \tau)] \frac{\partial}{\partial \tau}\left[\frac{r^2(\tau)}{2}\right] d\tau \tag{8.85}$$

The procedures for deriving the solutions of Eq. (8.85) are identical to those presented earlier (Su 2007); with an equation of anomalous infiltration, $I(t - \tau)$, used in Eq. (8.85), Eq. (8.85) would be able to incorporate anomalous infiltration during the integrated radial water movement.

8. Summary

This chapter discussed fKW models for overland flow, solute movement and possible erosion on hillslopes as coupled processes on slopes of different geometries. Detailed discussions of solutions have also been provided as examples of applications of the fKW model for overland flow. fKW is a non-linear fDE and its solutions have been presented for specific initial and boundary conditions, and asymptotic solutions on the depth of water and the discharge overland were developed for large times. An fKW model for solute transport has also been suggested, but no further analysis has been carried out. Data from laboratory and field are needed to verify some of the concepts suggested in this chapter.

Analyses have also been outlined for water flow in catchment networks and integrated water movement during irrigation.

Chapter 9

Fractional Partial Differential Equations for Groundwater Flow

1. Introduction

Water flow in saturated media such as aquifers and water-logged soils is a very important topic; it serves as a vital resource for many sectors, is crucial for human consumption and has a significant role in the environment. Groundwater hydrology and hydrogeology or geohydrology are the fields which deal with water flow in saturated porous media such as aquifers and in special cases with soils which experience variable saturation under water-logged conditions. The level of saturation in pores is a crucial difference which distinguishes soil water physics and groundwater hydrology; materials presented in Chapters 5, 6 and 7 are more related to this chapter besides the introductory Chapters 3 and 4.

Management, protection and assessment of groundwater require quantitative knowledge of storage, flow, distribution and physico-chemical and biological properties of water in saturated strata. These requirements have prompted interested individuals to develop knowledge and relevant methods for understanding groundwater flow. The first step towards the quantification of flow in porous media by Darcy (1856) marked the beginning of a quantitative era. Following Darcy's pioneering experiments, a rigorous mathematical formulation in the form of partial differential equations (PDEs) was presented and analyzed by Boussinesq (1904). Since Darcy and Boussinesq, numerous models and methodologies have appeared and extensive research is ongoing, as is evident from continuing publications in various forms.

Aquifers, depending on the type of pressure applied on them, are usually classified into two types: confined aquifers and unconfined aquifers.

A confined aquifer exists between two confining impervious (or low permeable) layers. When the hydraulic pressure of the aquifer is high enough for water to reach the surface of the ground (to well up or flow freely from a well), such an aquifer is also called an artesian, which takes after the name of the province of Artois in France, where this type of aquifer was first named (de Marsily 1986). A confined aquifer, which exists between two impervious geological formations without a free surface, is also called a pressure aquifer (Bear 1979, Bras 1990). The pressure on such an aquifer depends on whether external water is being added to it (such as an aquifer with an inclined topography), or water in the aquifer is being extracted from it (such as through pumping). In the former case, the pressure is positive, whereas it may be negative (at least for a period of time when the recharge rate of the aquifer is less than the pumping rate) in the latter case.

An unconfined aquifer, also called a phreatic aquifer, is one which has a free surface or water table with the unsaturated zone as the upper boundary. The word 'phreatic' originated from the Greek word *phrear* and *phreat* meaning '*a well*'. As such, an unconfined aquifer is under atmospheric pressure to which the aquifer is exposed through the pores of the unsaturated zone. For this reason, unconfined aquifers are also called water-table aquifers (Hantush 1964b).

In reality, any 'impervious' formation of geological origin has a certain level of permeability which allows water to move slowly. de Marsily (1986) 'arbitrarily' set a limit of hydraulic conductivity K as the criterion for separating confined and unconfined aquifers; this limit is $K = 10^{-9}$ *meter/s* or $K = 0.0036$ *mm/h*.

The arbitrary separation of confined and unconfined aquifers can be problematic, depending on the issue of interest. For example, for natural groundwater flow without major pollutants, this separation is feasible and is useful for handling measurements and quantification of groundwater hydrology; however, for designing and monitoring radioactive wastes stored in long-term underground facilities, the limit of hydraulic conductivity ($K = 0.0036$ *mm/h*) would be very high, hence creating a serious problem if leakage is possible and/or storage is designed for centuries or millennia.

Another reality is that both confined and unconfined aquifers may have connections with external formations at certain locations where a higher permeability forming a leaky aquifer exists; in many cases, portions of an aquifer are confined only while the rest of the aquifer has a free surface, such as an artesian aquifer on a slope. In some cases, the same aquifer can change its type from unconfined to confined, following engineering operations (de Marsily 1986).

Since the pioneering works of Darcy (1856) and Boussinesq (1904), methods for quantification of flow in porous media, specifically in groundwater hydraulics and hydrology, have been well developed. The dominant methods are based on integer calculus and are known as the governing equations of groundwater flow, derived from the conservation of mass and the moment equation from Darcy's law (Eagleson 1970, Bear 1972, 1979, Bras 1990). Depending on the choice of terminologies used for the water body (such as potential, density, hydraulic head or piezometric head), these governing equations appear in slightly different forms, particularly when different assumptions are made, leading to the respective derivations. In this chapter, we briefly outline the governing equations of groundwater flow for the two types of aquifers, along with their fractional counterparts, fractional partial differential equations (fPDEs) developed in recent decades. Some of their applications using limited data from field as examples have also been summarized in this chapter.

This chapter is only concerned with the following major issues in groundwater flow:

1. Isothermal flow in confined, heterogeneous and homogeneous aquifers.
2. Isothermal flow in unconfined, heterogeneous and homogeneous aquifers.

3. Isothermal radial flow in confined and unconfined aquifers, in terms of hydraulics of wells, including hydraulics of flow in aquifers on a sloping base and hydraulics of wells on a hillslope.

4. Earth tide and barometric tide effects on groundwater, etc.

5. fPDEs in different forms for groundwater flow in different types of aquifers, and other issues.

2. Governing Equations for Isothermal Groundwater Flow in Confined Aquifers

Referring to Eq. (3.49), consider a saturated media in which there is no change in the saturation in pores, that is, $\dfrac{\partial s}{\partial t} = 0$, along with Terzaghi's theory which states that $\Delta\sigma_z = -\Delta p$. The governing equation (Eq. 3.49) (Eagleson 1970) for flow of an incompressible fluid ($\beta = 0$) with a constant density, in a confined, heterogeneous, isothermal, vertically compressible medium, on a horizontal base is simplified as:

$$\alpha \frac{\partial p}{\partial t} = -\nabla \cdot \mathbf{q} \tag{9.1}$$

where p is the pressure head (see Chapter 3 for details), $\mathbf{q}$ is the flux vector and α is the vertical compressibility of the granular skeleton of the porous medium.

With the definitions of hydraulic head (which has three components) and piezometric head (which coincides with hydraulic head by neglecting kinetic head) (de Marsily 1986), Eq. (9.1) can be written in terms of hydraulic head h in confined aquifers as (Eagleson 1970):

$$S_e \frac{\partial h}{\partial t} = -\nabla \cdot \mathbf{q} \tag{9.2}$$

where S_e is denotes storage coefficient or storativity. When Darcy's law is expressed in the flux vector:

$$\mathbf{q} = -K_x \frac{\partial h}{\partial x} - K_y \frac{\partial h}{\partial y} - K_z \frac{\partial h}{\partial z} \tag{9.3}$$

where K_x, K_y, and K_z are the saturated hydraulic conductivities in the three directions, Eq. (9.2) can be expanded as (de Marsily 1986):

$$S_e \frac{\partial h}{\partial t} = \frac{\partial}{\partial x}\left(K_x \frac{\partial h}{\partial x}\right) + \frac{\partial}{\partial y}\left(K_y \frac{\partial h}{\partial y}\right) + \frac{\partial}{\partial z}\left(K_z \frac{\partial h}{\partial z}\right) \tag{9.4}$$

For practical purposes in field hydraulics and hydrology, some assumptions can be made to simplify Eq. (9.4). One simplification is the well-known Dupuit hypothesis that assumes that there is no vertical flow in the free surface of flow, which was clearly defined for unconfined aquifers. With the Dupuit hypothesis applied to confined aquifers, integrating Eq. (9.4) in the vertical direction (Bear 1979, de Marsily 1986) yields the following two-dimensional flow equation when a source term, R, is included for recharge (+) or withdrawal (–) (with leakage included):

$$S_e \frac{\partial h}{\partial t} = \frac{\partial}{\partial x}\left(T_x \frac{\partial h}{\partial x}\right) + \frac{\partial}{\partial y}\left(T_y \frac{\partial h}{\partial y}\right) \pm R - (F_t + F_b) \tag{9.5}$$

where F_t and F_b are the water exchange rates between the confined aquifer and the upper (F_t) and the lower (F_b) confining layers, and transmissivity in x direction is given as:

$$T_x = bK_x \tag{9.6}$$

Similarly, transmissivity in y direction is expressed as:

$$T_y = bK_y \tag{9.7}$$

with b as the thickness of the aquifer. When both transmissivities are constant, Eq. (9.5) further is simplified as:

$$S_e \frac{\partial h}{\partial t} = T_x \frac{\partial^2 h}{\partial x^2} + T_y \frac{\partial^2 h}{\partial y^2} \pm R - (F_t + F_b) \tag{9.8}$$

3. Governing Equation for Groundwater Flow in Unconfined Aquifers

The general equation of groundwater flow in unconfined aquifers (with constant water density) is Eq. (3.56), which is a consequence of conservation of mass:

$$\phi_e \frac{\partial h}{\partial t} = \frac{1}{\rho g}\frac{\partial p}{\partial t} = -\left(\frac{\partial q_x}{\partial x} + \frac{\partial q_y}{\partial y} + \frac{\partial q_z}{\partial z}\right) \tag{9.9}$$

where ϕ_e represents specific yield or drainage porosity (de Marsily 1986).

For unconfined flow, the upper surface of the aquifer is exposed to the atmosphere through pores in the soil. Equation (9.9) can be simplified for flow in unconfined, non-consolidating media, with the Dupuit hypothesis of no vertical flow in aquifers. For this type of flow, Boussinesq (1904) first presented a PDE, now known as the Boussinesq equation (BE) for groundwater flow in an unconfined aquifer. When the rate of recharge, R, is included, the BE is written as (Boussinesq 1904, Bear 1979, Bras 1990):

$$\frac{\partial h}{\partial t} = \frac{1}{2\phi_e}\left[\frac{\partial}{\partial x}\left(K_x \frac{\partial h^2}{\partial x}\right) + \frac{\partial}{\partial y}\left(K_y \frac{\partial h^2}{\partial y}\right)\right] + \frac{R}{\phi_e} \tag{9.10}$$

where piezometric head h is used. Equation (9.10) can also be written as:

$$\phi_e \frac{\partial h}{\partial t} = \frac{\partial}{\partial x}\left(K_x h \frac{\partial h}{\partial x}\right) + \frac{\partial}{\partial y}\left(K_y h \frac{\partial h}{\partial y}\right) + R \tag{9.11}$$

For homogeneous media, $K_x = K_y$; with recharge R, Eq. (9.10) is written as (Eagleson 1970):

$$\frac{\partial h}{\partial t} = \frac{K}{2\phi_e}\left(\frac{\partial^2 h^2}{\partial x^2} + \frac{\partial^2 h^2}{\partial y^2}\right) + \frac{R}{\phi_e} \tag{9.12}$$

Aquifers in natural conditions often have a certain slope in their underlying bases. When the slope gradient of the aquifer base, along with recharge rate, is incorporated in the one-dimensional form of Eq. (9.11), the following equation results (Werner 1957):

$$\phi_e \frac{\partial h}{\partial t} = K\left[h\frac{\partial^2 h}{\partial x^2} + \left(\frac{\partial h}{\partial x}\right)^2 - \omega\frac{\partial h}{\partial x}\right] + R \tag{9.13}$$

where $\omega = \tan \Lambda$ is the slope gradient of the aquifer, with an angle Λ on an impervious base.

Equation (9.13) can be written as follows to accommodate the variability of saturated hydraulic conductivity K:

$$\phi_e \frac{\partial h}{\partial t} = \frac{\partial}{\partial x}\left[Kh\left(\frac{\partial h}{\partial x} - \omega\right)\right] + R \tag{9.14}$$

which is a consequence of conservation of mass given as:

$$\phi_e \frac{\partial h}{\partial t} = -\frac{\partial q}{\partial x} + R \tag{9.15}$$

with q as mass flux per unit cross-sectional area (Werner 1953, 1957):

$$q = Kh\left(\omega - \frac{\partial h}{\partial x}\right) \tag{9.16}$$

Although the complete governing equations are more rigorous, field hydraulics practitioners often encounter many inaccuracies which make the rigorous mathematics impractical. One example of the simplified BE is the linearized form of Eq. (9.14), which can be used to derive the rate of time-dependent recharge by infiltrating water or any other form of increment in aquifers (Su 1994). In addition to the one-dimensional BE for groundwater flow in unconfined aquifers on a sloping base, Eq. (9.13), the two-dimensional BE can also be used (Bear 1979, Mahdavi 2015).

4. Unified Concepts and Equations for Groundwater Flow in Confined and Unconfined Aquifers

When defined in terms of mass flux, the equations of flow in both confined and unconfined aquifers are identical; however, when expressed in terms of hydraulic head, there is a difference evident by comparing Eq. (9.4) for confined aquifers with Eq. (9.10) for unconfined aquifers with Eq. (9.10). Referring to section 3.2 in

Chapter 3, based on Bernoulli's theorem (1738), which relates *elevation, pressure* and *velocity* along the given flow line of a fluid in frictionless flow, with potential Φ, another concept for unifying these two forms of equations of flow in the two types of aquifers is *hydraulic potential* (ψ). The relationship between hydraulic potential ψ and piezometric head h is expressed as (Hubbert 1940, Bras 1990):

$$\psi = h \tag{9.17}$$

for confined aquifers; for unconfined aquifers, the following relationship holds:

$$\psi = \frac{h^2}{2} \tag{9.18}$$

With the use of mass flux or hydraulic potential, the mathematical structures of the governing equations for confined and unconfined aquifers are identical, except for the different definitions of parameters ϕ_e, S_e and transmissivity. This similarity is very convenient for analyzing groundwater flow in both types of aquifers when one governing equation with one set of parameters is defined specifically for each type of aquifer.

4.1 Unified Equation for Groundwater Flow

In two dimensions, the equations of flow in both confined and unconfined aquifers can be unified by defining hydraulic potential ψ (Bras 1990). Hence, the following unified equation applies to both types of aquifers:

$$C_1 \frac{\partial \psi}{\partial t} = \frac{\partial^2 \psi}{\partial x^2} + \frac{\partial^2 \psi}{\partial y^2} + \frac{R}{C_2} \tag{9.19}$$

where the two parameters C_1 and C_2 are defined differently for the two types of aquifers:

1. For confined aquifers:

$$C_1 = \frac{S_c}{Kb} = \frac{S_c}{T} \tag{9.20}$$

$$C_2 = T \tag{9.21}$$

Here, S_c denotes storativity, b is the thickness of the aquifer and T represents transmissivity.

2. For unconfined aquifers:

$$C_1 = \frac{\phi_e}{Kh_0} = \frac{\phi_e}{T} \tag{9.22}$$

$$C_2 = K \tag{9.23}$$

where h_0 is the representative height of the aquifer, with which the original Boussinesq equation is linearized for unconfined aquifers, based on the Dupuit assumption (also see Bear 1979).

4.2 Heterogeneous Confined Aquifers

In this case, K is a function of depth z in one dimension, expressed as $K(z)$, and there are two methods in practice for its determination. The first method is the weighted mean hydraulic conductivity for confined aquifers, $\bar{K}_c$, recommended to replace the conventional hydraulic conductivity K for a heterogeneous confined aquifer (Bear 1972) in Eq. (9.4):

$$\bar{K}_c = \frac{1}{b}\int_0^b K(z)dz \tag{9.24}$$

where b is the thickness of the aquifer. Subsequently, the mean transmissivity for confined aquifers, $\bar{T}_c$, is:

$$\bar{T}_c = \bar{K}_c h \tag{9.25}$$

Another method for determining transmissivity in a heterogeneous confined aquifer is to calculate transmissivity integrated over the depth of the aquifer (Bras 1990):

$$\bar{T}_c = \int_{z_1}^{z_2} K(z)dz \tag{9.26}$$

where z_1 and z_2 are the heights of the aquifer at the two points of interest ($z_1 = 0$ if point 1 is used as the reference datum).

4.3 Heterogeneous Unconfined Aquifers

For heterogeneous unconfined aquifers, the weighted mean hydraulic conductivity, $\bar{K}_u$, is expressed as (Bear 1972):

$$\bar{K}_u = \frac{1}{h}\int_0^h K(z)dz \tag{9.27}$$

and the mean transmissivity is given by (Bras 1990):

$$\bar{T}_u = \int_{z_1}^{h} K(z)dz \tag{9.28}$$

where the upper integral limit is the free surface of the aquifer (rather than a fixed height) and the lower limit z_1 in Eq. (9.28) can be set to zero (as the reference datum).

It is thus evident that for both confined and unconfined homogeneous aquifers, the same unified governing equation, Eq. (9.19), applies when hydraulic potential is used as the key variable. For heterogeneous aquifers, the simplified governing equations appear identical in structure, but the two mean parameters, $\bar{T}$ and $\bar{K}$, are determined slightly differently, with the differences lying in the upper limit of the integral and the denominator for weighing the integral.

5. Radial Flow and Hydraulics of Wells in Confined and Unconfined Aquifers

5.1 Introduction to Radial Flow in Aquifers

Hydraulics of wells is one of the most important topics in groundwater hydraulics and hydrology as well as in the petroleum industry. There are extensive publications available on this topic (Hantush 1964c, Bear 1972, 1979, de Marsily 1986, Barenblatt et al. 1990, Bras 1990); related mathematical analyses of flow in radial direction can be found in the works of Carslaw and Jaeger (1959) and Evans et al. (1999).

The flow of water can be analyzed by mathematical equations formulated in different coordinates; cylindrical coordinates seem to be a better choice of coordinates for radial flows, for they are more suited to the radial flow patterns. To aid in the transformation from Cartesian coordinates to cylindrical coordinates, Fig. 9.1 shows the connection between the two coordinate systems, and the terminologies used in the transformation (Bourne and Kendall 1977).

In cylindrical coordinate system, as shown in Fig. 9.1, the position of a point P in the flow is determined in polar coordinates, r and θ, with respect to its projection on the plane of x and y, and the vertical coordinate z; here, θ is the polar angle of the imaginary pole pivoted at the center of the well turning horizontally around the zx-plane (or zy-plane) and r is the radial distance from the center of the well (Carslaw and Jaeger 1959, Bourne and Kendall 1977). Cartesian coordinates defined by (x, y, z) and cylindrical coordinates defined by (r, θ, z) are related by the following identities (Carslaw and Jaeger 1959, Bourne and Kendall 1977):

$$\left. \begin{aligned} x &= r\cos\theta \\ y &= r\sin\theta \\ z &= z \end{aligned} \right\} \tag{9.29}$$

In the following sections, different equations have been discussed for investigating flow in radial direction.

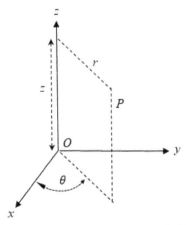

Fig. 9.1. Transformation of Cartesian coordinates (x, y, z) to cylindrical coordinates (r, θ, z).

5.2 Radial Flow in Homogeneous Confined Aquifers

5.2.1 Radial Flow in Homogeneous Confined Aquifers

With cylindrical coordinates, the equation of flow in homogeneous confined aquifers can be written as (Carslaw and Jaeger 1959):

$$\frac{\partial h}{\partial t} = \frac{K}{S_c}\left[\frac{1}{r}\frac{\partial}{\partial r}\left(r\frac{\partial h}{\partial r}\right) + \frac{1}{r^2}\frac{\partial^2 h}{\partial \theta^2} + \frac{\partial^2 h}{\partial z^2}\right] \tag{9.30}$$

which describes three components of flow, with flow in radial direction, the horizontally concentric direction and in the vertical direction measured by

$$\frac{K}{S_c r}\frac{\partial}{\partial r}\left(r\frac{\partial h}{\partial r}\right), \quad \frac{K}{S_c r^2}\frac{\partial^2 h}{\partial \theta^2} \quad \text{and} \quad \frac{K}{S_c}\frac{\partial^2 h}{\partial z^2} \text{ respectively.}$$

Equation (9.30) can be simplified to form different equations of radial flow subject to different flow conditions, as outlined below.

Case 1: Purely horizontally radial flow

When the flow of a line source takes place (identical to the case of heat flow with the z axis heated or cooled), with the initial condition and the boundary conditions independent of θ and z, indicative of flow of a line source from or towards a well, Eq. (9.30) is simplified as (Carslaw and Jaeger 1959):

$$\frac{\partial h}{\partial t} = \frac{K}{S_c}\frac{1}{r}\frac{\partial}{\partial r}\left(r\frac{\partial h}{\partial r}\right) \tag{9.31}$$

Case 2: Horizontally radial, concentric flow

When the initial condition and the boundary conditions are independent of z and the flow takes place perpendicular to the z axis, Eq. (9.30) gets simplified as (Carslaw and Jaeger 1959):

$$\frac{\partial h}{\partial t} = \frac{K}{S_c}\left[\frac{1}{r}\frac{\partial}{\partial r}\left(r\frac{\partial h}{\partial r}\right) + \frac{1}{r^2}\frac{\partial^2 h}{\partial \theta^2}\right] \tag{9.32}$$

As stated above, the flow described by $\dfrac{K}{S_c r^2}\dfrac{\partial^2 h}{\partial \theta^2}$ is concentric, similar to circulation around the center of the well. If there is only concentric flow, as in the case of circulation in a well, but no radial flow, Eq. (9.32) can be further simplified as:

$$\frac{\partial h}{\partial t} = \frac{K}{S_c r^2}\frac{\partial^2 h}{\partial \theta^2} \tag{9.33}$$

Case 3: Axis-symmetrical radial and vertical flow

This kind of flow takes place concurrently in both radial and vertical directions. It corresponds to the case when the initial condition and the boundary conditions are independent of θ and the flow takes place through the imaginary planes through the z axis. In this case, Eq. (9.30) becomes (Carslaw and Jaeger 1959):

$$\frac{\partial h}{\partial t} = \frac{K}{S_c}\left[\frac{1}{r}\frac{\partial}{\partial r}\left(r\frac{\partial h}{\partial r}\right) + \frac{\partial^2 h}{\partial z^2}\right] \tag{9.34}$$

Both Eqs. (9.31) and (9.34) have been widely applied to investigate flow in wells; however, Eq. (9.32) has been comparatively less investigated in published reports.

5.2.2 Radial Flow in Heterogeneous Confined Aquifers

For flow in heterogeneous aquifers, the variation of hydraulic conductivity K is a major concern, besides possible changes in other parameters. Several cases have been briefly discussed below:

1. For the complete equation of radial flow in Eq. (9.30), the variation of K needs to be considered in terms of r, θ and z.

2. For radial flow of a line source in Case 1, which corresponds to the situation described by Eq. (9.31) as the simplest scenario, hydraulic conductivity K for radial flow can be evaluated through the methods described in section 4.2.

3. For radial flow perpendicular to the z axis, described by Eq. (9.32), the methods described in section 4.2 can again be used to evaluate the hydraulic conductivity.

4. The equation of flow for axis-symmetrical radial flow which takes place in concurrent radial and vertical directions in heterogeneous aquifers is written as (Yeh and Chang 2013):

$$S_c\frac{\partial h}{\partial t} = \frac{K_r}{r}\frac{\partial}{\partial r}\left(r\frac{\partial h}{\partial r}\right) + K_z\frac{\partial^2 h}{\partial z^2} \tag{9.35}$$

where K_r and K_z are the hydraulic conductivities in horizontal radial and vertical directions respectively. It is evident that Eq. (9.34) is a special form of Eq. (9.35).

5.3 The Unified Equation for Radial Flow in Confined and Unconfined Aquifers

5.3.1 Radial Flow in Homogeneous Aquifers

Similar to Eq. (9.30), the unified equation of flow in Eq. (9.19) for confined and unconfined aquifers can be transformed into cylindrical coordinates (Bras 1990):

$$C_1\frac{\partial \psi}{\partial t} = \frac{1}{r}\frac{\partial}{\partial r}\left(r\frac{\partial \psi}{\partial r}\right) + \frac{1}{r^2}\frac{\partial^2 \psi}{\partial \theta^2} + \frac{\partial^2 \psi}{\partial z^2} + \frac{R}{C_2} \tag{9.36}$$

When Eq. (9.36) is used, distinction must be between confined and unconfined aquifers:

1. For confined aquifers, the parameters in Eq. (9.36) are specified for potential ψ by Eq. (9.17), with C_1 given by Eq. (9.20), C_2 by Eq. (9.21), and other parameters by Eqs. (9.24) to (9.26).

2. For unconfined aquifers, the parameters in Eq. (9.36) are specified for potential ψ in Eq. (9.18), with C_1 given by Eq. (9.22), C_2 by Eq. (9.23), and other parameters by Eqs. (9.27) and (9.28).

Similar to the generalized equation and Eq. (9.30) for flow in confined aquifers, Eq. (9.36) can be simplified into different forms as counterparts of Eqs. (9.31) to (9.35). The key difference lies in how the terms $\dfrac{1}{r^2}\dfrac{\partial^2 \psi}{\partial \theta^2}$ and $\dfrac{\partial^2 \psi}{\partial z^2}$ are retained. In the next section, only one case has been considered; other simplifications similar to Eqs. (9.31) to (9.34) follow likewise.

5.3.2 *Axis-symmetrical Radial Flow and Vertical Flow*

In this case, by neglecting the term $\dfrac{1}{r^2}\dfrac{\partial^2 \psi}{\partial \theta^2}$, the unified Eq. (9.36) for flow in the two types of aquifers is:

$$S\frac{\partial \psi}{\partial t} = \frac{K_r}{r}\frac{\partial}{\partial r}\left(r\frac{\partial \psi}{\partial r}\right) + K_z\frac{\partial^2 \psi}{\partial z^2} + \frac{R}{C_2} \tag{9.37}$$

where K_r and K_z are the hydraulic conductivities in horizontal radial and vertical directions respectively; furthermore, the following conditions apply:

1. For confined aquifers, Eq. (9.37) is specified with potential by Eq. (9.17), with Eq. (9.20) for C_1, Eq. (9.21) for C_2, and $S = S_c$.

2. For unconfined aquifers, Eq. (9.37) is specified with the potential by Eq. (9.18), with Eq. (9.22) for C_1, Eq. (9.23) for C_2, and $S = \phi_e$.

For axis-symmetrical flow with a vertical component in aquifers in homogeneous media, $K_r = K_z = K$. Hence, Eq. (9.37) can be modified as:

$$C_1\frac{\partial \psi}{\partial t} = K\left[\frac{1}{r}\frac{\partial}{\partial r}\left(r\frac{\partial \psi}{\partial r}\right) + \frac{\partial \psi}{\partial z}\right] + \frac{R}{C_2} \tag{9.38}$$

5.4 *Equations of Flow in Aquifers on a Sloping Impervious Base, in Cylindrical Coordinates*

Most reports in literature concerning flow in wells deal with aquifers on a horizontal base; however, realistic groundwater flow in many cases takes place when there is a gradient on the base of the aquifer. For wells installed into an aquifer which overlies a sloping base, the governing equation should incorporate the slope of the base on which the aquifer moves. Hantush (1962a, b, c, 1964a, b, c) extensively investigated flow in wells in terms of drawdown in aquifers on a sloping base and proposed two governing equations for unconfined and confined aquifers as summarized below.

5.4.1 Equation of Flow in Wells in an Unconfined Aquifer on a Sloping Impervious Base

Hantush (1964b) used polar coordinates to formulate the governing equation of flow in wells installed in an unconfined aquifer on a sloping impervious base. The equation is written in terms of drawdown:

$$Z = h_0^2 - h^2 \tag{9.39}$$

where h_0 is the initial water level in the well at time $t = 0$ and h is the water level after $t > 0$. When leakage is considered, the equation is written as:

$$\frac{1}{\phi_e}\frac{\partial Z}{\partial t} = \frac{1}{r}\frac{\partial}{\partial r}\left(r\frac{\partial Z}{\partial r}\right) + \frac{2}{\bar{\beta}}\left(\cos\theta\,\frac{\partial Z}{\partial r} - \sin\theta\,\frac{\partial Z}{r\partial\theta}\right) + \frac{\partial^2 Z}{r^2\partial\theta^2} - \frac{Z}{\bar{B}^2} \tag{9.40}$$

with

$$\bar{\beta} = \frac{2\bar{h}}{\tan\theta} \tag{9.41}$$

and $\bar{h}$ as the weighted mean of hydraulic heads:

$$\bar{h} = \frac{1}{2}(h_0 + h) \tag{9.42}$$

θ is the polar angle with the pole at the center of the pumped well and $\bar{B}$ is the leakage factor of the semi-pervious base of the aquifer, expressed as:

$$\bar{B} = \frac{K\bar{h}b'}{K'} \tag{9.43}$$

where K and K' are the hydraulic conductivities of the aquifer and the leaky semi-pervious base respectively, and b' is the thickness of the semi-pervious base.

In the above definition, the approximations are set as $\bar{h} \approx h_0$ for $r > 1.5D_0$, and $\bar{h} = \frac{1}{2}(h_0 + h_w)$ for $r < 1.5D_0$, where h_w is the depth of water in the well.

For determining flow parameters using the pumping test method, the initial condition and the boundary conditions for Eq. (9.40) were given by Hantush (1964b):

$$\lim_{r\to 0}\left(\pi K r\frac{\partial Z}{\partial r}\right) = -Q_t$$

$$= \begin{cases} -Q_s f(t), & 0 < t < t_0 \\ 0 & t > t_0 \end{cases} \tag{9.44}$$

$$Z(r, \theta, 0) = 0, \qquad t = 0, r > 0 \tag{9.45}$$

$$Z(\infty, \theta, t) = 0, \qquad t > 0, r \to \infty \tag{9.46}$$

where Q_t is the variable discharge and Q_s is the constant discharge of the well during pumping, and $f(t)$ is a function of time.

With the following identity (Hantush 1964b):

$$Z = s(r,t)\exp\left(-\frac{r}{\beta}\cos\theta\right)$$

(9.47)

Equation (9.40) and the IC and the BCs can be transformed into:

$$\frac{1}{\phi_e}\frac{\partial s}{\partial t} = \frac{1}{r}\frac{\partial}{\partial r}\left(r\frac{\partial s}{\partial r}\right) - \frac{s}{B^2}$$

(9.48)

$$\pi K r_w \frac{\partial s(r_w,t)}{\partial r} = -Q_t$$

$$= \begin{cases} -Q_s f(t), & 0 < t < t_0 \\ 0 & t > t_0 \end{cases}$$

(9.49)

$$s(r, 0) = 0, \qquad t = 0, r > 0$$

(9.50)

$$s(\infty, t) = 0, \qquad t > 0, r \to \infty$$

(9.51)

where r_w is the effective radius of the well.

Equation (9.40) is based on the same author's two-dimensional equation of flow on a sloping base (Hantush 1964c). The conditions for which Eq. (9.40) was derived are: $\tan\theta < 0.02$ and $(h_0 - h_w) < 0.5h_0$, where h_w is the depth of water in the well at the time when readings of the water table are taken (Hantush 1964b).

For the above analyses, no consideration is taken into account for differences in flow direction, affecting flow patterns in the porous medium. With a two-dimensional BE, Mahdavi (2015) demonstrated groundwater flow with recharge for an anisotropic aquifer on a sloping base, which clearly showed the effects of anisotropy on flow patterns.

5.4.2 Equation of Flow in Wells in a Confined Aquifer on a Sloping Impervious Base

In addition to Eq. (9.40) for unconfined aquifers, Hantush (1964b) also presented the governing equation of flow in wells in confined aquifers on a sloping impervious base; the equation is written in terms of drawdown, in polar coordinates, as:

$$\frac{1}{S_c}\frac{\partial s(r,t)}{\partial t} = \frac{1}{r}\frac{\partial}{\partial r}\left(r\frac{\partial s(r,t)}{\partial r}\right)$$

$$-\frac{2}{a}\left(\cos\theta\frac{\partial s(r,t)}{\partial r} - \sin\theta\frac{\partial s(r,t)}{r\partial\theta}\right) + \frac{\partial^2 s(r,t)}{r^2\partial\theta^2}$$

(9.52)

where a is a constant in the exponential function for aquifer thickness expressed as:

$$b = b_0\exp\left(-\frac{2}{a}r\cos\theta\right)$$

(9.53)

Here, b_0 is the thickness of the aquifer at $r = 0$, and the IC and the BCs are written as:

$$\lim_{r \to 0}\left(2\pi K b_0 r \frac{\partial s(r_w,t)}{\partial r} \right) = -Q_t$$

$$= \begin{cases} -Q_s f(t), & 0 < t < t_0 \\ 0 & t > t_0 \end{cases}$$

(9.54)

$$s(r, 0) = 0 \tag{9.55}$$

$$s(\infty, t) = 0 \tag{9.56}$$

5.5 Useful Transformations for Solutions of the Equation for Radial Flow

Methods for solving the equation of radial flow can be found in many handbooks and texts, which often contain the well function and Bessel functions. Furthermore, the following transformations can be used conveniently to transform the equation of flow in radial coordinates for the purpose of solving the equation:

1. The Carslaw-Jaeger transform

Carslaw and Jaeger (1959) used the following transformation:

$$u = \psi r \tag{9.57}$$

which can transform the three-dimensional form of Eq. (9.31) into the standard diffusion equation in radial coordinates, i.e.,

from

$$\left. \frac{\partial \psi}{\partial t} = K\left(\frac{\partial^2 \psi}{\partial r^2} + \frac{2}{r}\frac{\partial \psi}{\partial r} \right) = K\frac{1}{r^2}\frac{\partial}{\partial r}\left(r^2 \frac{\partial \psi}{\partial r} \right) \right\}$$

to

$$\frac{\partial u}{\partial t} = K\frac{\partial^2 u}{\partial r^2}$$

(9.58)

The resultant diffusion equation with the unknown function u is relatively easier to solve, compared to its three-dimensional counterpart in a radial coordinate.

2. The Passioura-Cowan transform

Passioura and Cowan (1968) introduced the following identity:

$$R = \ln\left(\frac{r}{r_1} \right) \tag{9.59}$$

where r_1 is a constant, which can transform Eq. (9.31) into the following equation:

$$\frac{\partial \psi}{\partial t} = K\frac{e^{-2r}}{r_1^2}\frac{\partial^2 \psi}{\partial r^2} \tag{9.60}$$

3. Philip's transformations

In elaborate developments towards solving advection-dispersion equations in different dimensions, Philip (1994) presented several transformations for the conversion of advection-dispersion equations in cylindrical radial coordinates into their counterparts in Cartesian coordinates.

4. The Qi-Liu transformation

With this transformation, time-fractional PDEs for axis-symmetrical flow and transport in radial coordinates can be transformed to their counterparts in Cartesian coordinates (Qi and Liu 2010). The transformed fPDEs in Cartesian coordinates are easier to handle and integer-based PDEs can also be solved as special cases of fPDEs.

5.6 Other Issues in Hydraulics of Wells

A number of practical issues in well hydraulics have been discussed in literature (Hantush 1964c, Bear 1972, 1979, de Marsily 1986, Barenblatt et al. 1990, Bras 1990, Yeh and Chang 2013). In this section, we only briefly discuss some of those issues which affect the interpretation of data from observations of aquifers such as pumping tests and slug tests.

5.6.1 Effects of Finite Well Radius and Wellbore Storage

When a well is drilled into an aquifer, confined or unconfined, the free surface of the bulk water body in the well is in many ways different from the water in the aquifer strata: the response of the bulk water in the well to changes in barometric pressure, air temperature, and movements of water in the well are very different from the anomalous movement of water in aquifers. The radius of the well and the volume of water stored in it also affect how information from the well is interpreted (de Marsily 1986). This issue is also addressed in terms of a capture zone, the area affected by water extraction or recharge.

5.6.2 Skin Effect

Drilling a well with a mechanical device, constructing a well screen and stabilizing the well screen with granular materials create mechanical disturbances in the natural formation of an aquifer. The layer of the aquifer close to the well, often called the skin, has different hydraulic properties compared to those of the aquifer formation, and the skin affects the flow of water in aquifers which can musk natural flow in aquifers. The skin effect can be a problem for interpreting data derived from pumping tests using wells, unless the data is derived from non-destructive techniques (Binley et al. 2015).

5.6.3 Interference of Flow by Wells

In many hydrological investigations, only a small number of wells are used and investigators do not seem to be concerned about the interference between wells if water is being extracted from a large number of nearby wells. In circumstances where many wells are used, the interference of nearby wells with flow patterns is an interesting topic.

Barenblatt et al. (1990) investigated the interference of wells in an array of wells which formed a well field in a two-dimensional plane (x, y). For wells with a distance $x = 2a$, where a represents half of the distance between two wells on the x axis, Barenblatt et al. (1990) showed that in terms of interference of wells in the y axis, outside of the strip, for $|y| \leq a$, flow can be regarded as one-dimensional; inside the strip, for $|y| > a$, the flow is two-dimensional and the effect of wells as point sinks or sources (depending on whether the wells are being pumped or recharged) cannot be ignored.

Hantush (1967) investigated special cases of interference of flow by a well installed in different combinations of aquifers overlaid as layers. One case considers flow to a well that cuts through two confined aquifers, and the second case deals with a well installed in an unconfined aquifer which overlays a confined aquifer. These cases are particularly important for regions where different types of aquifers are present and heterogeneity appears, or in cases where aquifers are separated by different substrata.

5.6.4 Partial Well Penetration

A well is called a partially penetrating well when the length of the well screen is less than the thickness of the aquifer in which the well is constructed. In partially penetrating wells, the flow patterns are different from those in fully penetrating wells in that a vertical component of flow often exists during pumping or recharge, which reduces or exerts extra pressure on the aquifer.

6. Earth Tides and Barometric Effects on Groundwater

6.1 Effects of Inland Earth Tides, Atmospheric Tides and Barometric Pressure

The foregoing discussions and mathematical formulations are based on the Navier-Stokes equations, which do not have provisions to explain numerous external natural stresses such as barometric pressure and earth tides (Spane 2002) besides external effects which can be treated as changes in boundary conditions, such as tidal or river-state fluctuations, etc. This is the case in mainstream literature on groundwater hydrology (Eagleson 1970, Bear 1979, Bras 1990), wherein effects of the periodic forces of earth tides and barometric pressure are implicitly ignored. Only in some special cases are the tidal effects on aquifers noticed and managed, due to seawater intrusion into aquifers in coastal regions, which could cause significant problems by increasing aquifer salinity and soil salinity, and affect ecology and land use in coastal regions.

The effects of atmospheric pressure on aquifers are commonly acknowledged but only a little guidance and a few operational procedures are available (Rasmussen and Crawford 1996); even today, the orthotidal effect, which is the gravitational effect on river water bodies through groundwater tides, is regarded as new (Briciu 2018). The review by Bredehoeff (1967) showed that the effect of earth tides on water wells was recognized and investigated 140 years ago by Klonne (1880). Since then, many investigations have been carried into earth tides and barometric effects on inland water bodies (on surface as well as subsurface). The application of the response of water levels in wells can be used for determining flow properties of the aquifers (Rojstaczer 1988) and the revelation of the changes in water quality (such as the electrical conductivity of water in inland rivers) (Briciu et al. 2018).

Both earth tides and atmospheric tides have periodic and random components, with gravity as the periodic force and weather-induced barometric changes, etc. as random forces. With the aid of wavelet analysis, Briciu (2018) identified 11 components of gravitational tides in inland rivers, with a 0.95 confidence level, which function indirectly through groundwater and atmospheric tides (see Table 9.1).

The traditional implicit convention in groundwater hydrology without earth tidal and barometric effects applies to basic definitions and mathematical models such as those presented in Chapter 3 and previous sections of this chapter. To explicitly incorporate these effects, Spane (2002) suggested the inclusion of barometric head h_a as a component of the total hydraulic head h:

$$h = h_o + h_a \tag{9.61}$$

where h_o is the observed hydraulic head (above the reference datum). With the inclusion of tidal effects, the total hydraulic head should also include tidal head h_t; hence, Eq. (9.61) can be further updated to be:

$$h = h_o + h_t + h_a \tag{9.62}$$

The observed hydraulic head (or potential) includes the pressure head at the time of measurement, with a given atmospheric pressure. This observed head can be called the apparent hydraulic head (with a specific datum defined), with velocity head (or potential) or other terms such as chemical head (or potential), etc. ignored. With Eq. (9.62), it follows that the total hydraulic head (or potential) is a sum of the observed/apparent hydraulic head h_o, the periodic earth tidal head h, and barometric

Table 9.1. Details of earth tides identified in inland rivers, as indirect implications of groundwater tides (data after Briciu 2018).

Sl. No.	Name of the tide	Symbol	Period (hour)
1	Principal lunar declinational diurnal	O1	25.819
2	Lunisolar diurnal	PI1	24.132
3	Principal solar declinational diurnal	P1	24.065
4	Solar diurnal	S1	24.0
5	Lunisolar declinational diurnal	K1	23.934
6	Lunisolar diurnal	PSI1	23.869
7	Lunisolar synodic fortnightly	MSf	14.76
8	Principal lunar semidiurnal	M2	12.42
9	Larger solar elliptic	T2	12.016
10	Principal solar semidiurnal	S2	12.0
11	Lunisolar semidiurnal	K2	11.967
	Mean period (hours)		18.511
	Mean amplitude of semidiurnal tides ranges from 0.7 mm to 96.9 mm		

head h_a, with h_t and h_a both containing periodic gravitational and random pressures in the tidal and barometric tides.

6.2 Coastal Waves and Saltwater Intrusion of Fresh Aquifers

In addition to the inland earth tides resulting from gravity and their impact on groundwater, rivers and other water bodies such as lakes and coastal waves have significant impact on coastal morphology and shape coastal waters in terms of chemistry and physics. Examples of these effects are seawater intrusion in coastal aquifers (Bear et al. 1999) and shaping of coastal morphology resulting from sediment transport along coastal regions at a regional scale (Elvin and Su 1998).

With the influence of inland earth tides, atmospheric tides and ocean waves in coastal regions, the hydraulics of groundwater in coastal regions is more complex. To account for these forms of tides and waves, the total head, incorporating each of these tides in Eq. (9.62), can be used, in addition to treating ocean waves as a periodic boundary condition.

Apart from seawater intrusion in fresh coastal aquifers, saltwater in inland water bodies can also affect inland fresh aquifers; examples include the disposal of irrigation drainage water and industrial wastewaters, which have higher concentrations of salts of various kinds, resulting in the eventual contamination of aquifers by leachates from these disposals.

7. Other Factors Related to Model Construction for Groundwater Flow

In Chapter 3, a brief discussion was presented on water flow in soils and aquifers as coupled multiple processes which include heat transfer, solute movement (diffusion and convection), gas movement, electrical effects, the thermal-diffusion process known as the Soret effect, the diffusion-thermo process known as the Dufour effect (Mortimer and Eyring 1980), and chemical and biochemical reactions with redox potentials that influence mass and energy transfer, elasticity, strain and load, etc. These processes are inter-related and multiphase in nature. Depending on the processes dominating the multiple processes coupled with flow of water in aquifers, any specific quantitative analysis of groundwater flow requires a mathematical model or its computerized code. Furthermore, data are needed for model calibration with suitable methodologies.

The foregoing discussion relates to the quantification of groundwater flow based on physical processes. The most profound impact on aquifers that is not explicitly addressed comes from the excessive extraction of groundwater for irrigation, and domestic and industrial uses, accounting for more than a third of Earth's accessible renewable freshwater (Schwarzenbach et al. 2006). The source term (either recharge or extraction) can be included in all the formulations.

Extraction of water from aquifers in many parts of the world over many centuries has depleted some aquifers; these extractions are still continuing in major aquifers of the world with which ecosystems and societies are associated (Taylor 2014). To this end, it is obvious that, to ensure a sustainable environment, it is essential to slow the population growth worldwide, and find ways to reduce groundwater extraction and achieve sustainable use of the groundwater resource.

8. fPDEs for Isothermal Groundwater Flow in Unconfined Aquifers

fPDEs for groundwater flow are grouped into two categories, one for unconfined aquifers and the second for confined aquifers.

For flow in unconfined aquifers, the classical Boussinesq equation (Boussinesq 1904) based on integer calculus has been in use extensively for over a century (Werner 1953, 1957, Bear 1972, Su 1994, Barenblatt et al. 2000, Abdellaoui et al. 2015, Mahdavi 2015, Telyakovskiy et al. 2016). Since the 1990s, various forms of fPDEs have been presented as alternatives to account for the more detailed stochastic motion of water in porous media.

Two important developments in the application of the CTRW theory, which led to fPDEs for groundwater flow, were: the one attributed to Compte (1997), who presented generic fPDEs (Eqs. (4.15) and (4.18)) which could be used for modelling a number of processes (including groundwater flow and solute transport), and the one accredited to He (1998), who presented different forms of fPDEs for water flow in aquifers, specifically parameterized for anomalous Darcy flow.

Another important development in modelling water flow in porous media was made by Wheatcraft and Meerschaert (2008), who presented a space-fractional equation of mass conservation for flow in porous media (Eq. (4.22)) in which the spatial element size Δx appears with the order of fractional derivatives λ, in the form of $\Delta x^{1-\lambda}$, to account for the scale effect resulting from non-local flow processes ($\Delta x^{1-\lambda} \rightarrow 1$ as $\lambda \rightarrow 1$).

Other examples of fPDEs for groundwater flow include those by Atangana and his colleagues (Atangana and Bildik 2013, Atangana 2014, 2016, 2018, Atangana and Vermeulen 2014, Alkahtani and Atangana 2016, Atangana and Alkahtani 2016, Atangana and Baleanu 2016, Djida et al. 2016), Mehdinejadiani et al. (2013), Zhuang et al. (2014), Su et al. (2015) and Su (2017a). A set of space-time fPDEs or the fractional Boussinesq equation (Su 2017a) has features to accommodate various properties of porous media, such as heterogeneity and distributed orders to account for multifractals; in particular, the two-term distributed-order fPDEs for flow in large and small pores in aquifers have been applied to radial flow (Su et al. 2015).

These developments on water flow in aquifers complement the applications of fPDEs and other concepts in fractional calculus for solute movement in groundwater as investigated by Lenormand (1992), Compte (1997) and Benson (1998) and in many subsequent investigations (Benson et al. 2000a, b, Schumer et al. 2003a, b). Reviews on these topics can be found in the works of Berkowitz et al. (2006), Zhang et al. (2009) and Benson et al. (2013).

In this section, we briefly summarize the different forms of fPDEs for both the types of aquifers, formulated in different coordinates, before more complicated forms of fPDEs are presented in later parts of the chapter.

8.1 The Continuous-Time Random Walk Theory and Flow in Unconfined Aquifers

In Chapter 4, a brief review showed the direct link between the continuous-time random walk (CTRW) theory and the fPDE for flow and transport processes, which are also applicable to groundwater flow in porous media. Water flow in unsaturated soils can be successfully described using the CTRW theory (Su 2014), and flow in saturated media should have similar properties as the level of saturation in porous media is a key parameter as the switch. Given that solute movement in porous media has been successfully modelled using the CTRW theory (Compte 1997, Meerschaert et al. 2002, Meerschaert 2012), water flow that entrains solute in the medium can be consistently interpreted using the CTRW concept. These logical connections and successful analyses prompted the extension of the approach to water flow in unconfined aquifers (Su 2017a).

8.2 Fractional Boussinesq Equations Derived from Conservation of Mass

The fractional Boussinesq equations (fBEs) can be derived with the aid of conservation of mass (Mehdinejadiani et al. 2013) or the CTRW theory (Su 2017a). The derivation can be achieved by incorporating fractional hydraulic gradient, fractional derivatives of hydraulic gradient, or a combination of both in a fashion similar to fPDEs for water flow in soils as presented in Chapter 5 and for groundwater flow (He 1998).

8.3 Different Forms of Fractional Boussinesq Equations and Their Implications

8.3.1 Space-time Fractional Boussinesq Equations for Groundwater Flow on a Slopping Base

From the set of fBEs for flow in unconfined aquifers under different conditions (Su 2017a), the simplest fBE is the one-dimensional counterpart of Eq. (9.15), written as:

$$\phi_e \frac{\partial^\beta h}{\partial t^\beta} = -\frac{\partial^\eta q}{\partial x^\eta} \tag{9.63}$$

where q is given by Eq. (9.16) and ϕ_e denotes drainage porosity or specific yield for the unconfined aquifer.

Fractional flux has two features, as Wheatcraft and Meerschaert (2008) explained, including the scale-invariant property that eliminates scale effects on parameters and a fractional divergence term in the fractional equation of conservation of mass.

When the aquifer has a sloping impervious base, Eq. (9.63) is written as (Su 2017a):

$$\phi_e \frac{\partial^\beta h}{\partial t^\beta} = \frac{\partial^\eta}{\partial x^\eta} \left(Kh \frac{\partial h}{\partial x} \right) - \frac{\partial^\eta}{\partial x^\eta} (\omega Kh) \tag{9.64}$$

where $\omega = \tan \Lambda$ is the slope gradient of the impervious base (see Eqs. (9.13) and (9.16)).

8.3.2 Simplified Fractional Boussinesq Equations for Homogeneous Unconfined Aquifers on a Slopping Base

For homogeneous aquifers with a constant K, Eq. (9.64) gets simplified as (Su 2017a):

$$\phi_e \frac{\partial^\beta h}{\partial t^\beta} = K \frac{\partial^\eta}{\partial x^\eta}\left[h\left(\frac{\partial h}{\partial x} - \omega\right)\right] \tag{9.65}$$

which is still a non-linear fPDE, even though K is constant.

8.3.3 Other Forms of Simplified fBEs in One Dimension

Equation (9.64) is a non-linear fPDE with the term Kh possessing more important properties, compared to its linear counterpart. In hydrological applications, certain simplifications can be made in Eq. (9.64). In the following cases, simplifications have been achieved for homogeneous aquifers with K being a constant; for heterogeneous aquifers, simplifications can be made with the use of transmissivity or the weighted average of K.

It should be noted that the expansion of the term $\dfrac{\partial^\eta}{\partial x^\eta}\left(Kh\dfrac{\partial h}{\partial x}\right)$ in Eq. (9.64) cannot be simply performed using the Leibniz chain rule for integer calculus; it requires a more complex Leibniz chain rule, a finite series for the product of integer derivatives and fractional derivatives of the two terms, Kh and $\dfrac{\partial h}{\partial x}$ (Podlubny 1999).

With the term Kh represented by transmissivity $\bar{T}_u$, Eq. (9.64) is written as follows for heterogeneous aquifers:

$$\phi_e \frac{\partial^\beta h}{\partial t^\beta} = \bar{T}_u \frac{\partial^\lambda h}{\partial x^\lambda} - \omega\bar{K} \frac{\partial^\eta h}{\partial x^\eta} \tag{9.66}$$

with $\lambda = \eta + 1$.

Despite the fact that the Boussinesq equation was originally developed for unconfined aquifers, with averaged hydraulic conductivity and transmissivity, the one-dimensional Boussinesq equation has an identical mathematical structure for flow in confined aquifers; the only difference lies in the parameters defined for the two types of aquifers (see sections 4.1 and 4.3).

Equation (9.66) is a linear fPDE and published reports which document analytical solutions of linear fPDEs are available (Jiang et al. 2012a, Su 2017a).

8.3.4 Two-dimensional fBEs for Homogeneous Unconfined Aquifers

With the Dupuit assumption, for unconfined aquifers with vertical equipotential surfaces (implying no vertical flow), the full three-dimensional flow equation in the form of the integer PDE can be integrated along the vertical coordinate to yield its two-dimensional form (Bear 1979, de Marsily 1986). For fPDEs, that operation is more complicated, hence, the two-dimensional fBE based on Eq. (9.10) can be simply written as:

$$\phi_e \frac{\partial^\beta h}{\partial t^\beta} = \frac{1}{2}\frac{\partial}{\partial x}\left(K_x \frac{\partial^{\eta_x} h^2}{\partial x^{\eta_x}}\right) + \frac{1}{2}\frac{\partial}{\partial y}\left(K_y \frac{\partial^{\eta_y} h^2}{\partial y^{\eta_y}}\right) \tag{9.67}$$

based on derivatives of the fractional gradient, and

$$\phi_e \frac{\partial^\beta h}{\partial t^\beta} = \frac{1}{2} \frac{\partial^{\eta_x}}{\partial x^{\eta_x}} \left(K_x \frac{\partial h^2}{\partial x} \right) + \frac{1}{2} \frac{\partial^{\eta_y}}{\partial y^{\eta_y}} \left(K_y \frac{\partial h^2}{\partial y} \right) \tag{9.68}$$

based on fractional derivatives of the gradient.

8.4 Dimensions of the Parameters in Fractional Boussinesq Equations

The dimensions or units of parameters in fPDEs are important to ensure that the interpretation of data is physically meaningful.

One approach is the example of the time-fractional diffusion equation for which Kilbas et al. (2006) suggested to introduce a new fractional diffusion coefficient D_f expressed as:

$$D_f = D\tau^{1-\beta} \tag{9.69}$$

where β is the order of the time-fractional diffusion equation, D is the classical diffusion coefficient with dimension $[L^2 T^{-1}]$ and τ is the new time constant parameter which accommodates the new dimensions.

Another approach is to follow Compte (1997) who introduced new parameters which ensured the physical explanation of the fPDEs. Compte's approach to dimensions can be applied to make sure the dimensions in the fPDEs are physically correct (see section 3.3 in Chapter 4).

8.5 Relationship between the Orders of Fractional Derivatives in Time and Space and Flow Patterns

Referring to section 3.4 in Chapter 4, based on the analysis by Zaslavsky (2002), the following relationships hold for anomalous diffusion in unconfined aquifers represented by Eq. (9.66) with $\omega = 0$ (or Eq. (9.63) for β and η):

$$\frac{\beta}{\lambda} = \frac{\ln L_s}{\ln L_t} = \frac{\mu}{2} \tag{9.70}$$

which connects the orders of spatio-temporal fractional derivatives, λ and β, and the spatio-temporal scaling measures for fractals, L_s and L_t. It also means that the transport exponent is given as:

$$\mu = \frac{2 \ln L_s}{\ln L_t} \tag{9.71}$$

or

$$\mu = \frac{2\beta}{\lambda} \tag{9.72}$$

which can be used to identify the type of diffusion: $\mu < 1$ is for sub-diffusion, $\mu = 1$ for classical diffusion and $\mu > 1$ for super-diffusion (Zaslavsky 2002).

9. fPDEs for Isothermal Groundwater Flow in Confined Aquifers

Parallel to the integer-based PDEs in section 2 for groundwater in confined aquifers, the fractional counterparts of these PDEs can be developed with the aid of the CTRW theory or from conservation of mass. The fractional counterpart of Eq. (9.2) can be written as:

$$S_e \frac{\partial^\beta h}{\partial t^\beta} = -\nabla^\eta \cdot \mathbf{q} \tag{9.73}$$

The above equation may be expanded as:

$$S_e \frac{\partial^\beta h}{\partial t^\beta} = \frac{\partial^{\eta_x}}{\partial x^{\eta_x}} \left(K_x \frac{\partial h}{\partial x} \right) + \frac{\partial^{\eta_y}}{\partial y^{\eta_y}} \left(K_y \frac{\partial h}{\partial y} \right) + \frac{\partial^{\eta_z}}{\partial z^{\eta_z}} \left(K_z \frac{\partial h}{\partial z} \right) \tag{9.74}$$

where η_x, η_y and η_z are the orders of spatial fractional derivatives and may be equal or different.

The fractional derivatives of flux in Eq. (9.74) can also be replaced by derivatives of fractional Darcy flux (He 1998) so that an fPDE alternative to Eq. (9.74) can be written as follows:

$$S_e \frac{\partial^\beta h}{\partial t^\beta} = \frac{\partial}{\partial x} \left(K_x \frac{\partial^{\eta_x} h}{\partial x^{\eta_x}} \right) + \frac{\partial}{\partial y} \left(K_y \frac{\partial^{\eta_y} h}{\partial y^{\eta_y}} \right) + \frac{\partial}{\partial z} \left(K_z \frac{\partial^{\eta_z} h}{\partial z^{\eta_z}} \right) \tag{9.75}$$

Following He (1998), a combination of the two forms of fractional derivatives in Eqs. (9.74) and (9.75) can be used as:

$$S_e \frac{\partial^\beta h}{\partial t^\beta} = \frac{\partial^{\alpha_1}}{\partial x^{\alpha_1}} \left(K_x \frac{\partial^{\alpha_2} h}{\partial x^{\alpha_2}} \right) + \frac{\partial^{\gamma_1}}{\partial y^{\gamma_1}} \left(K_y \frac{\partial^{\gamma_2} h}{\partial y^{\gamma_2}} \right) + \frac{\partial^{\eta_1}}{\partial z^{\eta_1}} \left(K_z \frac{\partial^{\eta_2} h}{\partial z^{\eta_2}} \right) \tag{9.76}$$

The one-dimensional form of Eq. (9.74) with advection for flow in a confined aquifer on a sloping base can be written as:

$$S_e \frac{\partial^\beta h}{\partial t^\beta} = \frac{\partial^{\eta_x}}{\partial x^{\eta_x}} \left(K_x \frac{\partial h}{\partial x} \right) - \omega K_x \frac{\partial^{\eta_x}}{\partial x^{\eta_x}} \tag{9.77}$$

where $\omega = \tan \Lambda$ is the slope of the impervious base of the aquifer, with Λ as the angle of the slope. One-dimensional forms of Eqs. (9.75) and (9.76) are also simpler by dropping the third terms on the right-hand sides of the equations and adding an advection term.

10. Distributed-order fPDEs in Cartesian Coordinates

As discussed in Chapters 4 and 5, the fractional orders in the fPDEs are not necessarily fractions or constants, they can be variables, functions of time, space and/or h or other parameters or variables (Samko 1995, Lorenzo and Hartley 2002).

Similar to integer PDEs, fPDEs such as Eq. (9.75) can also be simplified by integration over the vertical direction (Bear 1979) to reduce the three-dimensional fPDE to its two-dimensional form; however, integration of Eqs. (9.74) and (9.76) with respect to depth z is more complicated.

10.1 Two-term Distributed-order fPDE for Groundwater Flow in Confined Aquifers

The mathematical representation of physical processes of water movement in porous media is certainly more accurate when more terms are used in temporal fractional derivatives. However, the challenge with a large number of fractional terms in the formulation is the increasing difficulty of evaluating the parameters β_i. To be consistent with the widely-used mobile-immobile model of water flow and solute transport in porous media (saturated and unsaturated), and with the success in its extension as fPDEs (Schumer et al. 2003b, Su et al. 2015), we adapt our terminology for groundwater flow (Su et al. 2015) to expand the temporal fractional derivatives in Eq. (9.74) as:

$$S_{e2}\frac{\partial^{\beta_2}h}{\partial t^{\beta_2}}+S_{e1}\frac{\partial^{\beta_1}h}{\partial t^{\beta_1}}=\frac{\partial^{\eta_x}}{\partial x^{\eta_x}}\left(K_x\frac{\partial h}{\partial x}\right)$$
$$+\frac{\partial^{\eta_y}}{\partial y^{\eta_y}}\left(K_y\frac{\partial h}{\partial y}\right)+\frac{\partial^{\eta_z}}{\partial z^{\eta_z}}\left(K_z\frac{\partial h}{\partial z}\right) \tag{9.78}$$

which, with convection included, can be written as:

$$S_{e2}\frac{\partial^{\beta_2}h}{\partial t^{\beta_2}}+S_{e1}\frac{\partial^{\beta_1}h}{\partial t^{\beta_1}}=\frac{\partial^{\eta_x}}{\partial x^{\eta_x}}\left(K_x\frac{\partial h}{\partial x}\right)$$
$$+\frac{\partial^{\eta_y}}{\partial y^{\eta_y}}\left(K_y\frac{\partial h}{\partial y}\right)+\frac{\partial^{\eta_z}}{\partial z^{\eta_z}}\left(K_z\frac{\partial h}{\partial z}\right)-V_x-V_y-V_z \tag{9.79}$$

where S_{e2} and S_{e1} are the storativities of immobile and mobile zones respectively; $\beta_2 > \beta_1$, with β_2 for large pores and β_1 for small pores.

Similarly, the two-term distributed-order time-fractional derivatives can also be applied to Eqs. (9.75) and (9.76) as follows:

$$S_{e2}\frac{\partial^{\beta_2}h}{\partial t^{\beta_2}}+S_{e1}\frac{\partial^{\beta_1}h}{\partial t^{\beta_1}}=\frac{\partial}{\partial x}\left(K_x\frac{\partial^{\eta_x}h}{\partial x^{\eta_x}}\right)$$
$$+\frac{\partial}{\partial y}\left(K_y\frac{\partial^{\eta_y}h}{\partial y^{\eta_y}}\right)+\frac{\partial}{\partial z}\left(K_z\frac{\partial^{\eta_z}h}{\partial z^{\eta_z}}\right) \tag{9.80}$$

and

$$S_{e2}\frac{\partial^{\beta_2}h}{\partial t^{\beta_2}}+S_{e1}\frac{\partial^{\beta_1}h}{\partial t^{\beta_1}}=\frac{\partial^{\alpha_1}}{\partial x^{\alpha_1}}\left(K_x\frac{\partial^{\alpha_2}h}{\partial x^{\alpha_2}}\right)$$
$$+\frac{\partial^{\gamma_1}}{\partial y^{\gamma_1}}\left(K_y\frac{\partial^{\gamma_2}h}{\partial y^{\gamma_2}}\right)+\frac{\partial^{\eta_1}}{\partial z^{\eta_1}}\left(K_z\frac{\partial^{\eta_2}h}{\partial z^{\eta_2}}\right) \tag{9.81}$$

The simplest two-term distributed-order fPDE for groundwater flow in confined aquifers on a sloping base can be written as:

$$S_{e2} \frac{\partial^{\beta_2} h}{\partial t^{\beta_2}} + S_{e1} \frac{\partial^{\beta_1} h}{\partial t^{\beta_1}} = \bar{K}_f \frac{\partial^\lambda h}{\partial x^\lambda} - \omega \bar{K}_f \frac{\partial^\eta h}{\partial x^\eta} \tag{9.82}$$

which will be analyzed further in the next section.

10.2 Two-term Distributed-order fPDE for Groundwater Flow in Unconfined Aquifers

In line with the discussions for confined aquifers with mobile and immobile zones in section 10.1, fPDEs for unconfined aquifers can be formulated by modifying Eq. (9.67) or (9.68). In its equivalent form in terms of hydraulic head, Eq. (9.67) in two dimensions can be modified to be the two-dimensional fBE:

$$\phi_{em} \frac{\partial^{\beta_2} h}{\partial t^{\beta_2}} + \phi_{eim} \frac{\partial^{\beta_1} h}{\partial t^{\beta_1}} = \frac{1}{2} \frac{\partial^{\eta_x}}{\partial x^{\eta_x}} \left(K_x \frac{\partial h^2}{\partial x} \right) + \frac{1}{2} \frac{\partial^{\eta_y}}{\partial y^{\eta_y}} \left(K_y \frac{\partial h^2}{\partial y} \right) + \frac{R}{\phi_e} \tag{9.83}$$

which is a result of the fractional derivatives of the classical Darcy flux, with ϕ_{em} and ϕ_{eim} being the drainage porosities (or specific yields) in mobile and immobile zones respectively.

Similar to Eqs. (9.80) and (9.81) for confined aquifers, a three-dimensional distributed-order fPDE can also be applied to unconfined aquifers without applying the Dupuit hypothesis (Eagleson 1970):

$$\begin{aligned} \phi_{em} \frac{\partial^{\beta_2} h}{\partial t^{\beta_2}} + \phi_{eim} \frac{\partial^{\beta_1} h}{\partial t^{\beta_1}} &= \frac{1}{2} \frac{\partial}{\partial x} \left(K_x \frac{\partial^{\eta_x} h^2}{\partial x^{\eta_x}} \right) + \frac{1}{2} \frac{\partial}{\partial y} \left(K_y \frac{\partial^{\eta_y} h^2}{\partial y^{\eta_y}} \right) \\ &+ \frac{1}{2} \frac{\partial}{\partial z} \left(K_z \frac{\partial^{\eta_z} h^2}{\partial z^{\eta_z}} \right) + \frac{R}{\phi_e} \end{aligned} \tag{9.84}$$

which is based on fractional flux. The third formulation, similar to the approach by He (1998), is based on joint fractional derivatives:

$$\begin{aligned} \phi_{em} \frac{\partial^{\beta_2} h}{\partial t^{\beta_2}} + \phi_{eim} \frac{\partial^{\beta_1} h}{\partial t^{\beta_1}} &= \frac{1}{2} \frac{\partial^{\alpha_1}}{\partial x^{\alpha_1}} \left(K_x \frac{\partial^{\alpha_2} h^2}{\partial x^{\alpha_2}} \right) + \frac{1}{2} \frac{\partial^{\gamma_1}}{\partial y^{\gamma_1}} \left(K_y \frac{\partial^{\gamma_2} h^2}{\partial y^{\gamma_2}} \right) \\ &+ \frac{1}{2} \frac{\partial^{\eta_1}}{\partial z^{\eta_1}} \left(K_z \frac{\partial^{\eta_2} h^2}{\partial z^{\eta_2}} \right) + \frac{R}{\phi_e} \end{aligned} \tag{9.85}$$

When recharge and the slope of the aquifer base incorporated, Eq. (9.64) for heterogeneous media in one dimension can be written as follows (Su 2017a):

$$\phi_{em} \frac{\partial^{\beta_2} h}{\partial t^{\beta_2}} + \phi_{eim} \frac{\partial^{\beta_1} h}{\partial t^{\beta_1}} = \frac{\partial^{\eta_x}}{\partial x^{\eta_x}} \left(K_x h \frac{\partial h}{\partial x} \right) - \omega K_x \frac{\partial^\eta h}{\partial x^\eta} + \frac{R}{\phi_e} \tag{9.86}$$

For heterogeneous media, with the weighted mean transmissivity and conductivity, Eq. (9.86) becomes:

$$\phi_{em}\frac{\partial^{\beta_2}h}{\partial t^{\beta_2}}+\phi_{eim}\frac{\partial^{\beta_1}h}{\partial t^{\beta_1}}=\bar{T}_u\frac{\partial^{\lambda}h}{\partial x^{\lambda}}-\omega\bar{K}_u\frac{\partial^{\eta}h}{\partial x^{\eta}}+\frac{R}{\phi_e} \tag{9.87}$$

with $\lambda = \eta_x + 1$, and $\bar{T}_u$ and $\bar{K}_u$ determined using Eqs. (9.27) and (9.28).

The applications of symmetric fractional derivatives (SFD) in modelling solute movement were given by Meerschaert et al. (1999), Benson et al. (2000a), Schumer et al. (2003a) and Jiang et al. (2012a, b). Zhang et al. (2009) and Benson et al. (2013) provided extensive reviews and analyses of SFD-based models, and tested some of them using field data. The analyses by Zhang et al. (2009) and Benson et al. (2013) are mainly for solute transport, also providing an indirect picture of water flow in porous media as solute movement is coupled with water flow.

Zaslavsky (2002) analyzed SFD-based fPDEs in great detail and showed that the orders of fractional derivatives in the time-space fPDE are related to the exponents of the two probability density functions in the CTRW theory, and "the critical exponents that characterize the fractal structures of space-time" (Zaslavsky 2002).

Following Kilbas et al. (2006), who introduced a new parameter for the fractional diffusion equation, four parameters in Eq. (9.87) have been used here:

$$b_i = \phi_{eim}\tau_1^{\beta_1-1} \tag{9.88}$$

$$b_m = \phi_{em}\tau_2^{\beta_2-1} \tag{9.89}$$

$$\bar{T}_f = \bar{T}_u\tau_t^{\lambda-2} \tag{9.90}$$

$$\bar{K}_f = \bar{K}_u\tau_k^{\eta-1} \tag{9.91}$$

where τ_1, τ_2, τ_t and τ_k are parameters for dimension corrections, and ϕ_{eim} and ϕ_{em} are the drainage porosities of immobile and mobile zones respectively.

In practice, it is less convenient to define and measure the parameters in Eqs. (9.88) to (9.91); as a less precise approximation, b_i and b_m can be treated as lumped parameters to accommodate the changes in dimensions introduced by fPDEs.

With dimensional modifications to the parameters, Eq. (9.87) can be updated as:

$$b_m\frac{\partial^{\beta_2}h}{\partial t^{\beta_2}}+b_i\frac{\partial^{\beta_1}h}{\partial t^{\beta_1}}=\bar{T}_f\frac{\partial^{\lambda}h}{\partial x^{\lambda}}-\omega\bar{K}_f\frac{\partial^{\eta}h}{\partial x^{\eta}} \tag{9.92}$$

As stated earlier, the approach taken by Compte (1997) is also a valuable option to accommodate the dimensions. Due to the possible inconsistency between the dimensions of the fPDE and its solutions in different forms, particularly approximate solutions, it is strongly recommended that the dimensions of the parameters in the final solutions should be verified to ensure their physical relevance.

10.3 Moment Equations of Distributed-order fBEs for Aquifers

As demonstrated by Saichev and Zaslavsky (1997) and Zaslavsky (2002), and briefly presented in Chapters 4 and 5, the moment of an fPDE derived from a Dirac delta function input (Saichev and Zaslavsky 1997) is the assemble mean, which represents ensemble averaging of distance or displacement, x, over repeated observations; this moment can be used to estimate the model parameters. Hence, we briefly discuss the moments of Eq. (9.82) for confined aquifers and Eq. (9.92) for unconfined aquifers, and transport exponent, which can be used to define flow patterns.

Based on scaling parameters and moment equations for diffusion in Chapter 5, for the special case of $\lambda = 2$ and $\eta = 1$, the moment equations for distributed-order fPDEs are given as follows for confined aquifers and unconfined aquifers respectively (Chechkin et al. 2002, Sandev et al. 2015):

$$\left\langle x^2 \right\rangle = \frac{2\Gamma(1+\lambda)\bar{K}_c}{b_m} t^{\beta_2} E_{\beta_2-\beta_1,\beta_2+1}\left(-\frac{b_i}{b_m}t^{\beta_2-\beta_1}\right) \tag{9.93}$$

$$\left\langle x^2 \right\rangle = \frac{2\Gamma(1+\lambda)\bar{T}_f}{b_m} t^{\beta_2} E_{\beta_2-\beta_1,\beta_2+1}\left(-\frac{b_i}{b_m}t^{\beta_2-\beta_1}\right) \tag{9.94}$$

with $E_{\beta_2-\beta_1,\beta_2+1}\left(-\dfrac{b_i}{b_m}t^{\beta_2-\beta_1}\right)$ being the Mittag-Leffler function.

Because of the similarity in the moment equations, Eqs. (9.93) and (9.94), let us examine Eq. (9.93) only for the distributed-order fPDE in Eq. (9.82) for confined aquifers. Equation (9.93) has the following asymptotic properties for $t \to 0$ and $t \to \infty$ (Chechkin et al. 2002):

$$\left\langle x^2 \right\rangle \approx \frac{\Gamma(1+\lambda)\bar{K}_c}{\Gamma(1+\beta_2)b_m} t^{\beta_2} \propto t^{\beta_2} \qquad \text{for } t \to 0 \tag{9.95}$$

$$\left\langle x^2 \right\rangle \approx \frac{\Gamma(1+\lambda)\bar{K}_c}{\Gamma(1+\beta_2)b_m} t^{\beta_1} \propto t^{\beta_1} \qquad \text{for } t \to \infty \tag{9.96}$$

The above asymptotic solutions imply that large pores are important at early stages of water flow, while small pores play a crucial role at large times.

For uniform media with no distinction between mobile and immobile zones, Eq. (9.93) develops the following asymptotic results (Chechkin et al. 2002):

$$\left\langle x^2 \right\rangle \approx \Gamma(1+\lambda)\bar{K}_c t \ln\left(\frac{1}{t}\right) \qquad \text{for } t \to 0 \tag{9.97}$$

$$\left\langle x^2 \right\rangle \approx \Gamma(1+\lambda)\bar{K}_c \ln t \qquad \text{for } t \to \infty \tag{9.98}$$

where $\ln(1/t)$ is natural logarithm of $1/t$.

The asymptotic results in Eqs. (9.95) to (9.98) for confined aquifers can also be applied to Eq. (9.94) for unconfined aquifers by replacing $\bar{K}_c$ with $\bar{T}_f$.

10.4 Multi-term Distributed-order fPDEs for Unconfined Aquifers

The two-term distributed-order fPDEs discussed above are useful not only for analyzing flow in large and small pores, but also for explaining the concept of backwater effects in hydraulics, widely known at a large spatial scale. At the micro-scale, backwater effects on flow can be explained by backward fractional derivatives (BFD) in space.

The background of the SFD was discussed in Chapters 2 and 4. Bochner (1949) used backward and forward fractional linear operators, and Saichev and Zaslavsky (1997) described the 'wandering processes' using the term SFDs to take into account the backwater effect in particle motion. By definition, fractional derivatives incorporate those two components—BFDs and forward fractional derivatives (FFDs)—when the symmetric case is considered (Gorenflo and Mainardi 2001, Umarov and Gorenflo 2005b). As the symmetrical case is considered for the derivation of the fWDE based on the CTRW theory in section 3.2 of Chapter 4, the fPDEs we use here consider FFDs and BFDs.

The two-term distributed-order fPDEs discussed above consider flow in two levels of pores. For flow in multifractal pores, the flow of fluid through pores of different sizes can be represented by multi-term fractional time derivatives. The multi-term time-fractional derivatives (Jiang et al. 2012a) used here take the following form:

$$P(D_t)h(x,t) = \sum_{j=1}^{n} b_j D_t^{\beta_j} h(x,t) \tag{9.99}$$

where D_t^β and $D_t^{\beta_j}$ are Caputo fractional derivatives, with $0 \le \beta_n < ... \beta_1 < \beta \le 1$ as fractional orders of the time-fractional diffusion equation, or $0 \le \beta_n < ... \beta_1 < \beta \le 2$ as fractional orders of the time-fractional wave equation, with $j = 1,2,...m$. In fact, $0 \le \beta_n < ... \beta_1 < \beta \le 2$ is more generic when the multi-term fractional diffusion-wave equation is used. b_j ($j = 1,2,...n$) are the coefficients of the time-fractional derivatives, accounting for proportional contributions from each fractional time derivative to the sum of fractional derivatives.

Combining Eq. (9.99) with backward and forward fractional derivatives, using Saichev and Zaslavsky's notation, yields:

$$\sum_{j=1}^{n} b_j D_t^{\beta_j} h(x,t) = K \frac{\partial^\eta}{\partial |x|^\eta} \left(\frac{\partial h^2}{\partial |x|} \right) - \omega K \frac{\partial^\eta h}{\partial |x|^\eta} \tag{9.100}$$

for a homogeneous unconfined aquifer; for a heterogeneous unconfined aquifer, when $K(z)$ varies with depth, the equation reads as:

$$\sum_{j=1}^{n} b_j D_t^{\beta_j} h(x,t) = \frac{\partial^\eta}{\partial |x|^\eta} \left[K(z) \frac{\partial h^2}{\partial |x|} \right] - \omega \frac{\partial^\eta}{\partial |x|^\eta} \left[K(z)h \right] \tag{9.101}$$

In most of our discussions in this book, the sign $||$ for representing backward and forward fractional derivatives is not used, yet its meaning is retained.

When multi-term fPDEs are used, the amended dimensions are as follows:

$$
\left.
\begin{aligned}
b_1 &= \frac{\phi_1}{\phi}\tau^{\beta_1-1} \\[2ex]
b_2 &= \frac{\phi_2}{\phi}\tau^{\beta_2-1} \\[1ex]
&\quad\ldots\ldots \\[1ex]
b_n &= \frac{\phi_n}{\phi}\tau^{\beta_n-1}
\end{aligned}
\right\}
\tag{9.102}
$$

where $\phi_1, \phi_2, \ldots, \phi_n$ are the porosities at $1,2,\ldots n$ levels of pores respectively.

Equations (9.100) and (9.101) are distributed-order fBEs incorporating the SFDs for unconfined aquifers, as an extension of the earlier models in section 3 to multi-term time-fractional derivatives. The forward and the backward fractional components of derivatives account for forward and backward fractional components of flow if the motion of water parcels is regarded as a wandering process (Saichev and Zaslavsky 1997).

Readers may find more information on backward and forward fractional derivatives in the works of Saichev and Zaslavsky (1997), Gorenflo and Mainardi (1998a, 1998b, 2001), Benson et al. (2000b), Mainardi et al. (2001) and Umarov and Gorenflo (2005b), who detail the definitions and the properties of these derivatives.

In theory, the multi-term definition is an ideal way for modelling flow and particle motion in multifractal media with an unlimited level of micro-structures, a key characteristic of fractal media. The challenge is lies in parameter determination for flow in aquifers.

For examples of the applications of these concepts, the reader may refer to relevant literature, such as the two-term mobile-immobile model of solute transport (Schumer et al. 2003b), water flow in soils (Su 2012, 2014), and well hydraulics (Su et al. 2015).

10.5 Multi-term Distributed-order fPDEs for Confined Aquifers

The formulation of multi-term distributed-order fPDEs for confined aquifers is identical to the model for unconfined aquifers (section 10.4), with relevant definitions and parameters for confined aquifers.

11. fPDEs for Hydraulics of Anomalous Radial Flow in Wells on a Horizontal Base

As discussed in section 4, the mathematical structures of the models for confined and unconfined aquifers can be identical, except for the definitions of potential according to Eqs. (9.17) and (9.18), and their parameters. With the appropriate choice of parameters, such as aquifer diffusivity (which incorporates hydraulic conductivity, hydraulic head, and either of *specific yield for unconfined aquifers* and *storativity for confined aquifers*), the model for groundwater flow in confined and unconfined aquifers in radial coordinates is structurally similar.

11.1 Distributed-order Equation for Radial Groundwater Flow in a Unified Form

Referring to sections 4 and 5 for the unified equation of flow in aquifers, in radial coordinates, in terms of hydraulic potential ψ, a two-term temporal distributed-order fPDE can be formulated as follows:

$$b_1 \frac{\partial^{\beta_1} \psi}{\partial t^{\beta_1}} + b_2 \frac{\partial^{\beta_2} \psi}{\partial t^{\beta_2}} = \frac{D_f}{r} \frac{\partial}{\partial r}\left(r \frac{\partial \psi}{\partial r}\right) \tag{9.103}$$

where, for confined aquifers, hydraulic potential is $\psi = h$ in Eq. (9.17), and for unconfined aquifers, it is $\psi = \frac{h^2}{2}$ in Eq. (9.18).

For radial flow, such as water flow to or from wells in confined aquifers, a two-term distributed-order fPDE (Su et al. 2015) has been presented, which is a distributed-order fDWE of the form:

$$b_1 \frac{\partial^{\beta_1} h}{\partial t^{\beta_1}} + b_2 \frac{\partial^{\beta_2} h}{\partial t^{\beta_2}} = \frac{D_f}{r} \frac{\partial}{\partial r}\left(r \frac{\partial h}{\partial r}\right) \tag{9.104}$$

where $D_f = \frac{T}{S_e}$ (see also Eq. (9.19)), and $\beta_1 < 1$ and $\beta_2 < 1$ are the orders of fractional derivatives for immobile and mobile zones respectively; $b_1 + b_2 = 1$, $b_1 = \frac{\phi_{im}}{\phi} \tau^{\beta_1 - 1}$ and $b_2 = \frac{\phi_m}{\phi} \tau^{\beta_2 - 1}$, with ϕ_{im}, ϕ_m and ϕ being the porosities of the immobile zone, the mobile zone and total porosity respectively; this follows the dimensional amendment by Kilbas et al. (2006). D_f denotes fractional diffusivity, which also takes two forms for unconfined and confined aquifers, like their conventional counterparts in Eq. (9.19).

For radial flow in unconfined aquifers, as a counterpart of Eq. (9.104), the following fPDE is used:

$$b_1 \frac{\partial^{\beta_1} h^2}{\partial t^{\beta_1}} + b_2 \frac{\partial^{\beta_2} h^2}{\partial t^{\beta_2}} = \frac{D_f}{r} \frac{\partial}{\partial r}\left(r \frac{\partial h^2}{\partial r}\right) \tag{9.105}$$

where

$$D_f = \frac{T}{\phi_e} = \frac{1}{C_1} \tag{9.106}$$

is from Eq. (9.19). More information on integer PDEs for flow in aquifers can be found in the works of Bear (1979) and Bras (1990).

11.2 Applications of fDWEs to Determine Parameters of Flow in Confined Aquifers

The procedures for determining the parameters in Eq. (9.104) were reported by Su et al. (2015). These procedures have been briefly summarized in this section.

In traditional practice in groundwater hydrology, there are many methods for deriving flow parameters by observing the water level in a well. One of these methods is *pumping test*, which involves continuous pumping of water into or from a well and measuring the changes in the water level of the well to determine the flow parameters. Another method is *slug test*, which involves injection into or extraction from a well a small quantity of water and observing changes in the water level to determine the flow parameters.

11.2.1 Solutions for Evaluating Parameters Using Pumping Test

With pumping test, Eq. (9.104) is solved here subject to a constant rate Q in an infinite aquifer, which corresponds to the following IC and BCs discussed by Sternberg (1969):

$$h(r, t) = 0, \qquad\qquad t = 0 \qquad\qquad (9.107)$$

$$h(r, t) = 0, \qquad\qquad r = \infty \qquad\qquad (9.108)$$

$$R\frac{dh}{dr}\bigg| = -\frac{Q}{2\pi T_f}, \qquad\qquad r = R \qquad\qquad (9.109)$$

where T_f is the transmissivity of the aquifer, defined for the fDWE, R is the radius of the well and Q is the volumetric flow rate during pumping from the well (for injection, the right hand side of Eq. (9.109) is positive).

11.2.1.1 Solutions and Approximation for Pumping Test

The complete solutions for this case were given by Su et al. (2015). The approximate solution for water flow in confined aquifers with mobile and immobile zones takes the following form (Su et al. 2015):

$$h = \frac{\gamma Q}{2\pi b_2 T} t^{\beta_2 - 1} E_{\beta_2 - \beta_1, \beta_2}\left(-\frac{b_1}{b_2} t^{\beta_2 - \beta_1}\right) \qquad\qquad (9.110)$$

where $\pi = 3.1416$, $\gamma = 0.5772$ is Euler's number and $E_{\beta_2 - \beta_1, \beta_2}\left(-\frac{b_1}{b_2} t^{\beta_2 - \beta_1}\right)$ is the two-parameter Mittag-Leffler function expressed as:

$$E_{\beta_2 - \beta_1, \beta_2}\left(-\frac{b_1}{b_2} t^{\beta_2 - \beta_1}\right) = \sum_{k=0}^{\infty} \frac{1}{\Gamma[(\beta_2 - \beta_1)k + \beta_2]}\left(-\frac{b_1}{b_2} t^{\beta_2 - \beta_1}\right)^k \qquad (9.111)$$

where $\Gamma[(\beta_2 - \beta_1)k + \beta_2]$ is the gamma function.

The Mittag-Leffler function in Eq. (9.110) can be further simplified for practical applications by retaining only limited terms. By retaining the first two leading terms in Eq. (9.110), the approximate solution is written as:

$$h = \frac{\gamma Q t^{\beta_2 - 1}}{2\pi b_2 T}\left[\frac{1}{\Gamma[\beta_2]} - \frac{b_1}{b_2 \Gamma[2\beta_2 - \beta_1]} t^{\beta_2 - \beta_1}\right] \qquad\qquad (9.112)$$

By retaining the first term only, Eq. (9.112) yields the one-term approximation as:

$$h = \frac{\gamma Q t^{\beta_2-1}}{2\pi b_2\, T\, \Gamma[\beta_2]} \tag{9.113}$$

Equation (9.113) implies that the simplest approximation does not consider water flow in immobile zones in an aquifer, which can be used for interpreting short-term data from aquifer tests. As $\beta_2 \leq 1$, Eqs. (9.112) and (9.113) describe water level recessions.

11.2.1.2 Asymptotic Results of the Solution for Pumping Test

It has been seen that Eq. (9.110) is the generalized Prabhakar function (GPF) in Chapter 2, which can be written as:

$$h = \frac{\gamma Q}{2\pi b_2 T}\, e_{\beta_2-\beta_1,\beta_2}\left(t; -\frac{b_1}{b_2}\right) \tag{9.114}$$

In addition to the approximate solutions with the leading terms in Eq. (9.112), the asymptotic trends of the solution in Eq. (9.110) or Eq. (9.114) are worthwhile exploring. Two aspects of these asymptotic solutions have been briefly discussed below.

Case 1. Asymptotic solution with $0 < \beta_2 - \beta_1 < \beta_2 \leq 1.0$ for confined aquifers with mobile and immobile zones

Following the procedures of Mainardi and Garrappa (2015), the asymptotic result of the GPF solution in Eq. (9.114) for large times is:

$$h \sim \frac{\gamma Q}{2\pi\Gamma[\beta_1]b_2 T}\, t^{\beta_1-1}, \qquad t \rightarrow \infty \tag{9.115}$$

which means that at large times, both small and large pores are functioning which is manifested by the parameters for the two types of pores.

Case 2. Asymptotic solution with $0 < \beta_2 - \beta_1 = \beta_2 \leq 1.0$ for uniform aquifers

This condition applies to aquifers without immobile zones, that is, $\beta_1 = 0$, $b_2 = 1$ and $\beta_2 = \beta$. Subsequently, Eq. (9.114) yields the following asymptotic solution:

$$h \sim \frac{\gamma\beta Q}{2\pi\Gamma(1-\beta)T}\, t^{-\beta-1}, \qquad t \rightarrow \infty \tag{9.116}$$

In both the cases as expressed in Eqs. (9.115) and (9.116), the water level decreases with time. The difference lies in the fact that in aquifers *with* mobile and immobile pores (the first case), at large times ($t \rightarrow \infty$), the order of fractional derivatives for small pores and the porosity for large pores (b_2) are present.

11.2.2 Solutions for Evaluating Parameters Using Slug Test

For slug test, a small quantity of water is either added to or removed from the well. The following IC and BCs are used to describe this situation (Cooper et al. 1967, Su et al. 2015):

$$h(r, t) = h(t), \qquad\qquad r = r_s + 0, \qquad t > 0 \qquad\qquad (9.117)$$

$$h(r, t) = 0, \qquad\qquad r = \infty, \qquad t > 0 \qquad\qquad (9.118)$$

$$2\pi r_s T_f \frac{\partial h(r,t)}{\partial r} = \pi r_c^2 \frac{\partial h(r,t)}{\partial t}, \qquad r = r_s + 0, \qquad t > 0 \qquad\qquad (9.119)$$

$$h(r, t) = 0, \qquad\qquad r > r_s, \qquad t = 0 \qquad\qquad (9.120)$$

$$h(0,t) = \frac{V}{\pi r_c^2} = 1, \qquad\qquad r = 0, \qquad t > 0 \qquad\qquad (9.121)$$

where r_s is the effective radius and r_c is the radius of the casing in the interval (or depth) over which the water level changes.

The detailed derivation of solutions of Eq. (9.104) subject to the conditions in Eqs. (9.117) to (9.121) for slug test has been presented by Su et al. (2015). Here, the solutions are briefly summarized.

The full solution of Eq. (9.104) subject to the conditions in Eqs. (9.117) to (9.121) is very complex and of the form (Su et al. 2015):

$$h = \frac{r_s t^{\beta_2 - 1}}{b_2 T_f r} \sum_{k=0}^{\infty} \left[-\sqrt{\frac{n_e b_1}{b_2 T_f}} \left(\frac{r_s}{r_c} \right)^2 \right]^k \frac{2 t^{\beta_2 k/2}}{r} E_{\beta_2 - \beta_1, \beta_2 (1+k/2)}^{k+1} \left(-\frac{b_1 t^{\beta_2 - \beta_1}}{b_2} \right) \qquad (9.122)$$

11.2.3 Approximate Solutions

By retaining only the first term in the summation (with $k = 0$), Eq. (9.122) is simplified as:

$$h = \frac{2 r_s}{b_2 T_f r^2} e_{\beta_2 - \beta_1, \beta_2} \left(t; \frac{b_1}{b_2} \right) \qquad\qquad (9.123)$$

where

$$e_{\beta_2 - \beta_1, \beta_2} \left(t; \frac{b_1}{b_2} \right) = t^{\beta_2 - 1} E_{\beta_2 - \beta_1, \beta_2} \left(-\frac{b_1}{b_2} t^{\beta_2 - \beta_1} \right) \qquad\qquad (9.124)$$

is the GPF.

11.2.4 Asymptotic Results of the Approximate Solutions

As demonstrated in Chapter 6, the GPF in Eq. (9.123) has important asymptotic solutions which were originally presented in detail by Mainardi and Garrappa (2015) for $t \to \infty$.

In this case, there are two possibilities where asymptotes develop, depending on the combination of parameters, if we extend the upper limit from 1.0 to 2.0 (Mainardi and Garrappa 2015).

Case 1

$$0 < \beta_2 - \beta_1 < \beta_2 \le 2.0 \qquad (9.125)$$

Under this condition, the GPF in Eq. (9.123) develops the following asymptote:

$$h \sim \frac{2r_s}{b_2 T_f r^2} \frac{t^{\beta_1 - 1}}{\Gamma(\beta_1)}, \qquad t \to \infty \qquad (9.126)$$

Case 2

$$0 < \beta_2 - \beta_1 = \beta_2 \le 2.0 \qquad (9.127)$$

Under this condition, the GPF in Eq. (9.123) develops the following asymptote:

$$h \sim \frac{2\Gamma(1 - \beta_2) r_s}{\beta_2 b_2 T_f r^2} t^{-\beta_2 - 1}, \qquad t \to \infty \qquad (9.128)$$

The condition in Eq. (9.127) means that the pores are uniform, i.e., there are no immobile pores in the media, which means that $\beta_1 = 0$. Chechkin et al. (2002) presented an asymptote of an equation for MSD identical to Eq. (9.123) as time approaches the onset, i.e., Eq. (9.123) in this case yields ($\tau = 1$ in Chechkin et al.):

$$h \sim \frac{2r_s}{\Gamma(\beta_2 + 1) b_2 T_f r^2} t^{\beta_2}, \qquad t \to 0 \qquad (9.129)$$

With the asymptotic analyses above, it is clear that the solution with $k = 0$ in Eq. (9.122) is, in fact, the only physically meaningful solution considering the conditions in Eqs. (9.125) and (9.129).

11.2.5 Approximation to the Large-time Solution

In this case, one of the simplified forms for Eq. (9.122) is given as follows (Su et al. (2015):

$$h = \frac{2r_s t^{\beta_2 - 1}}{b_2 T_f r^2} \left\{ \begin{pmatrix} \dfrac{1}{\Gamma(2 - \beta_1)} - \dfrac{b_1 t^{\beta_2 - \beta_1}}{\Gamma(\beta_2 - 2\beta_1 + 2) b_2} \end{pmatrix} \\ - \sqrt{\dfrac{n_e b_1}{b_2 T_f}} \left(\dfrac{r_s}{r_c} \right)^2 t^{\beta_2/2} \left(\dfrac{1}{\Gamma(3\beta_2/2)} - \dfrac{2b_1 t^{\beta_2 - \beta_1}}{\Gamma[(5\beta_2/2) - \beta_1] b_2} \right) \right\} \qquad (9.130)$$

The simplest solution is the one retaining the first term in Eq. (9.131) (Su et al. 2015):

$$h = A_0 t^{\beta_2 - 1} \qquad (9.131)$$

with

$$A_0 = \frac{2r_s}{\Gamma(2 - \beta_1) b_2 T_f r^2} \qquad (9.132)$$

As $\beta_2 < 1$, water level described by Eq. (9.132) decreases as a power function of time. For different approximations of Eq. (9.122), dimensions vary, which can be lumped in one of the parameters.

11.2.6 Solutions at Small Time

The exact solution for slug tests and its approximate solution are also detailed in Su et al. (2015), and the one-term approximation to the small-time solution for slug tests is of the form (Su et al. 2015):

$$h = \frac{2T_f^{1/2}}{\Gamma(\beta_2)b_2}\left(\frac{r_s}{r}\right)^{3/2} t^{\beta_2-1} \tag{9.133}$$

11.3 Other Forms of Fractional Equations for Radial Groundwater Flow

11.3.1 Time-fractional PDE for Radial Flow

Over the past few years, several versions of fPDEs have appeared (Park et al. 2010, Qi and Liu 2010, Raghavan 2012, Su et al. 2015, Raghavan and Chen 2019, Razminia et al. 2015a, b, 2019). The fPDE for radial water flow as studied by Raghavan (2012) is of the form:

$$\phi_e c \frac{\partial^\beta p(r,t)}{\partial t^\beta} = \frac{1}{r^{n-1}} \frac{\partial}{\partial r}\left(K(r)r^{n-1}\frac{\partial p(r,t)}{\partial r}\right) \tag{9.134}$$

where $p(r, t)$ represents hydraulic head, r denotes radial distance, β is the order of fractional derivatives, n is the number of dimensions (for plane radial flow to or from a well, $n = 2$), ϕ_e stands for specific yield or drainage porosity for unconfined aquifers and storativity for confined aquifers, c denotes compressibility, and $K(r)$ is the scale-dependent conductivity expressed as:

$$K(r) = K_0 r^{-\theta_r} \tag{9.135}$$

where K_0 and θ_r are parameters.

Equation (9.134) is basically an extension of the generalized diffusion equation for radial flow in a heterogeneous porous medium (O'Shaunghnessy and Procaccia 1985, Metzler et al. 1994), based on the works of Gefen et al. (1983). Metzler et al. (1994) investigated the ranges of parameters; for instance, $\theta_r = d + a - 2$, where d is the fractal dimension embedded in the integer Euclidian space and a is the empirical parameter in the conductivity (O'Shaughnessy and Procaccia 1985). It has been observed from the above analysis that Eq. (9.135) incorporates scale parameters in the radial direction.

11.3.2 Time- and Space-fractional PDEs for Radial Flow in a Unified Form

Atangana and Vermeulen (2014) presented different forms of equations for radial flow in groundwater, which include time-fractional, space-fractional and time-space-fractional PDEs. Their time-fractional PDE is similar to Eq. (9.134) with a constant K and $n = 2$. However, their space-fractional PDEs are questionable because the expansion of fractional derivatives of compound functions does not follow simple

Leibniz rules; instead, the fractional expansion is a more complicated series (Podlubny 1999).

Considering axis-symmetrical radial and vertical groundwater flow in heterogeneous aquifers with large and small pores (mobile and immobile zones), a more generic extension of Eq. (9.34) (see also Bear 1979) can be written in cylindrical coordinates as follows:

$$b_1 \frac{\partial^{\beta_1} h^2(r,t)}{\partial t^{\beta_1}} + b_2 \frac{\partial^{\beta_2} h^2(r,t)}{\partial t^{\beta_2}} = \frac{1}{r}\frac{\partial^\alpha}{\partial r^\alpha}\left(K_r r \frac{\partial^\eta h^2(r,t)}{\partial r^\eta}\right) + K_z \frac{\partial^2 h^2(r,t)}{\partial z^2} \tag{9.136}$$

for unconfined aquifers; the equation for confined aquifers reads as:

$$b_1 \frac{\partial^{\beta_1} h(r,t)}{\partial t^{\beta_1}} + b_2 \frac{\partial^{\beta_2} h(r,t)}{\partial t^{\beta_2}} = \frac{1}{r}\frac{\partial^\alpha}{\partial r^\alpha}\left(K_r r \frac{\partial^\eta h(r,t)}{\partial r^\eta}\right) + K_z \frac{\partial^2 h(r,t)}{\partial z^2} \tag{9.137}$$

In both cases, z is the vertical coordinate.

11.3.3 Radial Composite Flows in Aquifers

Razminia et al. (2015a) proposed a system of fPDEs for radial flow in groundwater, taking into account wellbore and skin effects. This system of fPDEs, by assigning different parameters for the skin and the rest of the aquifer, takes the following form:

$$\frac{\partial^\beta p_1}{\partial t^\beta} = \frac{1}{r^\Theta}\frac{\partial^2 p_1}{\partial r^2} + \frac{\zeta}{r^{\delta+1}}\frac{\partial p_1}{\partial r}, \qquad \text{for } r \le R_D \tag{9.138}$$

and

$$\frac{\partial^\beta p_2}{\partial t^\beta} = \frac{1}{r^\Theta}\frac{\partial^2 p_2}{\partial r^2} + \frac{\zeta}{r^{\delta+1}}\frac{\partial p_2}{\partial r}, \qquad \text{for } r > R_D \tag{9.139}$$

where p_1 and p_2 are the pressures (strictly, the piezometric heads) in the skin and the aquifer respectively, R_D is the radius of the skin and r is the radial distance. The other parameters are defined as follows:

$$\zeta = d_m - \delta - 1 \tag{9.140}$$

where d_m is the mass fractal dimension, and

$$\beta = \frac{2}{2+\delta} \tag{9.141}$$

denotes the conductivity index, with $\zeta \ge 0$ and $0 < \beta \le 1$.

12. Exchange of Water between Mobile and Immobile Zones

The major difference between a saturated and an unsaturated porous medium is the level of water content in the medium. As for water flow in unsaturated soils (Su 2012, 2014), a mathematical relationship for water movement at different levels of pores in aquifers can be similarly formulated. The modified relationship (Su et al. 2015, Su 2017a) for water exchange between the two zones, which differs from the traditional counterpart by the order of fractional derivative, β_3, is written as:

$$\frac{\partial^{\beta_3} h_{im}}{\partial t^{\beta_3}} = A h_{im} + B \tag{9.142}$$

with

$$A = -\frac{\omega}{b_1} \tag{9.143}$$

$$B = \frac{\omega}{b_1} h_m \tag{9.144}$$

where h_m and h_{im} are the hydraulic heads in the mobile and the immobile zone respectively, and ω is the water exchange rate between the two zones.

Similar to flow in unsaturated soils, the order of fractional derivative, β_3, for water exchange between mobile and immobile zones during the draining of saturated aquifers can have similar quantities as β_1 and β_2. For generality, we retain β_3 in the formulation.

Solutions of the problem defined by Eqs. (9.142) to (9.144) are similar in mathematical terms for water exchange between mobile and immobile zones in unsaturated zones subject to two types of boundary conditions at the interface (Su 2012) and can be derived by choosing appropriate terminologies from the solutions by Su (2012) and also in section 6 in Chapter 6.

12.1 Solutions for Water Exchange between Large and Small Pores

The procedures for solving Eq. (9.142) are similar to the methods presented in Chapter 6 for water exchange between small and large pores in soils. In the present case of $h_m = h_{m0}$ (h_{m0} being the constant value of h_m measured on the surface of mobile zones), $\dfrac{\partial^{\beta_3} h_{im}}{\partial t^{\beta_3}}$ can be replaced by $\dfrac{d^{\beta_3} h_{im}}{dt^{\beta_3}}$ for solutions.

We examine water exchange between mobile and immobile zones with two types of ICs at the interface of mobile and immobile zones:

IC 1:

$$\frac{d^{\beta_3} h_{im}}{dt^{\beta_3}} = v_k, \qquad\qquad k = 1,2, \ldots,n, \qquad\qquad t = 0, \tag{9.145}$$

where v_k is the fractional rate of change of water in the immobile zone.

IC 2:

$$h_{im} = h_{im0}, \qquad\qquad t = 0, \tag{9.146}$$

where h_{im0} is the initial value of h_{im}.

12.1.1 Exact Solutions for Water Exchange at the Interface when the Initial Rate of Change in Hydraulic Head is Known

In this case, Eq. (9.142) to be solved with the condition in Eq. (9.145) is similar to the one for distributed-order movement of water in mobile and immobile pores (Su 2012); the solution is of the form:

$$h_{im} = \sum_{k=1}^{n} v_k t^{\beta_3-k} E_{\beta_3,\beta_3-k+1}\left(-\frac{\omega}{b_1}t^{\beta_3}\right) + \int_0^t (t-\tau)^{\beta_3-1} E_{\beta_3,\beta_3}\left[-\frac{\omega}{b_1}(t-\tau)^{\beta_3}\right]\frac{\omega h_m}{b_1}d\tau$$

$$(9.147)$$

where $E_{\beta_3,\beta_3-k+1}\left(-\dfrac{\omega}{b_1}t^{\beta_3}\right)$ and $E_{\beta_3,\beta_3}\left[-\dfrac{\omega}{b_1}(t-\tau)^{\beta_3}\right]$ are Wiman functions or two-parameter Mittag-Leffler functions (MLFs).

As we consider a surface with $h = h_0$ (a constant in the mobile zone), Eq. (9.147) can be integrated (Kilbas et al. 2004) to give:

$$h_{im} = \sum_{k=1}^{n} v_k t^{\beta_3-k} E_{\beta_3,\beta_3-k+1}\left(-\frac{\omega}{b_1}t^{\beta_3}\right) + \frac{\alpha h_0}{b_{im}}t^{\beta_3} E_{\beta_3,\beta_3+1}\left(-\frac{\omega}{b_1}t^{\beta_3}\right)$$

$$(9.148)$$

where $E_{\beta_3,\beta_3+1}\left(-\dfrac{\omega}{b_1}t^{\beta_3}\right)$ is the Wiman function.

In Chapters 6 and 7, a similar model was applied to investigate exchanges of water and solute between large and small pores, which showed that the solution in Eq. (9.148), with $k = 1$, was the only physically relevant solution. Hence, in the following section, we only consider the case $k = 1$, which results in a physically relevant solution of Eq. (9.142) subject to the condition in Eq. (9.145):

$$h_{im} = v_1 t^{\beta_3-1} E_{\beta_3,\beta_3}\left(-\frac{\omega}{b_1}t^{\beta_3}\right) + \frac{\alpha h_0}{b_{im}}t^{\beta_3} E_{\beta_3,\beta_3+1}\left(-\frac{\omega}{b_1}t^{\beta_3}\right)$$

$$(9.149)$$

12.1.2 Asymptotic Solutions when the Initial Rate of Change of Hydraulic Head is Known

1. For $t \to \infty$:

Similar to the analysis in Chapter 6, the asymptotic solution of Eq. (9.149) as $t \to \infty$ can be derived by following the procedures of Mainardi and Garrappa (2015):

$$h_{im} \sim \left[\frac{\beta_3 v_1}{\Gamma(1-\beta_3)t^{\beta_3+1}} + h_0\right]\frac{\omega}{b_{im}}, \qquad t \to \infty \qquad (9.150)$$

2. For $t \to 0$:

The asymptotic solution under this condition is given as (Friedrich 1991):

$$h_{im} \sim \left[\frac{v_1}{t\Gamma(\beta_3)} + \frac{h_0}{\Gamma(\beta_3+1)}\right]\frac{\omega}{b_{im}}t^{\beta_3}, \qquad t \to 0 \qquad (9.151)$$

which indicates that as time approaches the onset, the component of hydraulic head in the immobile zone associated with the rate of change decreases, whereas the component associated with the initial hydraulic head increases due to $\beta_3 \le 1$.

12.2 Solutions for Water Exchange when the Initial Hydraulic Head is Known

12.2.1 Exact Solutions

In this case, Eq. (9.142) is solved with Eq. (9.146) and the solution is given as (Kilbas et al. 2006):

$$h_{im} = h_{im0} E_{\beta_3}\left(-\frac{\omega}{b_1}t^{\beta_3}\right) + \int_0^t (t-\tau)^{\beta_3-1} E_{\beta_3,\beta_3}\left[-\frac{\omega}{b_1}(t-\tau)^{\beta_3}\right]\frac{\omega h_m}{b_1}d\tau \tag{9.152}$$

where $E_{\beta_3}\left(-\frac{\omega}{b_1}t^{\beta_3}\right)$ is the MLF and $E_{\beta_3,\beta_3}\left[-\frac{\omega}{b_1}(t-\tau)^{\beta_3}\right]$ is the Wiman function.

With a constant hydraulic head ($h = h_0$), on the interface of mobile and immobile pores of the aquifer, Eq. (9.152) is integrated (Kilbas et al. 2004) to yield:

$$h_{im} = h_{im0} E_{\beta_3}\left(-\frac{\omega}{b_1}t^{\beta_3}\right) + \frac{\alpha h_0}{b_{im}}t^{\beta_3} E_{\beta_3,\beta_3+1}\left(-\frac{\omega}{b_1}t^{\beta_3}\right) \tag{9.153}$$

where $E_{\beta_3}\left(-\frac{\omega}{b_1}t^{\beta_3}\right)$ is the MLF and $E_{\beta_3,\beta_3+1}\left(-\frac{\omega}{b_1}t^{\beta_3}\right)$ is the Wiman function.

12.2.2 Asymptotic Solutions

12.2.2.1 Asymptotic Solutions when the Hydraulic Head at the Interface of Mobile and Immobile Zones is Known at Large Time $t \to \infty$

Here, again, there are two cases of asymptotes for Eq. (9.153):
The asymptotic solution in this case is expressed as:

$$h_{im} \sim \left[\frac{h_{im0}\beta_3 t^{-\beta_3-1}}{\Gamma(1-\beta_3)} + h_0\right]\frac{\omega}{b_{im}}, \qquad t \to \infty \tag{9.154}$$

12.2.2.2 Asymptotic Solutions when the Hydraulic Head at the Interface of Mobile and Immobile Zones is Known at Small Time $t \to 0$

The asymptotic solution under this condition is written as:

$$h_{im} \sim \left(h_{im0} + h_0 t^{\beta_3}\right)\frac{\omega}{b_{im}}, \qquad t \to 0 \tag{9.155}$$

13. Example: Solutions of fPDEs for Groundwater Flow in Aquifers Subject to Boundary Conditions of the First Kind

13.1 Exact Solutions

In this section, the solutions of Eq. (9.92) for groundwater flow in heterogeneous unconfined aquifers are presented. These solutions can be easily extended to Eq. (9.82) for confined aquifers by replacing $\bar{T}_f$ with $\bar{K}_f$. The solution is subject to the following initial condition and boundary conditions:

$$h(x, t) = h(x, 0), \qquad\qquad t = 0, \qquad\qquad 0 \le x \le L \qquad\qquad (9.156)$$

$$h(x, t) = h(0, t), \qquad\qquad t > 0, \qquad\qquad x = 0 \qquad\qquad (9.157)$$

$$h(x, t) = h(L, t), \qquad\qquad t > 0, \qquad\qquad x = L \qquad\qquad (9.158)$$

For water flow in an aquifer with two levels of pores, the large-small (or mobile-immobile) porosity model is a reasonable choice. Hence, the solution by Jiang et al. (2012a) can be modified to accommodate this situation:

$$h(x,t) = \frac{[h(L,t) - h(0,t)]x}{L} + h(0,t) + \sum_{n=1}^{\infty} h_{n1}(0) h_{n0}(t) \sin\left(\frac{n\pi}{L}x\right) \qquad (9.159)$$

where

$$h_{n1}(0) = \frac{2}{L} \int_0^L \left[h(x,0) - h(0,0) - \frac{[h(L,0) - h(0,0)]x}{L} \right] \sin\left(\frac{n\pi x}{L}\right) dx, \quad n = 1,2,\dots \quad (9.160)$$

$$h_{n0}(t) = 1 - k_n t^{\beta_2} G_\mu^n(t) \qquad\qquad (9.161)$$

where $G_\mu^n(t)$ is the two-variable Mittag-Leffler function (2MLF) (Bhalekar and Daftardar-Gejji 2013):

$$G_\mu^n(t) = E_{(\beta_2 - \beta_1, \beta_2), 1 + \beta_2}\left(-\frac{b_i}{b_m} t^{\beta_2 - \beta_1}, -k_n t^{\beta_2} \right) =$$

$$\sum_{j=0}^{\infty} \sum_{i=0}^{j} \binom{j}{i} \frac{\left(-\frac{b_i}{b_m} t^{\beta_2 - \beta_1} \right)^i \left(-k_n t^{\beta_2} \right)^{j-i}}{\Gamma[(1 + \beta_2) + (\beta_2 - \beta_1)i + \beta_2(j - i)]} \qquad (9.162)$$

where

$$\binom{j}{i} = \begin{cases} \dfrac{j!}{i!(j-i)!} & \text{for } 0 \le i \le j \\ 0 & \text{for } 0 \le j < i \end{cases} \qquad\qquad (9.163)$$

is the binomial coefficient (Bronshtein and Semendyayev 1979), and

$$k_n = \frac{1}{b_m}\left[\overline{T}_f \lambda_n^\lambda - \omega \overline{K}_f \lambda_n^\eta \right], \qquad\qquad n = 1,2,\dots \qquad (9.164)$$

with

$$\lambda_n = \frac{n\pi}{L}, \qquad\qquad n = 1,2,\dots \qquad (9.165)$$

The 2MLF is a special form of the multinomial Mittag-Leffler function (MMLF) (Hadid and Luchko 1996, Luchko and Gorenflo 1999, Luchko 2011). For further details on MMLF, the reader is referred to the works of Saxena et al. (2011) and Li et al. (2015).

Equation (9.159) is very complex; for practical applications, it can be simplified by using different methods as discussed below.

13.2 Simplified Solutions

For $j = i$, Eq. (9.162) becomes the Wiman function or two-parameter Mittag-Leffler function:

$$G_\mu^n(t) = E_{\beta_1, 1+\beta_2}\left(-\frac{b_i}{b_m} t^{\beta_2 - \beta_1}\right)$$

(9.166)

Hence, Eq. (9.161) involves the GPF in the following form:

$$h_{n0}(t) = 1 - k_n e_{\alpha,\beta}\left(t; \frac{b_i}{b_m}\right)$$

(9.167)

with

$$e_{\alpha,\beta}\left(t; \frac{b_i}{b_m}\right) = t^{\beta_2} E_{\beta_1, 1+\beta_2}\left(-\frac{b_i}{b_m} t^{\beta_2 - \beta_1}\right)$$

(9.168)

which is the GPF.

13.3 Particular Solutions for Groundwater Levels with Known Values for IC and BCs

The solution presented by Jiang et al. (2012b) can be completed by specifying particular values for IC and BCs: $h(0, 0)$, $h(x, 0)$ and $h(L, 0)$. To complete the integration in Eq. (9.160), the specified initial value of hydraulic head, $h(x, 0)$, and the constant values at the two boundaries, $h(0, 0)$ and $h(L, 0)$, are used. Eq. (9.160) can thus be written as:

$$h_{n1}(0) = \frac{2}{L}\int_0^L (a_1 + a_2 x)\sin\left(\frac{n\pi}{L}x\right)dx$$

(9.169)

with

$$a_1 = h(x, 0) - h(0, 0)$$

(9.170)

$$a_2 = -\frac{[h(L,0) - h(0,0)]}{L}$$

(9.171)

Equation (9.169) can be subsequently integrated (Gradshteyn and Ryzhik 1994) to yield:

$$h_{n1}(0) = \frac{2}{n\pi}\left[\frac{a_2 L}{n\pi}\sin\left(\frac{n\pi x}{L}\right) - (a_1 + a_2 x)\cos\left(\frac{n\pi x}{L}\right)\right]$$

(9.172)

With Eq. (9.172), the solution of Eq. (9.92) is now written as:

$$h(x,t) = h(0,t) + \frac{[h(L,t)-h(0,t)]x}{L} +$$

$$\sum_{n=1}^{\infty} \frac{2\sin\left(\frac{n\pi x}{L}\right)}{n\pi}\left[\frac{a_2 L\sin\left(\frac{n\pi x}{L}\right)}{n\pi} - (a_1 + a_2 x)\cos\left(\frac{n\pi x}{L}\right)\right]\left[1 - k_n e_{\alpha,\beta}\left(t;\frac{b_i}{b_m}\right)\right]$$

(9.173)

It is clear from Eq. (9.173) that hydraulic head at the lower boundary ($x = L$) is expressed as:

$$h(x,t) = h(L,t)$$

(9.174)

For $t = 0$, Eq. (9.173) yields the initial hydraulic head in different locations as follows:

$$h(x,0) = h(0,0) + \frac{[h(L,t)-h(0,0)]x}{L} +$$

$$\sum_{n=1}^{\infty} \frac{2}{n\pi}\sin\left(\frac{n\pi x}{L}\right)\left[\frac{a_2 L}{n\pi}\sin\left(\frac{n\pi x}{L}\right) - (a_1 + a_2 x)\cos\left(\frac{n\pi x}{L}\right)\right]$$

(9.175)

13.4 Asymptotic Solutions of Simplified and Particular Solutions

The temporal part of Eq. (9.173) is the GPF, which has asymptotic results for $t \to \infty$. By extending the upper limit in the condition set from 1.0 to 2.0, as investigated by Mainardi and Garrappa (2015), which requires $n = 1$ in Eq. (9.173):

$$0 < \beta_1 < \beta_2 + 1 \le 2.0$$

(9.176)

the GPF has the following asymptotic form:

$$e_{\alpha,\beta}(t;\rho) = t^{\beta_2} E_{\beta_1,1+\beta_2}\left(-\frac{b_i}{b_m}t^{\beta_2-\beta_1}\right) \sim \frac{t^{\beta_2-\beta_1-1}}{\Gamma[\beta_2-\beta_1]}$$

(9.177)

The asymptotic solution of Eq. (9.173), with $n = 1$, is thus written as:

$$h(x,t) \sim h(0,t) + \frac{[h(L,t)-h(0,t)]x}{L} +$$

$$\frac{2\sin\left(\frac{\pi x}{L}\right)}{\pi}\left[\frac{a_2 L\sin\left(\frac{\pi x}{L}\right)}{\pi} - (a_1 + a_2 x)\cos\left(\frac{\pi x}{L}\right)\right]\left[1 - k_1\frac{t^{\beta_2-\beta_1-1}}{\Gamma[\beta_2-\beta_1]}\right]$$

(9.178)

13.5 The Physically Relevant Solution for Groundwater Levels in an Unconfined Aquifer

It can be inferred from the asymptotic analysis that the only physically relevant term in the solution, Eq. (9.173), is the first term (for $n = 1$). Thus, Eq. (9.173) can be simplified as:

$$h(x,t) = h(0,t) + \frac{[h(L,t) - h(0,t)]x}{L} +$$

$$\frac{1}{\pi}\left\{\frac{a_2 L}{\pi}\left[1 - \cos\left(\frac{2\pi x}{L}\right)\right] - (a_1 + a_2 x)\sin\left(\frac{2\pi x}{L}\right)\right\}\left[1 - k_1 e_{\alpha,\beta}\left(t; \frac{b_i}{b_m}\right)\right] \quad (9.179)$$

with

$$k_1 = \frac{1}{b_m}\left[\overline{T}_f \lambda_1^\lambda - \omega \overline{K}_f \lambda_1^\eta\right] \quad (9.180)$$

In the above solutions, the 2MLF can also be approximated by retaining limited terms. Retaining only two leading terms in the 2MLF in Eq. (9.179) ($j = 0,1$ and $i = 0,1$) results in:

$$E_{(\beta_2 - \beta_1, \beta_2),1 + \beta_2}\left(-\frac{b_i}{b_m}t^{\beta_2 - \beta_1}, -k_1 t^{\beta_2}\right)$$

$$= \frac{1}{\Gamma[1 + \beta_2]} + \frac{b_i t^{-\beta_1}}{k_1 b_m \Gamma[1 + \beta_2 - \beta_1]} - \frac{k_1 t^{\beta_2}}{\Gamma[1 + 2\beta_2]} - \frac{b_i t^{\beta_2 - \beta_1}}{b_m \Gamma[1 + 2\beta_2 - \beta_1]} \quad (9.181)$$

Hence, the approximate solution of Eq. (9.179) is:

$$h(x,t) = h(0,t) + a_0 x + \left\{\frac{a_2 L}{\pi}\left[1 - \cos\left(\frac{2\pi x}{L}\right)\right] - (a_1 + a_2 x)\sin\left(\frac{2\pi x}{L}\right)\right\}\frac{(1 - F_t)}{\pi} \quad (9.182)$$

where

$$a_0 = \frac{[h(L,t) - h(0,t)]}{L} \quad (9.183)$$

and

$$F_t = k_1 t^{\beta_2}\left(\frac{1}{\Gamma[1 + \beta_2]} + \frac{b_i t^{-\beta_1}}{k_1 b_m \Gamma[1 + \beta_2 - \beta_1]} - \frac{k_1 t^{\beta_2}}{\Gamma[1 + 2\beta_2]} - \frac{b_i t^{\beta_2 - \beta_1}}{b_m \Gamma[1 + 2\beta_2 - \beta_1]}\right) \quad (9.184)$$

Equation (9.182) can be used to determine the dynamics of the aquifer, given the initial and the boundary values of $h(x, t)$, namely, a_0, a_1 and a_2 in Eqs. (9.183), (9.170) and (9.171) respectively.

F_t in Eq. (9.184) should be dimensionless so that the dimension of hydraulic head $h(x, t)$ is $[L]$, which requires that k_1 has the dimension $[T^{-\beta_2}]$ and $\frac{b_i}{b_m}$, has the dimension $T^{\beta_1 - \beta_2}$, hence changing the definitions of b_i and b_m to be consistent with the dimensions as stated earlier. Due to infinite terms in the original multi-term solutions, the dimensions of other parameters such as k_2, k_3 and k_i also change.

13.6 Equations Derived from fPDEs for Groundwater Discharge from Aquifers

The above solutions of Eq. (9.92) are applicable to unconfined aquifers with weighted transmissivity. With the use of weighted hydraulic conductivity in place of weighted transmissivity in the diffusion term, the solutions are also applicable to confined aquifers described by Eq. (9.82).

Here, we use the simplified solution in Eq. (9.182) to derive the discharge recession curves (Su 2017a):

$$q = \bar{K}_f h(L, t)[\omega - \xi_t + 2\xi_0(1 - F_t)] \tag{9.185}$$

where F_t is given by Eq. (9.184), and

$$\xi_t = \frac{[h(L,t) - h(0,t)]}{L} \tag{9.186}$$

is the average gradient of the aquifer; the average gradient of the aquifer at the initial time $t = 0$ is expressed as:

$$\xi_0 = \frac{[h(L,0) - h(0,0)]}{L} = -a_2 \tag{9.187}$$

The unit discharge q has a dimension $L^2 T^{-1}$ and is expressed in units such as m^2/h. The total discharge from an aquifer is given as:

$$Q = qw \tag{9.188}$$

where w is the width of the aquifer.

In the following situations, Eq. (9.185) can be further simplified:

Case 1: Aquifer on a horizontal base:

In this case, $\omega = 0$. Hence, Eq. (9.185) becomes

$$q = \bar{K}_f h(L, t)[2\xi_0(1 - F_t) - \xi_t] \tag{9.189}$$

If $\xi_0 = \xi_t$ or $\xi_0 \approx \xi_t$, then both the quantities can be denoted as ξ. Subsequently, Eq. (9.185) becomes

$$q = \bar{K}_f h(L, t)\xi(1 - 2F_t) \tag{9.190}$$

Case 2: Initial discharge at $t = 0$.

In this case, Eq. (9.189) with $F_t = 0$ yields

$$q = \bar{K}_f h(L, t)(\omega + 2\xi_0 - \xi_t) \tag{9.191}$$

which implies that the initial discharge of the aquifer from the lower end of the aquifer depends on the slope angle ω and the gradients of the aquifer. For an aquifer on a horizontal base, $\omega = 0$, for which Eq. (9.191) becomes

$$q = \bar{K}_f h(L, t)(2\xi_0 - \xi_t) \tag{9.192}$$

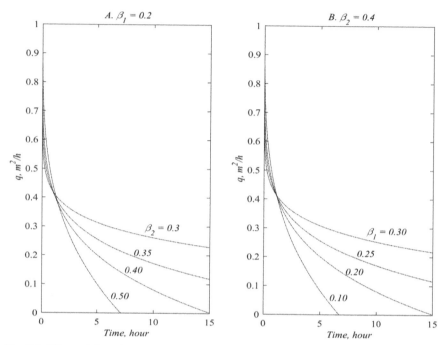

Fig. 9.2. Effects of the orders of fractional derivatives, β_2 and β_1, on recession curves given by Eq. (9.190). A: Fixed β_1 with variable β_2, indicating the facilitating role of β_2 in large pores. B: Fixed β_2 with variable β_1, indicating the trapping role of β_1 in small pores.

Also, if $\xi_0 = \xi_t$ or $\xi_0 \approx \xi_t$, ξ can be used for both the quantities. Eq. (9.192) can thus be written as:

$$q = \bar{K}_f h(L, t) \xi \qquad (9.193)$$

which is the statement of Darcy's law. The above simplifications are also a verification of the derivations based on fractional calculus.

As an example, Eq. (9.190) has been illustrated in Fig. 9.2 with the hypothetical parameters $\bar{K}_f h(L, t) \xi_0 = 1.0$, $b_i = 0.2$, $b_m = 0.8$ and $k_1 = 0.01$.

The illustration in Fig. 9.2 indicates that an increase in the porosity of large pores, represented by an increased b_m, facilitates groundwater drainage; however, an increase in the porosity of small pores slows down drainage due to decreasing large pores. These mathematical relationships are physically intuitive.

14. Groundwater Flow as a Multiphase Flow

In section 7, it was briefly stated that groundwater flow is not an isolated process but is, in fact, coupled with many other physical and chemical phenomena referred to as multiphase flow (Miller et al. 1998). At this stage, fPDEs have been successfully developed and applied to water flow and solute transport in aquifers; however, only limited work has been reported on coupled flow in porous media (Natale and Salusti 1996, Garra 2011, Garra and Salusti 2013, Garra et al. 2015).

14.1 Coupled Water Flow and Heat Transfer

Garra (2011) presented a coupled system of fPDEs for concurrent fluid flow and heat transfer in saturated media, assuming a local thermal equilibrium between the fluid and the porous medium; this system of fPDEs is of the form:

$$\frac{\partial^{\mu} T}{\partial t^{\mu}} = k \frac{\partial^2 T}{\partial z^2} + \beta_h \frac{\partial T}{\partial z} \frac{\partial p}{\partial z} + \gamma \left(\frac{\partial p}{\partial z} \right)^2 \tag{9.194}$$

$$\frac{\partial^{\beta} p}{\partial t^{\beta}} - \alpha_h \frac{\partial^{\beta} T}{\partial t^{\beta}} = D \frac{\partial^2 p}{\partial z^2} \tag{9.195}$$

where p denotes fluid pressure, T is the temperature of the fluid-porous medium system, k is the thermal diffusivity due to diffusion (Garra and Salusti 2013), μ and β are the orders of fractional derivatives, α_h is the thermal expansion coefficient, β_h is the average thermal diffusivity, D stands for fluid diffusivity, and γ represents average dissipative diffusivity due to media-fluid friction which is expressed as:

$$\gamma = \frac{\beta_h}{\rho c} \tag{9.196}$$

where ρ denotes fluid density and c is the fluid's heat capacity. Garra and Salusti (2013) and Garra et al. (2015) presented slightly different versions of Eqs. (9.194) and (9.195).

14.2 Other Coupled Water Flow and Transport Processes

In addition to the thermal effect, as de Marsily (1986) stated, other factors such as chemical gradient, electric gradient, etc. exert influence on water flow, generating reciprocal effects as other mechanisms, similar to the reciprocal Soret and Dufour effects. de Marsily (1986) proposed a more generic Darcy's law in the following form:

$$V = - K_h \operatorname{grad} h - K_E \operatorname{grad} E - K_c \operatorname{grad} c - K_T \operatorname{grad} T \tag{9.197}$$

where V denotes velocity; grad h, grad E, grad c and grad T represent the hydraulic gradient, electrical gradient, solute gradient and thermal gradient respectively; and K_h, K_E, K_c and K_T are the respective coefficients of the corresponding gradients.

To complete Eq. (9.197) as a system of PDEs for multiple coupled processes, more PDEs for electrical gradient, solute gradient and thermal gradient are required, which echo Maxwell's equations as a group of equations. If an anomalous version of Eq. (9.197) is used, then the fractional models for other reciprocal processes are required to be consistent with Eq. (9.197).

15. Summary

Groundwater flow in aquifers is an integral part of hydrological processes which connects water movement in soils. With fPDE-based models presented for water movement in soils as unsaturated porous media, fPDEs and their applications are discussed in this chapter for flow in saturated media form a complete spectrum of models for flow in porous media.

Unlike water movement in soils as unsaturated media that is freely connected to dynamics of atmospheric pressure, water flow in confined aquifers is subject to impervious layers which distinguishes confined aquifers from unconfined aquifers. For this reason, models for these two types of aquifers are different. This chapter outlines some fPDE-based models and their applications, and further verifications of these models are needed to refine their structures for different aquifers.

Chapter 10

Fractional Partial Differential Equations for Solute Transport in Groundwater

1. Introduction

The hydraulics and hydrology of groundwater flow are fertile fields where creativity flourishes; examples include the contributions by Boussinesq (1904), which place the analysis of groundwater flow on a rigorous foundation. Solute movement in groundwater is another important topic besides groundwater itself, for its relationship with contamination of aquifers from various sources, especially when anthropogenic and natural sources of pollutants are increasingly frequent on various lands, in stream flows and along coastal regions.

While mathematical analysis of groundwater flow has been made rigorous by the contributions of Boussinesq (1904), the mathematical analysis of solute transport in groundwater started later and models for solute transport are being continuously explored as they seem less satisfactory than their counterparts for groundwater flow.

As briefly discussed in Chapter 4, the applications of mathematical models mostly involve integer-order partial differential equations (PDEs), often termed as the advection-dispersion equation (ADE) or the Fokker-Planck equation (FPE) (Fokker 1914, Planck 1917). The refinement of the ADE or FPE continued when it was applied to the topic of dispersion of solutes at a field scale, giving rise to the scale issue (Scheidegger 1954, 1961). The scale issue has been investigated since the 1950s (Lallemand-Barres and Peaudecerf 1978, Gelhar et al. 1992, Bear 1972); however, a few applications of fractional PDEs (or fPDEs) to solute transport in the subsurface in geosciences were also reported before the 1990s.

In the 1990s, several significant developments were made regarding the applications of fPDEs for modelling solute transport in the subsurface:

1. One of the most important developments in modelling solute transport in the subsurface was suggested by Lenormand (1992), who introduced fractional derivatives to describe flow in heterogeneous porous media and derived a fractional PDE as a 'general transport equation' in Eq. (4.14).

2. Further developments to be discussed in the next section were made by Compte (1997), who proposed both time-fractional and space-fractional PDEs for modelling environmental problems applicable to solute transport in the subsurface.

3. Other fruitful developments took place when Benson (1998) applied Zaslavsky's (1994a) symmetrical fPDEs to solute transport in aquifers at a laboratory scale. Benson's (1998) investigations were specifically designed for hydrological problems at a laboratory scale rather than purely mathematical or physical analysis of generic nature. The reports published by Benson and his colleagues in subsequent research in this field involve mathematical developments and verification using data from experiments, which stimulated closely related investigations such as the fractional conservation of mass in a saturated zone (Wheatcraft and Meerschaert 2008) and new models based on fPDEs (Herrick et al. 2002, Schumer et al. 2003a, b, Benson et al. 2004, 2013, Zhang et al. 2009).

Solute transport in soils has been extensively discussed in Chapter 7. While major structures of mathematical models for solute transport in soils are similar to those for aquifers in terms of diffusion, advection and other processes, sharing similar porous structures with varying properties known as regions of excluded volume (Cushman and Ginn 1993), there are important differences which need attention. Among these differences is the distinction between unconfined and confined aquifers, which have different definitions of hydraulic head that determines flow velocity, and the level of water saturation in porous media (soils being unsaturated or variably saturated and aquifers being more frequently saturated).

For confined aquifers, storativity as a measure of both pore space and hydraulic pressure is an important parameter, for it is dependent on the hydraulic pressure of water in the medium, and the compressibility of water and the porous medium under the hydraulic pressure.

For unconfined aquifers, specific yield or effective drainage porosity is the measure of pore space which is exposed to atmospheric or barometric pressure.

As de Marsily (1986) stated, the mechanisms operating in these two types of aquifers are different: unconfined aquifers feature movement on a free surface exposed to atmospheric pressure, while confined aquifers are subject to the compressibility of water, particle grains and geological strata as a whole, depending on the hydraulic pressure on the aquifers. The difference in the pore space parameter for porous media in these two types of aquifers has been described in succeeding paragraphs.

In unconfined aquifers, specific yield (also called drainage porosity or effective porosity) is $\phi = \phi_e$. de Marsily (1986) discussed it in great detail; in particular, he showed that the variation of the mass of water, M, moved by a change in the water table in an unconfined aquifer is:

$$\frac{\partial M}{\partial t} = \rho \phi_e \frac{\partial h}{\partial t} dxdy \tag{10.1}$$

where $dxdy$ is the elemental area.

In confined aquifers, the aquifer's storativity S_c is defined as the volume of water released from or added to an aquifer, ΔV, per unit horizontal area of the aquifer, ΔA, per unit decline in the piezometric head (averaged vertically) $\Delta \psi$ (Bear 1972):

$$\phi = S_c = \frac{\Delta V}{\Delta A \Delta \psi} \tag{10.2}$$

de Marsily (1986) presented the following quantitative relationship between storativity and other parameters as:

$$\phi = S_c = \rho \phi_t gb \left(\beta_l - \beta_s + \frac{\alpha}{\phi_t} \right) \tag{10.3}$$

where ρ is the density of water, ϕ_t is the total porosity, g is acceleration due gravity, b is the thickness of the aquifer, β_l is the compressibility coefficient of water, β_s is the compressibility coefficient of solid grains of the porous medium, and α is the compressibility coefficient of the strata.

In confined aquifers, the outflow of water from the aquifer has two consequences (de Marsily 1986), which have been incorporated in Eq. (10.3): (a) the decompression of water in the aquifer, and (b) the compaction of the porous medium. There is no issue with the decompression of water in unconfined aquifers because the aquifer is exposed to the atmosphere with the restraint of geological materials. Solute transport in these two types of aquifers, therefore, differ due to these differences.

Comparing Eqs. (10.1) and (10.3), it is evident that the compressibility of water and that of the solid grains of the porous medium play a role in distinguishing between the two types of aquifers. In the following sections, several categories of reported models for solute transport in groundwater in both types of aquifers have been summarized along with some new ideas and developments.

2. fPDE-based Models for Solute Transport in Different Dimensions

Since Lenormand (1992) proposed fPDEs as the 'general transport equation' relevant to solute transport in porous media (see Eq. (4.14) in Chapter 4), Compte (1997) demonstrated the connection between fractional advection-diffusion equations (fADEs) and continuous-time random walk (CTRW) theory by presenting time-fractional and space-fractional fADEs (Compte 1997) in three dimensions. Compte presented separate time-fractional and space-fractional PDEs, which can be regarded as special cases of the time-space fractional PDEs derived by Uchaikin and Saenko (2003) by including a velocity vector $v(x)$. After Compte's (1997) developments, multi-dimensional space-fractional PDEs were also derived by Meerschaert et al. (1999) and Umarov and Gorenflo (2005b), although with slightly different approaches.

When the velocity vector $v(x)$ is included, the extended model based on Eq. (4.38) derived by Uchaikin and Saenko (2003) can be interpreted as the shift jump size distribution $(f^*(x - v(x)t)) = f(x)$ in one dimension or $(f^*(\mathbf{x} - v(\mathbf{x})t)) = f(\mathbf{x})$ in multiple dimensions (Zhang et al. 2009). With the two transitional probability density functions for jump lengths and waiting times, as given in Eqs. (4.34) and (4.35) respectively, Eq. (4.37) can be extended as follows to include convection or advection:

$$\frac{\partial^{\beta} c(\mathbf{x},t)}{\partial t^{\beta}} = \mathbf{D}\nabla^{\lambda} c(\mathbf{x},t) - \nabla \cdot \left[v(\mathbf{x})c(\mathbf{x},t)\right] \tag{10.4}$$

where $c(\mathbf{x}, t)$ is the solute concentration (instead of probability density) and $\mathbf{D}$ is the diffusivity vector given by:

$$\mathbf{D} = \frac{\Gamma(n/2)\Gamma(1-\lambda/2)A}{2^{\lambda}\Gamma[(\lambda+n)/2]\Gamma(1-\beta)B} \tag{10.5}$$

with A and B being coefficients in the transitional probability density functions given by Eqs. (4.34) and (4.35) respectively.

Equation (10.5) in different dimensions reads as follows:

$$D = \frac{\pi^{1/2}\Gamma(1-\lambda/2)A}{2^{\lambda}\Gamma[(\lambda+1)/2]\Gamma(1-\beta)B}, \qquad n=1 \tag{10.6}$$

$$D = \frac{\Gamma(1-\lambda/2)A}{2^{\lambda}\Gamma[(\lambda+2)/2]\Gamma(1-\beta)B}, \qquad n=2 \tag{10.7}$$

$$D = \frac{\pi^{1/2}\Gamma(1-\lambda/2)A}{2^{\lambda+1}\Gamma[(\lambda+3)/2]\Gamma(1-\beta)B}, \qquad n=3 \tag{10.8}$$

The developments made by Uchaikin and Saenko (2003) indicate that diffusivity D or the diffusivity vector $\mathbf{D}$ depends on the orders of fractional derivatives in space and time, and the scaling parameters, A and B, in the probability distribution density functions for jump length, $P(X>x) = Ax^{-\lambda}$, and waiting time, $P(J>t) = Bt^{-\beta}$. Similar approaches have been used to develop a multi-dimensional space-fractional PDE for generic particle dispersion (Meerschaert et al. 1999) and multi-dimensional time-space fractional PDEs for applications in cell biology (Andries et al. 2006).

Umarov and Gorenflo (2005b) also demonstrated the derivation of a space-fractional PDE in different dimensions as a special case of Eq. (10.4), which is identical to Compte's (1997) space-fractional fPDE in Eq. (4.18).

It must be emphasized here that the models for solute transport based on fPDEs reported in literature so far do not explicitly consider the difference in the solute transport processes in confined and unconfined aquifers. As briefly discussed in section 1, when water moves with different mechanisms in confined and unconfined aquifers, its velocity is different and the dispersion coefficient as a function of flow velocity (Bear 1972) also differs.

3. Fractional Conservation of Mass

An important development in modelling water flow in porous media was made by Wheatcraft and Meerschaert (2008), who presented a space-fractional equation of mass conservation for flow in porous media (see Eq. (5.117)). In fractional conservation of mass of liquid presented by Wheatcraft and Meerschaert (2008), the spatial element size Δx appears together with the order of space fractional derivatives α, in the form of $\Delta x^{1-\alpha}$, to account for the scale effect due to non-local flow processes. It is clear that this scale parameter $\Delta x^{1-\alpha} \rightarrow 1$ as $\alpha \rightarrow 1$.

Olsen et al. (2016) further presented a two-sided mass conservation (see Eq. 5.120), which is more generic, with the one-sided conservation of mass (Wheatcraft and Meerschaert 2008) as a special case. Similar to fractional conservation of mass of water in porous media, fractional conservation of mass of solutes can be developed by following the same approach. For example, similar to Eq. (5.120) for water movement in soils, the two-sided time-space fPDE for the conservation of mass by Olsen et al. (2016) can be extended to solute transport in aquifers as:

$$\frac{2^{\lambda}\Gamma(\lambda+1)}{(\Delta x)^{\lambda}}\frac{\partial^{\beta}}{\partial t^{\beta}}(\Delta x_3 \phi c) = {}^{R}D_{x_i}^{\lambda}(Dc) - {}^{L}D_{x_i}^{\lambda}(Dc) - D_{x_i}^{\eta}(vc) \tag{10.9}$$

where ϕ is different for confined ($\phi = S_c$) and unconfined ($\phi = \phi_e$) aquifers.

4. Symmetrical fADE for Solute Transport

The space-time fractional PDE, based on Zaslavsky's symmetrical fractional derivatives, was given by Benson (1998) as:

$$\frac{\partial^{\beta}P}{\partial t^{\beta}} = \frac{\partial}{\partial x}(AP) + \frac{1}{2}(1-\omega)\frac{\partial^{\lambda}}{\partial(-x)^{\lambda}}(BP) + \frac{1}{2}(1+\omega)\frac{\partial^{\lambda}}{\partial x^{\lambda}}(BP) \tag{10.10}$$

where A and B are coefficients related to the convectional velocity v and dispersion coefficient D, and ω denotes skewness.

For solute particles undergoing random walks with finite variance and finite mean waiting times, the order of fractional derivatives in Eq. (10.10) is integral, that is, $\beta = 1$ and $\lambda = 2$, and Eq. (10.10) becomes the conventional ADE.

For convenience in hydrological applications and as a convention in hydrology and soil science, the probability of the solute particles, P, is replaced by the (expected) solute concentration $c = c(x, t)$ to yield the following equation:

$$\frac{\partial^{\beta}c}{\partial t^{\beta}} = \frac{(1-\beta)}{2}\frac{\partial^{\lambda}(Dc)}{\partial(-x)^{\lambda}} + \frac{(1+\beta)}{2}\frac{\partial^{\lambda}(Dc)}{\partial x^{\lambda}} - v\frac{\partial c}{\partial x} \tag{10.11}$$

For $\beta = 1$ and constant D, Eq. (10.11) can be written as (Benson et al. 2001):

$$\frac{\partial c}{\partial t} = pD\frac{\partial^{\lambda}c}{\partial x^{\lambda}} + qD\frac{\partial^{\lambda}c}{\partial(-x)^{\lambda}} - v\frac{\partial c}{\partial x} \tag{10.12}$$

with $p + q = 1$; the relative weights of p and q describe the skewness of the formulation.

With a solution subject to an instantaneous input to Eq. (10.12) for the Green function, Benson et al. (2001) derived the following expression for dispersion coefficient D:

$$D = W\left(\frac{J_c}{\phi}\right)^{\lambda}\tau^{\lambda-1}|\Gamma(1-\lambda)| \tag{10.13}$$

where J_c denotes solute gradient, ϕ represents porosity, W is the scale (width) parameter (similar to the standard deviation of a Gaussian), and τ is the transit time for solute particles. Moreover, the absolute value of the gamma function (denoted by $|\,|$) is used here.

The peak concentration c_p and the peak velocity v_p were given by Benson et al. (2001) as:

$$c_p = \frac{c_0}{(Dt)^{1/\lambda}} \tag{10.14}$$

where c_0 is the initial solute concentration, and

$$v_p = v + \frac{A}{\lambda}\left[D\cos(\pi\lambda/2)\right]^{1/\lambda} t^{(1/\lambda)-1} \tag{10.15}$$

where A is a constant which depends on the skewness and the order λ. The second part of Eq. (10.15), depending on the value of the order of spatial fractional derivatives ($0 < \lambda < 2$), determines the shapes of the peak.

Both Eqs. (10.14) and (10.15) provide methods for estimating the order of spatial fractional derivatives. With Eq. (10.14), the log-log plot of the peak solute concentration c_p against time yields $-1/\lambda$ as the slope. With Eq. (10.15), the log-log plot of the difference $(v_p - v)$ against time also gives the value of λ.

While applying Eq. (10.11) to particle transport, Herrick et al. (2002) found that velocity v and saturated hydraulic conductivity K are related through the following power function:

$$v = CK^m \tag{10.16}$$

where C is a constant; this expression is basically the non-linear form of Darcy's law with respect to saturated hydraulic conductivity.

5. fPDEs for Reactive Solute Transport with Sink and Source Terms

5.1 fPDEs for Solute Transport with Equilibrium Absorption

For reactive solute transport in aquifers, a retardation factor R should be included in the formulation. It is again emphasized here that the difference between confined and unconfined aquifers should be noted.

By simplifying van Genuchten and Alves' (1982) formulation, the fADE (with a retardation factor) by Atangana and Kilicman (2013) can be written as:

$$R\frac{\partial c}{\partial t} = D\frac{\partial^\lambda c}{\partial x^\lambda} - v\frac{\partial^\gamma c}{\partial x^\gamma} - \mu c \tag{10.17}$$

where λ and γ are orders of fractional derivatives and μ is the decay parameter expressed as (van Genuchten and Alves 1982):

$$\mu = \mu_w + \mu_s \rho_s k/\phi_e \tag{10.18}$$

Here, μ_w and μ_s are the rate constants for first-order solute decay in liquid and solid phases of the aquifer, ρ_s is the density of the porous medium, ϕ_e is the effective equivalent porosity and k is the absorption rate in the linear equilibrium absorption relationship of the form:

$$S = kc \tag{10.19}$$

where S is the absorbed solute concentration. Here, it should be noted that the original definition of effective porosity ϕ_e for unsaturated media should be replaced with *storativity* for *confined aquifers* and with *specific yield* (or *drainage porosity*) for *unconfined aquifers*. For saturated media, we can also call it the equivalent porosity for the sake of simplicity.

Likewise, the solute production parameters can also be considered in Eq. (10.17) with approaches similar to integer-based PDEs (van Genuchten and Alves 1982); alternative absorption relationships can also be used (Bear and Verruijt 1992).

For multiple solutes having certain levels of absorption in porous media, the retardation factor should be considered in other forms such as fractional models in time and/or space.

5.2 Sorption-desorption and Non-equilibrium Mass Transfer

Absorption of solutes is a very important issue for environmental processes as desorption releases the solutes absorbed in the sites of porous media into the fluid of the porous media, possibly leading to contamination. In addition to the equilibrium absorption in Eq. (10.19), in an extensive review, Sardin et al. (1991) analyzed the following equation for local mass balance in an adsorbed phase:

$$\frac{\partial S}{\partial t} = k_{des}\left(K_a c_{im} - S\right) \tag{10.20}$$

where S is the concentration of adsorbed solute, c_{im} is the concentration of solute in water in the immobile zone, k_{des} is the desorption rate, and K_a is the adsorption parameter. For saturated media such as an aquifer, the solute concentration in the aggregate, c_p, and that in immobile water are related as (Sardin et al. 1991):

$$c_p = \phi_{im} c_{im} + \rho_a S \tag{10.21}$$

Here, ϕ_{im} is porosity of the immobile zone and ρ_a is the bulk density of aggregates. Similar to Eq. (7.93) for anomalous diffusion in immobile zones in soils and Eq. (10.35) for immobile zones in aquifers, the desorption process should be consistent with adsorption in terms of anomalous diffusion such as a fractional version of Eq. (10.20):

$$\frac{\partial^{\beta_4} S}{\partial t^{\beta_4}} = k_{des}\left(K_a c_{im} - S\right) \tag{10.22}$$

where β_4 is the order of fractional derivatives. Solutions of Eq. (10.22) subject to two types of initial condition can be found in the works of Su (2012) and in Chapters 7 and 9.

6. fPDEs of Distributed Order for Solute Transport in Aquifers

As special cases of the multi-term fPDEs of distributed order discussed in section 3, when heterogeneous porous media are treated as mobile and immobile zones only, solute transport in these two zones can be separately handled with fPDEs of distributed order.

6.1 Multi-term fPDEs of Distributed Order for Solute Transport in Groundwater

In Chapter 4, two types of distributed-order fPDEs were briefly discussed. Here, we further the discussion in the context of solute transport in aquifers.

6.1.1 Multi-term Time-fractional PDEs of Distributed Order

Compared to two-term fPDEs, multi-term fPDEs appear in limited reports only and are important in theory for quantifying flow and transport in multiple levels of fractal pores. Jiang et al. (2012a, b) investigated the following multi-term time-space symmetrical fPDE of distributed order:

$$\sum_{j=1}^{n} b_j D_t^{\beta_j} c(x,t) = D \frac{\partial^{\lambda} c(x,t)}{\partial |x|^{\lambda}} - v \frac{\partial^{\eta}}{\partial |x|^{\eta}} c(x,t) \tag{10.23}$$

The convective term can be fractional (Jiang et al. 2012a, b, Atangana and Kilicman 2013) or non-fractional. For constant distributed orders with non-fractional convection ($\eta = 1$), Eq. (10.23) can be rewritten as:

$$\sum_{j=1}^{n} b_j D_t^{\beta_j} c(x,t) = D \frac{\partial^{\lambda} c(x,t)}{\partial x^{\lambda}} - v \frac{\partial}{\partial x} c(x,t) \tag{10.24}$$

A special case of Eq. (10.23) is the two-term fPDE of distributed order, which has been extensively discussed in literature and investigated in the context of solute transport in large and small pores (or mobile and immobile pores) later in this chapter.

6.1.2 Multi-term Space-fractional PDE of Distributed Order

The multi-term space-fractional PDE of distributed order by Umarov and Steinberg (2006), when diffusion coefficients D_k are included, can be written as:

$$\frac{\partial c(x,t)}{\partial t} = \sum_{k=1}^{m} D_k a_k D_0^{\lambda_k} c(x,t) \tag{10.25}$$

A convective term may be introduced or can be a special term in this summation and the fractional derivatives of distributed order in time can also be incorporated in Eq. (10.25), making it an fPDE of distributed order in time and space. This model introduced by Umarov and Steinberg (2006) requires further analysis and applications.

6.2 Two-term fPDEs of Distributed Order for Solute Transport in Groundwater

Schumer et al. (2003b) developed fPDEs for total solute concentration c_t, solute concentration in the mobile zone, c_m, and that in the immobile zone, c_{im}. For $0 < \beta \le 1$, the fPDEs for these three components can be written as (Schumer et al. 2003b, Benson et al. 2013):

$$\frac{\partial c_t}{\partial t} + \phi_{mim} \frac{\partial^{\beta} c_t}{\partial t^{\beta}} = D \frac{\partial^2 c_t}{\partial x^2} - v \frac{\partial c_t}{\partial x} \tag{10.26}$$

$$\frac{\partial c_m}{\partial t} + \phi_{mim} \frac{\partial^\beta c_m}{\partial t^\beta} = D \frac{\partial^2 c_m}{\partial x^2} - v \frac{\partial c_m}{\partial x} - \phi_{mim} c_{m,0}(x) \frac{t^{-\beta}}{\Gamma(1-\beta)} \tag{10.27}$$

and

$$\frac{\partial c_{im}}{\partial t} + \phi_{mim} \frac{\partial^\beta c_{im}}{\partial t^\beta} = D \frac{\partial^2 c_{im}}{\partial x^2} - v \frac{\partial c_{im}}{\partial x} + c_{m,0}(x) \frac{t^{-\beta}}{\Gamma(1-\beta)} \tag{10.28}$$

where $c_{m,0}(x) = c_m(x, 0)$ and $c_{im,0}(x) = c_{im}(x, 0) = 0$ are the initial mobile and immobile concentrations at $t = 0$, and the mobile-immobile capacity coefficient (Schumer et al. 2003b) is defined as:

$$\phi_{mim} = \frac{\phi_{im}}{\phi_m} \tag{10.29}$$

where ϕ_{im} and ϕ_m are the porosities of immobile and mobile zones respectively. Once again, the porosities in these models should be replaced by the relevant parameter (storativity for confined aquifers and effective drainage porosity or specific yield for unconfined aquifers).

With time-fractional PDEs of distributed order, the fPDEs for total solute and mobile solute concentrations take different forms for solute transport at early and late stages (Schumer et al. 2003b). Solute transport at late stages (i.e., at a large time) is characterized by solute exchange between mobile and immobile zones while solute transport at early stages is dominated by advection. For solute transport at a late stage, the distributed-order fPDEs converge to the following one-term fPDEs respectively (Schumer et al. 2003b):

$$\frac{\partial c_t}{\partial t} + \phi_{mim} \frac{\partial^\beta c_t}{\partial t^\beta} \approx \phi_{mim} \frac{\partial^\beta c_t}{\partial t^\beta}, \qquad t \to \infty \tag{10.30}$$

and

$$\frac{\partial c_m}{\partial t} + \phi_{mim} \frac{\partial^\beta c_m}{\partial t^\beta} \approx \phi_{mim} \frac{\partial^\beta c_m}{\partial t^\beta}, \qquad t \to \infty \tag{10.31}$$

With Eq. (10.27), Schumer et al. (2003b) showed one important property: the recession or decay of mobile solutes in the Laplace domain follows:

$$\tilde{M}_m(s) = \frac{M_{m,0}}{s + \phi_{mim} s^\beta} \tag{10.32}$$

which, following the inverse Laplace transform (Kilbas et al. 2004), yields the Mittag-Leffler function:

$$M_m(t) = M_{m,0} E_{1-\chi}(z) = M_{m,0} \sum_{k=0}^{\infty} \frac{1}{\Gamma[1 + (1-\beta)k]} \left(-\phi_{mim} t^{1-\beta} \right)^k \tag{10.33}$$

For large times, an asymptote develops from Eq. (10.32) (Schumer et al. 2003b, Haubold et al. 2011):

$$M_m(t) \sim M_{m,0} \frac{t^{\beta-1}}{\phi_{mim} \Gamma(\beta)}, \qquad 0 < \beta \leq 1 \tag{10.34}$$

which can be used to determine the value of β, given the measured ϕ_{mim} and the initial solute concentration in the mobile phase, $M_{m,0}$.

Schumer et al. (2003b) showed that when the order of fractional derivatives approaches unity, a factor $(1 + \phi_{mim})$ appears as retardation, where ϕ_{mim} is the mobile-immobile capacity coefficient. This form of retardation is due to solute trapping in immobile zones; therefore, it can be called physical retardation to distinguish from the retardation due to physico-chemical absorption R, as in Eq. (10.20), which can be called physico-chemical retardation.

7. Solute Transfer between Mobile and Immobile Zones

In addition to the transport processes involving diffusion and advection for the total solute concentration and its two components, other important processes are solute transfer between mobile and immobile zones, and adsorption into and desorption from surfaces. Since the dual porosity concept was proposed by Rubinshtein in 1948 and further applied by Barenblatt et al. in 1960, the following integer ADE has been widely used for unsaturated media (Nielsen et al. 1986, Sardin et al. 1991) as discussed in Chapter 7:

$$\theta_{im} \frac{\partial c_{im}}{\partial t} = k_m \left(c_m - c_{im} \right)$$
(10.35)

where k_m is the mass transfer coefficient and θ_{im} is the volume fraction of the immobile water in the immobile zone. In saturated media, $\theta_{im} = \phi_{im}$, where ϕ_{im} is the porosity in the immobile zones; Hence, for solute exchange between the two zones in the aquifer, Eq. (10.35) is modified as:

$$\frac{\partial c_{im}}{\partial t} = \varpi \left(c_m - c_{im} \right)$$
(10.36)

with

$$\varpi = \frac{k_m}{\phi_{im}}$$
(10.37)

Sardin et al. (1991) showed that the characteristic mass transfer time t_m is expressed as:

$$t_m = \frac{1}{\varpi} = \frac{\theta_{im}}{k_m}$$
(10.38)

which implies that the dimension of ϖ is $[1/T]$ when the concentrations are dimensionless relative values. For a given geometry, Sardin et al. (1991) showed that t_m is related to velocity as $t_m = \frac{L}{v}$, which can be combined with Eq. (10.38) to yield:

$$k_m = \frac{\phi_{im} v}{L}$$
(10.39)

which is a simple and convenient way of measuring the mass transfer coefficient k_m.

Two types of initial condition can be used to solve Eq. (10.36) (Su 2012):

1. $\quad c_{im} = c_{im}(x, 0) = c_{im0}$, $\qquad\qquad\qquad t = 0$ $\qquad\qquad\qquad$ (10.40)

2. $\quad \dfrac{dc_{im}}{dt} = v_{im}$, $\qquad\qquad\qquad\qquad\quad t = 0$ $\qquad\qquad\qquad$ (10.41)

where c_{im0} is the initial value of c_{im} and v_{im} is the rate of change of solute concentration in the immobile zone.

The solution of Eq. (10.36) for a constant initial solute concentration on the surface, $c_{im}(x, 0)$ in Eq. (10.40), is (Schumer et al. 2003b):

$$c_{im} = c_{im}(x,0)e^{-\omega t} + \int_0^t \omega e^{-\omega t} c_m(x, t-\tau)d\tau$$

$\qquad\qquad\qquad$ (10.42)

which can be written in a compact form as:

$$c_{im} = c_{im}(x, 0)e^{-\omega t} + c_m * \omega e^{-\omega t}$$

$\qquad\qquad\qquad$ (10.43)

where the asterisk denotes convolution. Other forms of solutions subject to the flux condition in Eq. (10.40) can also be derived as special forms of solutions in the works of Su (2012).

For the anomalous transport processes to be consistent, the mass exchange between mobile and immobile zones should also follow the anomalous process; hence, a fractional mass transfer model is used as follows (Su 2012) (see also Chapters 7 and 9):

$$\frac{\partial^{\beta_3} c_{im}}{\partial t^{\beta_3}} = \omega_f \left(c_m - c_{im} \right)$$

$\qquad\qquad\qquad$ (10.44)

where β_3 is the order of fractional derivatives. In Eq. (10.44), the dimension of ω_f is $[T^{-\beta_3}]$, which is different from that of ω in Eq. (10.36).

In Chapters 7 and 9, two models similar to Eq. (10.44) had been analyzed—for solute transport in soils and groundwater flow respectively. Those solutions and asymptotic results can be slightly modified for solute transport in aquifers, except for the definitions specified for aquifers to replace moisture ratio or content. The issues which should be considered here include the difference in solute transport in unconfined and confined aquifers, the difference in the flow velocities, and definitions of effective porosity for unconfined aquifers and storativity for confined aquifers.

8. fPDEs for Flux and Residential Solute Relationships

Similar to solutes in unsaturated soils discussed in Chapter 7, depending on the definitions used for quantifying solutes in porous media, concentration of solutes is widely used as one of the measured quantities. Solute concentration can be measured as flux concentration $c_f(x, t)$, defined either as the mass of solutes per unit of fluid discharge, or as residential concentration $c_{r,t}(x, t)$, defined as the total mass of solutes per unit of total fluid volumes in both mobile and immobile zones. Both these terms can be defined for the total quantity of solutes or the components (such as in mobile and immobile zones).

The total residential solute concentration $c_{r,t}(x, t)$ and its two components are simply related by the following expression:

$$c_{r,t}(x, t) = c_{r,m}(x, t) + c_{r,im}(x, t) \tag{10.45}$$

where $c_{r,m}(x, t)$ and $c_{r,im}(x, t)$ are mobile and immobile residential solute components respectively.

8.1 Relationships between Flux and Residential Solutes

In Chapter 7, the relationships between residential concentration $c_{r,t}(x, t)$ and flux concentration $c_f(x, t)$ were discussed in the context of solute transport in soils as unsaturated porous media. In saturated media, these relationships can be interpreted correctly only when a distinction is made between confined and unconfined aquifers.

For flux concentrations and their relationships with their residential counterparts, the equations discussed in Chapter 7 need to be modified slightly for saturated aquifers. These two quantities are related as (Kreft and Zuber 1978):

$$Q \int_0^t c_f(x,\tau)d\tau = \phi A \int_x^\infty c_r(\xi,t)d\xi \tag{10.46}$$

where Q is the volumetric flow rate of the fluid $[L^3 T^{-1}]$, A is the cross-sectional area and ϕ has two definitions with one for unconfined and one for confined aquifers. When the volumetric flow rate per unit area, $q = Q/A$, is used, Eq. (10.46) can be written as:

$$q \int_0^t c_f(x,\tau)d\tau = \phi \int_x^\infty c_r(\xi,t)d\xi \tag{10.47}$$

The two quantities are related through the following identity (Kreft and Zuber 1978, Parker and van Genuchten 1984):

$$c_f = c_r - \frac{D}{v} \frac{\partial c_r}{\partial x} \tag{10.48}$$

The transform in Eq. (10.48) used by Kreft and Zuber (1978) for particle transport is a standard procedure in solving the heat equation widely used in applied mathematics (Carslaw and Jaeger 1947, 1959). In Chapter 7, this transformation was discussed in association with solute transport in soils. In terms of the application of fPDEs to flux concentration and residential solute concentration, Zhang et al. (2006) outlined the relationships between the two types of solute concentrations.

8.2 Space-fractional Advection-Dispersion Equation

Zhang et al. (2006) showed that with the following fADE for total residential concentration (for simplicity $c_{r,t}(x, t) = c_{r,t}$):

$$\frac{\partial c_{r,t}}{\partial t} = D \frac{\partial^\lambda c_{r,t}}{\partial x^\lambda} - v \frac{\partial c_{r,t}}{\partial x} \tag{10.49}$$

the following relationship holds between flux concentration and residential concentration:

$$c_f(x,t) = c_{r,t}(x,t) + \frac{x - vt}{v\lambda t} c_{r,t}(x,t)$$

(10.50)

8.3 Time-fractional Advection Equation of Distributed Order

A special case of the distributed-order ADE for solute transport in mobile and immobile zones (Schumer et al. 2003b) is the following fPDE without the dispersion term when the flow velocity is very large so that dispersion can be neglected (Zhang et al. 2006):

$$\phi_{mim} \frac{\partial^\beta c_{r,t}}{\partial t^\beta} + \frac{\partial c_{r,t}}{\partial t} = -v \frac{\partial c_{r,t}}{\partial x}$$

(10.51)

When written in terms of flux concentration, the appropriate equation reads as (Zhang et al. 2006):

$$\phi_{mim} \frac{\partial^\beta c_f}{\partial t^\beta} + \frac{\partial c_f}{\partial t} = -v \frac{\partial c_f}{\partial x} + \phi_{mim} \frac{t^{-\beta} \delta(x)}{\Gamma(1-\beta)}$$

(10.52)

where $\delta(x)$ is the Dirac delta function. The relationship between flux concentration and residential concentration with advection is expressed as (Zhang et al. 2006):

$$c_f(x,t) = \frac{\beta x}{\phi_{mt}[vt - x(1-\beta)]} c_{r,t}(x,t)$$

(10.53)

where

$$\phi_{mt} = \frac{\phi_m}{\phi}$$

(10.54)

is the ratio of the equivalent porosity in the mobile zone, ϕ_m, and the total equivalent porosity, ϕ.

8.4 Time-space Fractional Advection-dispersion Equation

When both temporal and spatial fractional derivatives are considered for the total residential solute concentration (Zhang et al. 2006) for $0 < \beta \leq 1$, that is

$$\phi_{mim} \frac{\partial^\beta c_{r,t}}{\partial t^\beta} + \frac{\partial c_{r,t}}{\partial t} = D \frac{\partial^\lambda c_{r,t}(x,t)}{\partial x^\lambda} - v \frac{\partial c_{r,t}(x,t)}{\partial x},$$

(10.55)

flux and residential solute concentrations are related as (Zhang et al. 2006):

$$c_f(x,t) = \left(\frac{x - vt}{v\lambda t} + 1 \right) c_{r,m}(x,t) + \frac{\beta x}{v\lambda t} \phi_{mim} c_{r,im}(x,t)$$

(10.56)

The algebraic solutions in Eqs. (10.50), (10.53) and (10.56) imply that the relationship between flux concentration and residential solute concentration under different conditions is independent of dispersion, that is, D is absent from these relations.

8.5 Symmetrical fPDEs for Residential Solute Concentration

Symmetrical fPDEs briefly discussed in section 5 are further discussed here in the context of flux concentration and residential solute concentration.

It has been shown that the symmetrical fPDE discussed above can be derived from the CTRW theory (Gorenflo et al. 2007, Gorenflo and Mainardi 2005, 2012, Zhang et al. 2009). The fPDEs discussed here, which include distributed-order fPDEs, essentially include backward fractional derivatives (BFDs) and forward fractional derivatives (FFDs). The time-space fractional PDE of distributed order for the total residential solute can be written as (Zhang et al. 2009):

$$\phi_{mim}\frac{\partial^\beta c_{r,t}}{\partial t^\beta}+\frac{\partial c_{r,t}}{\partial t}=qD\frac{\partial^\lambda c_{r,t}}{\partial x^\lambda}+(1-q)D\frac{\partial^\lambda c_{r,t}}{\partial(-x)^\lambda}-v\frac{\partial c_{r,t}}{\partial x} \tag{10.57}$$

where q is the skewness parameter and $0 \leq q \leq 1$.

8.5.1 Backward fPDE and the Backwater Effect

For $q = 0$, Eq. (10.57) becomes the fADE with BFDs only:

$$\phi_{mim}\frac{\partial^\beta c_{r,t}}{\partial t^\beta}+\frac{\partial c_{r,t}}{\partial t}=D\frac{\partial^\lambda c_{r,t}}{\partial(-x)^\lambda}-v\frac{\partial c_{r,t}}{\partial x} \tag{10.58}$$

which can be further simplified as:

$$\frac{\partial^\beta c_{r,t}}{\partial t^\beta}=D\frac{\partial^\lambda c_{r,t}}{\partial(-x)^\lambda}-v\frac{\partial c_{r,t}}{\partial x} \tag{10.59}$$

without considering immobile zones.

With the backward fADE, although four characteristics were discussed by Zhang et al. (2009), the most important feature of this model is the microscopic backwater effect as a counterpart of the large-scale backwater effect in hydraulics (observed in flow towards an end of a channel or the confluence of two streams).

8.5.2 Forward fPDE

For $q = 1$, Eq. (10.57) becomes the fADE with FFD only:

$$\phi_{mim}\frac{\partial^\beta c_{r,t}}{\partial t^\beta}+\frac{\partial c_{r,t}}{\partial t}=D\frac{\partial^\lambda c_{r,t}}{\partial x^\lambda}-v\frac{\partial c_{r,t}}{\partial x} \tag{10.60}$$

which was discussed in previous sections in relation to solute transport in aquifers with immobile zones.

As discussed in Chapter 2, the FFDs model fast diffusion, while the BFDs model the backwater effect with the time fractional derivative modelling slow diffusion. These mechanisms function as a dynamic consequential compromise resulting in the observed transport phenomena which can be quantified by the transport exponent (Zaslavsky 2002).

9. fPDEs of Distributed Order and Their Asymptotic Solutions

In Chapter 5, some important properties of fPDEs for special cases were discussed for water movement in soils. Those relationships are also applicable to solute transport in groundwater when the variables and parameters are correctly designated.

In addition to the other forms of fPDEs and parameters, the distributed-order time-fractional PDE introduced by Caputo (1969) (Gorenflo and Mainardi 2005) in the form of

$$\int_0^1 b(\beta), D_*^\beta c(x,t) d\beta = \frac{\partial^2 c(x,t)}{\partial x^2}, \tag{10.61}$$

with

$c(x, 0) = \delta(x)$,
$b(\beta) \geq 0$, and
$b(\beta) = b_1 \delta(\beta - \beta_1) + b_2 \delta(\beta - \beta_2)$

has the following mean square displacement (MSD) (Chechkin et al. 2002, Sandev et al. 2015):

$$\langle x^2 \rangle = \frac{2D}{b_2} t^{\beta_2} E_{(\beta_2-\beta_1),(\beta_2+1)} \left(-\frac{b_1}{b_2} t^{\beta_2-\beta_1} \right) \tag{10.62}$$

where $E_{(\beta_2-\beta_1),(\beta_2+1)} \left(-\frac{b_1}{b_2} t^{\beta_2-\beta_1} \right)$ is the Mittag-Leffler function with $\beta_2 > \beta_1$.

Equation (10.62) has the following asymptotic properties for small time $t \to 0$ and large time $t \to \infty$ (Chechkin et al. 2002):

$$\left. \begin{array}{ll} \langle x^2 \rangle \approx \dfrac{2D}{b_2 \Gamma(\beta_2 +1)} t^{\beta_2} \propto t^{\beta_2}, & t \to 0 \\[3mm] \langle x^2 \rangle \approx \dfrac{2D}{b_1 \Gamma(\beta_1 +1)} t^{\beta_1} \propto t^{\beta_1}, & t \to \infty \end{array} \right\} \tag{10.63}$$

The above asymptotic solutions imply that large pores are functional at early stages of solute transport, whereas small pores are important at large times. Gorenflo and Mainardi (2005) called the second MSD for $t \to \infty$ in Eq. (10.63) *sub-diffusion with retardation*.

With a distribution function of the form $b(\beta) = 1$ with $0 \leq \beta \leq 1$, Eq. (10.62) develops the following asymptotic properties (Chechkin et al. 2002):

$$\langle x^2 \rangle \approx 2Dt \ln \frac{1}{t} \qquad \text{for } t \to 0 \tag{10.64}$$

and

$$\langle x^2 \rangle \approx 2D \ln t \qquad \text{for } t \to \infty \tag{10.65}$$

where ln is the natural logarithm. Eq. (10.64) develops *anomalous super fast diffusion* at small times while Eq. (10.65) evolves as *anomalous super slow diffusion* (or ultra slow diffusion) at large times.

10. Radial Anomalous Solute Transport in Groundwater

10.1 Introduction to Radial Solute Transport

For non-reactive solute transport in planar radial flow in aquifers (or two-dimensional flow), the solute transport equation is written as (Bear 1979, Philip 1994, Moench 1995):

$$\frac{\partial c}{\partial t} = \frac{1}{r}\left(rD_L\frac{\partial c}{\partial r}\right) - v(r)\frac{\partial c}{\partial r} \tag{10.66}$$

where $c = c(x, t)$ is the solute concentration, D_L is the horizontal dispersion coefficient, r is the radial distance, and $v(r)$ is the planar radial velocity given as (Philip 1994):

$$v(r) = \frac{k_r}{r} \tag{10.67}$$

Here the constant $k_r = \dfrac{Q}{2\pi b\phi}$ has the dimension $[L^2T^{-1}]$. Subsequently, Eq. (10.67) can be written as:

$$v(r) = \frac{Q}{2\pi b\phi r} \tag{10.68}$$

where Q is the volumetric flow rate, ϕ denotes porosity and b is the thickness of the aquifer, indicating that the aquifer in question is a confined aquifer (Bras 1990) (for unconfined aquifers, it is replaced with the hydraulic head h).

Equation (10.66) can be written as:

$$\frac{\partial c}{\partial t} = -\frac{1}{r}\frac{\partial}{\partial r}\left[rq(r,t)\right] \tag{10.69}$$

where the flux of solutes in a radial direction, $q(r, t)$, is expressed as (Benson et al. 2004):

$$q(r,t) = -D_L\frac{\partial c}{\partial r} + v(r)c(r,t) \tag{10.70}$$

with $v(r)$ given by Eq. (10.68).

In some cases, the vertical one-dimensional movement of solutes driven by the vertical gradient is an issue. The vertical dispersion of solutes in groundwater could be significant, depending on the flow conditions and the gradient of solute concentration. The presence of a vertical gradient in the solute concentration is different from the flow of water in aquifers in that the vertical flow of water could be assumed to be negligible when the Dupuit hypothesis is valid.

In cases where the vertical component of solute transport is included, the equation for axis-symmetrical solute transport in cylindrical coordinates (Carslaw and Jaeger 1959) should be used; it can be written as:

$$\frac{\partial c}{\partial t} = \frac{1}{r}\left(rD_L\frac{\partial c}{\partial r}\right) - v(r)\frac{\partial c}{\partial r} + D_z\frac{\partial^2 c}{\partial z^2} \tag{10.71}$$

where D_z is the diffusion coefficient in vertical direction. When the transport process is due to diffusion only, Eq. (10.71) is simplified as:

$$\frac{\partial c}{\partial t} = \frac{1}{r}\left(rD_L\frac{\partial c}{\partial r}\right) + D_z\frac{\partial^2 c}{\partial z^2} \tag{10.72}$$

which, for homogeneous media, is further simplified as:

$$\frac{\partial c}{\partial t} = D\left[\frac{1}{r}\left(r\frac{\partial c}{\partial r}\right) + \frac{\partial^2 c}{\partial z^2}\right] \tag{10.73}$$

For reactive solutes, the retardation factor R is included in Eq. (10.71) so that the equation governing the solute transport is:

$$R\frac{\partial c}{\partial t} = \frac{1}{r}\left(rD_L\frac{\partial c}{\partial r}\right) - v(r)\frac{\partial c}{\partial r} + D_z\frac{\partial^2 c}{\partial z^2} \tag{10.74}$$

Equation (10.66) is clearly a result of conservation of mass in a radial coordinate with axis-symmetry in Eq. (10.74) (Meerschaert and Tadjeran 2004) by neglecting the vertical component $D_z\dfrac{\partial^2 c}{\partial z^2}$ (see Ozdemir and Karadeniz 2008).

10.2 Space-fractional and Time-Space Fractional PDEs for Radial Solute Transport

Depending on how fractional derivatives are applied to Eqs. (10.71) to (10.74) and the flux in Eq. (10.70), different forms of fPDEs can be derived. An example of the applications of fractional derivatives was proposed by Benson et al. (2004), who applied fractional derivatives to (total) solute transport as follows:

$$\frac{\partial c}{\partial t} = -\frac{k_v}{r}\frac{\partial c}{\partial r} + \frac{Dk_v}{r}\frac{\partial^\lambda c}{\partial r^\lambda} + f(r,t) \tag{10.75}$$

Benson et al. (2004) further extended Eq. (10.57) to a distributed-order fADE for radial solute transport as:

$$\phi_{mim}\frac{\partial^\beta c}{\partial t^\beta} + \frac{\partial c}{\partial t} = -\frac{k_v}{r}\frac{\partial c}{\partial r} + \frac{Dk_v}{r}\frac{\partial^\lambda c}{\partial r^\lambda} - \phi_{mim}c_{m,0}\frac{t^{-\beta}}{\Gamma(1-\beta)} \tag{10.76}$$

where ϕ_{mim} is the capacity coefficient by Eq. (10.29) and $c_{m,0}$ is the initial value of solute concentration in the mobile zone.

Equation (10.75) can also be extended to the following form for (residential) mobile solutes, c_m:

$$\frac{\partial c_m}{\partial t} + \phi_{mim}\frac{\partial^\beta c_m}{\partial t^\beta} = -\frac{1}{r}\frac{\partial^{\lambda_1}}{\partial t^{\lambda_1}}[rq(r,t)] + f(r,t) \tag{10.77}$$

with

$$q(r,t) = -D_r \frac{\partial^{\lambda_2} c}{\partial r^{\lambda_2}} + v(r)c(r,t) \tag{10.78}$$

or

$$\phi_{im} \frac{\partial^{\beta_1} c_m}{\partial t^{\beta_1}} + \phi_m \frac{\partial^{\beta_2} c_m}{\partial t^{\beta_2}} = -\frac{1}{r} \frac{\partial^{\lambda_1}}{\partial t^{\lambda_1}} [rq(r,t)] + f(r,t) \tag{10.79}$$

When vertical dispersion is included, Eq. (10.79) is extended to the following form:

$$\phi_{im} \frac{\partial^{\beta_1} c_m}{\partial t^{\beta_1}} + \phi_m \frac{\partial^{\beta_2} c_m}{\partial t^{\beta_2}} = -\frac{1}{r} \frac{\partial^{\lambda_1}}{\partial t^{\lambda_1}} [rq(r,t)] + D_z \frac{\partial^{\lambda_3} c}{\partial z^{\lambda_3}} + f(r,t) \tag{10.80}$$

where $q(r, t)$ is given by Eq. (10.78) and fractional derivatives in the vertical direction are also present to represent vertical dispersion.

These fPDEs above can be applied to total solute concentration and immobile concentration.

11. Functional-order fPDEs

The orders of fractional derivatives, taken with respect to time or space, can be either a real number (integers and fractions) or a function. The fractional calculus of functional order was pioneered by Unterberger and Bokobza (1965a) and Višik and Èskin (1967) and followed by Samko and Ross (1993), Jacob and Leopold (1993), Lorenzo and Hartley (2002) and many others. Caputo (1969) developed the idea of distributed-order differential equations with a distribution function as the order of fractional derivatives (such as β for time and λ for space-fractional derivatives). The combination of the two ideas is equally important in terms of fPDEs of distributed functional order.

Following a brief discussion of distributed-order fPDEs in Chapters 5 onwards, a more generic fPDE of distributed order for transport of total residential solutes in mobile zones and immobile zones. In groundwater, $c_{r,t} = c_{r,t}(x, t)$, can be written as:

$$b(\beta_1) \frac{\partial^{\beta_1} c_{r,t}}{\partial t^{\beta_1}} + b(\beta_2) \frac{\partial^{\beta_2} c_{r,t}}{\partial t^{\beta_2}} = D \frac{\partial^{\lambda} c_{r,t}}{\partial |x|^{\lambda}} - v \frac{\partial c_{r,t}}{\partial |x|} \tag{10.81}$$

where symmetrical fractional derivatives are used in addition to two-term time-fractional derivatives accounting for two-scale transport in large and small pores or in mobile and immobile zones.

The relationships between residential solutes $c_{r,t}$ and flux solutes $c_f(x, t)$ can also be investigated in a fashion similar to the results outlined in section 9.

With variable-order fractional derivatives, different forms of functional-order fractional derivatives can be introduced in the models discussed above. One such extension is the generalization of the fPDE for solute transport in aquifers with functional orders as:

$$b(\beta_1) \frac{\partial^{\beta_1(x,t)} c_{r,t}}{\partial t^{\beta_1(x,t)}} + b(\beta_2) \frac{\partial^{\beta_2(x,t)} c_{r,t}}{\partial t^{\beta_2(x,t)}} = D \frac{\partial^{\lambda(x,t)} c_{r,t}}{\partial |x|^{\lambda(x,t)}} - v \frac{\partial c_{r,t}}{\partial |x|} \tag{10.82}$$

Equation (10.82) applies to total residential solute transport in porous media with mobile and immobile zones; similar formulations can be developed for flux solute concentration. Without the distinction between mobile and immobile zones, Eq. (10.82) is simplified which was investigated by Sun et al. (2014).

12. Multi-dimensional Symmetrical fPDEs with Variable and Functional Orders

The foregoing discussions in section 11 are primarily concerned with one-dimensional problems. In fact, physical processes in aquifers are more likely multi-dimensional so that multi-dimensional fPDEs are more appropriate. Multi-dimensional fPDEs derived from multi-dimensional CTRW concepts for environmental processes (including solute transport) have been reported in literature (Compte 1997, Meerschaert et al. 1999, Schumer et al. 2003a, Uchaikin and Saenko 2003, Umarov and Gorenflo 2005b). With the use of solute concentration instead of multi-dimensional probability densities, the extended model by Uchaikin and Saenko (2003) in Eq. (10.4) can be written as:

$$b(\beta_1)\frac{\partial^{\beta_1(x,t)}c_{r,t}}{\partial t^{\beta_1(x,t)}} + b(\beta_2)\frac{\partial^{\beta_2(x,t)}c_{r,t}}{\partial t^{\beta_2(x,t)}} = \mathbf{D}\Delta^{\lambda(x,t)}c(\mathbf{x},t) - \nabla\left[v(\mathbf{x})c(\mathbf{x},t)\right] \qquad (10.83)$$

Similar investigations on multi-dimensional fPDEs of distributed order can be found in the report by Zayernouri and Karniadakis (2015) and related studies.

Equation (10.83) applies to residential solute transport in two levels of pores in three dimensions. Solute transport in highly heterogeneous media is an interesting topic and fPDEs of variable order or functional order such as Eq. (10.83) can be used to represent transport in multifractal porous media. It may be possible that the orders of fractional derivatives are also functions of solute concentration.

13. Tempered Anomalous Solute Transport

Based on the assumption that sub-diffusion takes place as a result of extended delay in small or immobile pores characterized by an exponentially tempered power-law, instead of a power-law, Meerschaert et al. (2008) proposed the following tempered fPDE for solute transport in porous media for the waiting time probability distribution function in the CTRW theory:

$$\frac{\partial c_{r,t}}{\partial t} + \phi_{mim}e^{-\mu t}\frac{\partial^{\beta}c_{r,t}}{\partial t^{\beta}}\left(e^{-\mu t}c_{r,t}\right) = D\frac{\partial^2 c_{r,t}}{\partial x^2}$$
$$- v\frac{\partial c_{r,t}}{\partial x} - \phi_{mim}g(t)c_{r,t,0}(x,0)\delta(x) \qquad (10.84)$$

where $g(t)$ is the memory function given by:

$$g(t) = \int_t^{\infty}e^{-\mu t}\frac{\beta\tau^{-\beta-1}}{\Gamma(1-\beta)}d\tau \qquad (10.85)$$

with μ being the tempering parameter and e as the exponential function.

Zhang et al. (2015) found the following relationship:

$$\mu \propto \frac{D_*}{Z^2} \tag{10.86}$$

where D_* is the effective molecular diffusivity and Z is the effective thickness of the layers with low permeability (in which the tempering process takes place). Eq. (10.86) was derived from the Damköhler number D_a, which embodies μ as a parameter:

$$D_a = \frac{\mu a \phi_e}{KJ} \tag{10.87}$$

where J is the hydraulic gradient, K represents hydraulic conductivity, a is the characteristic length scale of the aquifer and ϕ_e is the effective equivalent porosity.

The fPDE in Eq. (10.84) is for residential solute concentration and its relationship with flux concentration can be found in previous sections.

14. Summary

This chapter briefly overviews fractional models for solute transport in aquifers, while also introducing some new ideas such as Eq. (10.83), etc. With the application of fPDEs to groundwater hydrology since Lenormand (1992) and Compte (1997), significant progress has been made by many researchers in developing and applying new models to improve our understanding of physical, physico-chemical and hydrological processes in aquifers. It is important to note that distinction should be made in the models for confined and unconfined aquifers as the hydraulics in these two types of aquifers is different, which determines the different flow velocities, hence resulting in distinctive water flow and solute transport mechanisms.

Chapter 11

Fractional Partial Differential Equations and their Applications to Poroviscoelastic Media and Geomechanics

Re-familiarization and Introduction for Hydrologists and Soil Scientists

1. Introduction

The Navier-Stokes equations (NSEs) briefly discussed in Chapter 3 are fundamental equations that govern flow of fluids (including flow of water in porous media). The NSEs accommodate viscoelasticity of porous media, or poroviscoelasticity, and the vertical compressibility of the granular skeleton of porous media and compressibility of water as far as water movement in soils and aquifers is concerned. Of course, high-order hydrodynamics or Burnett hydrodynamics (Burnett 1935) can provide a better description of water and related processes in deforming porous media, but the present form of simplified NSEs for flow in porous media remains a dominant concept for practitioners.

Viscoelasticity of porous media, or poroviscoelasticity, and compressibility of water link groundwater hydrology, soil physics and soil mechanics, which treat viscoelastic properties (or poroviscoelastic) and compressibility properties with different degrees of simplification.

In soil physics, the consideration of viscoelastic properties of porous media is compromised, where the NSEs are highly simplified and the basic equation of motion, known as Darcy's law, incorporates hydraulic conductivity K, which includes permeability k as a more empirical property of porous media, and the dynamic viscosity for water μ, (Bear 1972) in the following form:

$$K = \frac{k\rho g}{\mu} = \frac{kg}{v} \tag{11.1}$$

where kinetic viscosity v is related to dynamic viscosity as $\mu = v\rho$ with ρ being the density of water and g denotes acceleration due to gravity. When Darcy's law is combined with conservation of mass for water flow in unsaturated soils, a PDE known as the Richards equation (Richards 1931) can be derived, which includes Eqs. (5.9) and (5.12). However, the Richards equation accounts for neither the compressibility of soil and water nor the stress-strain relations of soils.

In groundwater hydrology, the compressibility of fluid and porous media are incorporated and less explicitly represented by storage coefficient (or storativity) for confined aquifers and specific yield (or drainage porosity) for unconfined aquifers (see Eqs. (9.4) and (9.9)). When the governing equations are simplified for homogeneous media and/or a further simplification of non-vertical flow in aquifers is used with the Dupuit hypothesis, transmissivity T also incorporates hydraulic conductivity, which, in turn, embraces the viscosity of water (dynamic viscosity μ or kinetic viscosity v).

In soil mechanics, which is more concerned with stress-strain relations, the central issues are volume change and associated dynamics of soils (elasticity, viscoelasticity or poroviscoelasticity of porous media).

It can be seen that in the spectrum from soil physics, groundwater hydrology and soil mechanics, more and more attention is being paid to viscosity, elasticity and poroviscoelasticity. With increasing consideration of details of flow in porous media in this spectrum, more detailed mathematical models are anticipated.

With varying levels of attention to the same topic of soil, different concepts and methods have gradually been developed by practitioners, leading to the formation of invisible boundaries between soil mechanics and soil physics in the 1930s, and the comment "In my opinion the separation of Soil Physics and Soil Mechanics, whatever the historical and practical reasons for it, impedes progress in both fields." (Philip 1974). The consequence of separating soil mechanics from soil physics discourages hydrologists and soil scientists from investigating such topics as stress, strain, elasticity, viscoelasticity or poroviscoelasticity even though hydrologists (particularly groundwater hydrologists) and soil scientists deal with soils and aquifers which are viscoelastic and poroviscoelastic materials.

Subsequently, the following questions naturally arise:

1. Hydrology and soil science: How can water movement, solute transport, energy transfer and gas transport, or multi-phase flow, in soils and aquifers be addressed together with poroviscoelasticity?

2. Geomechanics and soils mechanics: How can elasticity, viscoelasticity and poroelasticity be handled together with water movement, solute transport and energy transfer?

With an intention of re-familiarizing hydrologists and soil scientists with the developments in the concepts and methodologies in soil mechanics, this chapter briefly traces the evolution of the key concepts and methodologies of elasticity, viscoelasticity, poroelasticity, and the latest models based on fractional calculus, particularly, fractional differential equations (fDEs).

2. Basic Concepts Regarding Poroviscoelastic Materials, and Relationships between Them

2.1 Stress, Strain and Pore Fluid Pressure

Stress and strain are two concepts central to the mechanics of deformation. Depending on which theory is used to define them, the definitions can slightly differ. With the definitions for fluid-filled porous media given by Biot (1935, 1941) and restated by Detournay and Cheng (1993) considering poroelastic properties, these concepts have been described below.

2.1.1 Stress

Stress is the force applied to an object per unit area. In mechanics, stress is often represented by the symbol σ. Forces can have different components; hence, stress can also have different components such that the concepts of total stress and strain tensor are used. In many situations, stresses are investigated in different dimensions so that a stress tensor σ_{ij} is generally used, which denotes forces applied in the j direction, the normal to which is in the i direction (Detournay and Cheng 1993).

Findley et al. (1976) defined stress σ as the ratio of the force acting on a given area in a body to the area as the size of the area tends to zero. This definition is infinitesimal or microscopic, whereas the definition used by Detournay and Cheng (1993) is macroscopic.

2.1.2 Strain

Strain is the deformation of the material in response to the stress applied on a body; it is represented here by the symbol ε.

Findley et al. (1976) gave the following two definitions of strain, which is essentially a measure of the intensity of deformation at a point. The two types of strain can be distinguished as:

a. *linear strain*, that is, change in the length of a line segment (per unit length) as the length of the line segment becomes infinitesimal, and

b. *engineering shear strain*, that is, change in the angle at the point of intersection of the two lines originally at right angles.

2.1.3 Relation of Stress and Strain Components in Biot's Definitions

In an important treatise on poroelasticity, Biot (1941) introduced the following 4 components of stress and strain as concepts for both bulk dynamic and kinematic variables (Detournay and Cheng 1993): the *total stress tensor* σ_{ij}, *fluid pressure p*, *solid strain tensor* ε_{ij} and *variation of fluid volume* ζ.

With the causal reciprocal relation between stress and strain in poroelastic materials (an increase in pore pressure induces a change in the form of dilation of the porous medium, and compression of the porous medium causes a rise in pore pressure), Biot (1941) (see Detournay and Cheng 1993) decomposed the strain tensor σ_{ij} into two components: solid partial stress τ_{ij}, and fluid partial stress τ; the strain in response to these stresses is also decomposed into two components: solid strain tensor ε_{ij}, and fluid dilatation e. The quantities in these two sets of stress and strain are related through the following identities:

$$\sigma_{ij} = \tau_{ij} + \delta_{ij}\tau \tag{11.2}$$

$$p = -\frac{\tau}{\phi} \tag{11.3}$$

$$\zeta = \phi(\varepsilon - e) \tag{11.4}$$

where ϕ denotes porosity and δ_{ij} is the Kronecker delta function.

 With the above descriptions, deformation ε_{ij} and the variation in fluid pressure in the medium, ζ, following the application of stress, results in work done by the process (Detournay and Cheng 1993):

$$dW = \sigma_{ij}\varepsilon_{ij} + pd\zeta \tag{11.5}$$

where liquid pressure p is positive for saturated and negative for unsaturated media. This process in a poroelastic system has a constitutive relation known as the Euler condition:

$$\frac{\partial \varepsilon_{ij}}{\partial p} = \frac{\partial \zeta}{\partial \sigma_{ij}} \tag{11.6}$$

 The above definitions have been given for fluid-filled porous media in which the pressure of fluid is separate from the matrix; in that case, the 'fluid' which fills the medium can include both liquids (such as water) and gases (such as air) for natural soils. Hence, the total pressure in the porous medium is the sum of pressures resulting from both liquids and gases.

2.1.4 Elastic and Plastic

Deformation or dilatation can be classified as two types depending on whether it is reversible or not; the material is called elastic when deformation is reversible, and it is called plastic when deformation is irreversible (Diethelm and Freed 1999).

2.2 Overview of the Developments in Ideas and Methodologies for Elasticity

For the benefit of hydrologists and soil scientists in understanding the relevance of deformation and related concepts about porous media, a short overview of historical developments in elasticity and viscoelasticity is provided here.

 The elastic and viscous properties of materials of both rigid and flexible nature have been very important topics in mechanics, mathematics and experimental pursuits. There exists extensive literature which documents major landmarks in these fields, dating back to the 17th century. Among these reports, the reviews of Love (1892) on mathematical theory of elasticity, Gurtin (1973) on linear theory of elasticity, Detournay and Cheng (1993) on the fundamentals of poroelasticity, Truesdell and Noll (2004) on the non-linear theory of elasticity, and Kausel (2010) on the early history of soil-structure interactions are particularly detailed.

2.2.1 Developments in General Theory of Elasticity

Among the pre-modern literature on elasticity, the treatise of Love (1892) detailed the conception and historical developments of ideas on strain and stress of materials, showing that the concept of elasticity dates back to Galilei in 1638, who attempted the quantification of bending of a beam with one end fixed on a wall. Since then, elasticity and extended concepts such as viscoelasticity and poroviscoelasticity have been continuously investigated. Koeller (2010) and Mainardi (2010) traced the major developments of creep and relaxation for elastic materials which are also called linear materials with memory.

The following are the pioneers and notable contributors in this field (the details before 1892 are mainly from Love 1892):

1638: Galilei is "the first mathematician to consider the nature of the resistance of solids to rupture...", and his endeavor to quantify the bending of a beam with one end fixed to a wall, known as Galilei's problem, is the onset of mechanics of elastic materials (Love 1892).

1660: Hooke's law (published in 1676) is proposed, and "this law formed the basis of the mathematical theory of elasticity" (Love 1892). Hooke's discovery in 1660, which was published in 1676, is not limited to a 'spring' made of rigid materials, it is a more generic approach to any 'springy body', and 'tension' refers to general 'strain'.

1691–1705: Jacques Bernoulli is among the mathematicians who extended Galilei's problem, and his work is attributed as "The first investigation of any importance is that of elastic line" (Love 1892). Daniel Bernoulli in 1742 further investigates J. Bernoulli's problem (Sachkov 2008).

1742–1744: Euler publishes a book in 1744 (*Methodus inveniendi lineas curvas maximi minimive proprietate gaudentes*) and investigates an elastic rod known as the *Euler elastic* (Sachkov 2008) or *Euler elasticity*. Euler's additional investigation is a reply to D. Bernoulli's statement in 1742 that the elastic energy of a bent rod is proportional to the magnitude of the bent body (Sachkov 2008).

1776–1787: Coulomb's theory on flexure published in 1776 is regarded as most scientific of all early mathematical memoirs dealing with Galilei's problem, and correctly calculated the moment of elastic forces; he was the first to consider resistance of thin fibers due to torsion.

1820–1827: Navier is the first to investigate the general equations of equilibrium and vibration of elastic solids (1821), and presents the general equations of elasticity (1827). He was one of the pioneers in establishing hydraulics as a discipline in addition to be the first to develop the theory of suspension bridges (Univ. St. Andrews 2000).

1823–1841: Cauchy (1823, 1827a, b) lays the foundation of the general theory of elasticity and extends it to mathematical physics (Love 1892); he expresses six components of stress and strain, which also appeared in the early works of Green and Saint-Venant and, some of which are equivalent to the independent discoveries by Lamé and Clapeyron (1828) and Lamé (1841).

1828: Poisson investigates radial vibration of spheres and considers the theory of wave propagation in an unlimited isotropic elastic medium; he identifies two kinds of waves: waves of compression and waves of distortion, which propagate independently with different velocities.

1833: Neumann proposes thermoelastic equations, which were also investigated slightly later by Duhamel (1838), leading to the thermoelastic equations of Neumann and Duhamel (Love 1892), which were further investigated by Zener (1938, 1948).

1834: Vicat is the first to systematically investigate the phenomena of creep.

1837: Green's contribution is appreciated as: "the revolution which Green effected in the elements of elastic theory is comparable in importance with that produced by Navier's discovery of the general equations" (Love 1892).

1844: de Saint-Venant analyzes the components of finite strain and stress-strain relations (Love 1892). His name is associated with the set of equations for open channel flows in hydraulics.

1845: Stokes (1845, 1851) discovers viscosity and hypothesizes that "…in fluids the force of resistance is in simple proportion to the velocity of change of shape…" (Thomson (Lord Kelvin) 1890); he presents the equations of elasticity that deal with fluids from perfect fluids to perfectly elastic solids through plastic solids; he also discovers viscosity in the form of Stokes' law and the elastic property of matter in terms of stress-strain relations. Stokes was one of the key pioneers in fluid mechanics and his contributions along with those of others, particularly Navier, form the Navier-Stokes equations as the fundamental set of equations governing fluid flow.

1845: Thomas Young first considers shear as an elastic strain and defines the term today known as Young's modulus.

1855: Thomson (Lord Kelvin) investigates thermoelasticity.

1859: Lamé introduces the method of curvilinear coordinates into the study of physics and develops the whole theory on these methods.

1864: Clebsch develops a general theory of free vibrations in solids.

1868: Maxwell investigates elasticity; the model he investigated is now named after him.

1879: Thomson and Tait introduce the concept of strain-ellipsoid and develop the kinematic theory, which reaches its climax this time.

1882–1887: Voigt (1882) investigates a model of elasticity (now named after him). Stress-strain relations were further investigated by Voigt (1882), de Saint-Venant (Clebsch 1883), Thomson (1884) and Neumann (1885) among others.

1895: Thurston is apparently the first to propose a model on three stages of creep and study stress relaxation (Findley et al. 1976).

1904: Lamb presents the Lamb's problem—a fundamental solution for homogeneous half space subjected to a dynamic load on its surface (Kausel 2010).

1907: Volterra (1907, 1909, 1913, 1928, 1929, 1940, 1959) carries out extensive investigations into many issues on deformation and elasticity. He rigorously analyzes elasticity of materials using integro-differential equations, defines the term *hereditary linear elasticity (ereditarietà leneare elastica)* and develops a series of fundamental relationships.

1913–1934: Hatschek in 1913 and Philippoff in 1934 observe 'structure-viscosity'.

1921: Nutting (1921a, b) investigates deformation and proposes a new law which relates deformation with stress and time, and the stress (or the force applied) can be a function of temperature, similar to thermoelasticity as investigated by Neumann (1833), Duhamel (1838), Thomson (1855), Zener (1938) and Rabotnov (1948) (Rabotnov used fractional concepts).

1923: Terzaghi develops a series of theories on porous media, including the earliest theory accounting for the influence of fluids in pores on quasi-static deformation of soils; he proposes a model for one-dimensional consolidation and relationships between permeability and porosity, etc.

1935: Gemant investigates elasticity and introduces the concept of complex viscosity.

1938: Zener (1938, 1948), in a series of papers, develops models for elasticity and theromoelasticity, which are relevant to water flow and solute transport in porous media subject to heat gradients. A model developed by Zener is now named after him as well.

1939: Biot (1939 to 1973) develops theory and methodologies in a series of papers on stress-strain relations for porous media. The contributions of Terzaghi and Biot formed the key foundation of poroviscoelastic media (see Detournay and Cheng 1993).

1943: Blair and Coppen propose a time-fractional form of Nutting's model. Blair (1944) and Blair and Reiner (1951) further investigate this model.

1948: Gerasimov is arguably the first, according to Valério et al. (2014), to propose models on anomalous dynamics of the mechanical behavior of viscoelastic materials. He proposes a time-fractional creep-relaxation relation with his own fractional derivatives, similar to the definition of Caputo fractional derivatives (1967) and equivalent to the fractional model in Eq. (11.15) below; he also investigates with a set of fPDEs a viscoelastic problem involving the motion of a viscous fluid between moving surfaces (see also Podlubny 1999, Kilbas et al. 2010).

1948: Rabotnov, seemingly coinciding with Gerasimov, investigates thermal dependence of elasticity and proposes a multi-term differential equation (his Eq. (3.1)) for the stress-strain relation which generalizes all forms of linear models of elasticity to be outlined below.

1959: van der Knaap develops a theory of non-linear elasticity of porous media or non-linear poroelasticity.

1961: Slonimsky proposes a fractional deformation model (a linear combination of an integer integral equation and a fractional integral equation).

1966: Caputo proposes a multi-term time-fractional model for stress-strain relations (his Eq. (2)) and investigates elastic waves in spherical coordinates.

1971: Caputo and Mainardi (1971a, b) propose a time-fractional generic model (named in this book as the Caputo-Mainardi model), investigate dissipation properties and compare the modelled results with experimental data.

2.2.2 Developments in the Theory of Poroelasticity

Among the historical developments in elasticity, plasticity and viscoelasticity, the contributions of Terzaghi (1923) and Biot (1935, 1941) are the most important and relevant to flow in porous media. The property of porous media with elasticity is also termed as poroelasticity (Detournay and Cheng 1993). Poroelasticity was apparently first used by Geertsma (1966), according to Wang (2000), and is named poroviscoelasticity in some cases (considering flow of viscous liquids in elastic porous media) (Bemer et al. 2001). Both poroelasticity and poroviscoelasticity are the central issues to be discussed below.

Furthermore, the time-dependent deformation process described by Nutting (1921a, b) can be fundamentally treated as a creep-relaxation process with memory, which was initiated by Caputo and Mainardi (1971a, b). The general properties of thermoelastic relationships by Neumann (1833), Duhamel (1838) and Zener (1938), etc. constitute more complex yet more realistic processes of deformation and related mechanisms when poroviscoelastic materials are under variable conditions.

There are extensive literature reviews on poroelastic media, some of which are specialized to relate stress and strain for water flow and solute transport in poroelastic media, besides other approaches used for stress-strain relations (Love 1892, Bagley and Torvik 1983, Koeller 1984, 2010, Bemer et al. 2001, Liingaard et al. 2004, Kausel 2010, Mainardi 2012, Lai et al. 2016, Sun et al. 2016).

Soil physics, since its separation from soil mechanics in the 1930s, developed its own methodologies to deal with properties of soils. This approach in soil physics was discussed in Chapter 3, where an integral transform with the void ratio as the variable was used to convert the vertical Cartesian coordinate to a *material coordinate* for swelling soils.

The material coordinate represented by *m* was proposed in the 1960s (Raats 1965, Raats and Klute 1968, Smiles and Rosenthal 1968, Philip 1969b, Philip and Smiles 1969). With this approach for swelling-shrinking soils, the material coordinate is valid for vertically deforming soils only by neglecting horizontal expansion of soil (resulting from an increase in the water content of swelling soils). This simplification allows deformation to be analyzed with void ratio as the variable which depends on the moisture content of the soil. Similar to the development of ideas of elasticity and deformation, more complicated factors (such as the thermal effect and time-dependent or processes with memory, etc.) can also be introduced.

2.3 Linear, Non-linear and Semi-linear Poroelasticity and Poroviscoelasticity

Different methods have been gradually developed over time to quantify the relationship between stress and strain for poroelastic media with the following concepts as important reciprocal relations (Wang 2000):

Solid-to-fluid coupling: It occurs when a change in applied stress produces a change in fluid pressure or fluid mass.

Fluid-to-solid coupling: This occurs when a change in fluid pressure or fluid mass produces a change in the volume of the porous medium.

These two phenomena, as demonstrated by Mainardi (2010) and Mainardi and Spada (2011) for linear elasticity, can be quantitatively related as reciprocal relations in more generic forms based on fractional calculus.

2.3.1 Linear and Non-linear Problems

In geomechanics and related topics, three concepts are often important for theoretical developments and applications: *stress* σ, *strain* ε, and *free energy* associated with them, F_0. These quantities are related by the following relationship (Bemer et al. 2001, Eq. (16)):

$$\sigma = \frac{\partial F_0}{\partial \varepsilon} \tag{11.7}$$

The *free energy* F_0 associated with the relationship between changes in stress and strain is the sum of *classical potential linear elasticity W* and *non-linear potential H*; *linear elasticity* can be further decomposed into different components. When these relations are given under the isothermal condition, the term *isothermal free energy* is used (Biot 1973, 4926).

When such concepts as elasticity, poroelasticity and poroviscoelasticity are used for describing materials which are analyzed with models with *non-linear potential H* neglected, the concepts used for describing the properties are linear (such as *linear elasticity, linear poroelasticity* and *linear* poroviscoelasticity) (Bemer et al. 2001).

The use of linear and non-linear methods depends on the scale of the problem under consideration. Non-linear potential is less important at small scales but significant at large scales. Biot (1939) stated:

"It is well known that the classic Theory of Elasticity is restricted to small deformation and rotations and that this is the underlying reason for its linear character." "It was thought that a non-linear theory including terms of the first and second order only would yield the essential features due to large deformation which are not explained by a linear theory."

Biot (1973) further stated that *"The non-linear behavior is then mainly a consequence of local geometric effects such as changes in contact areas, crack closure, etc."* In addition to the classification of linear and non-linear approaches, Biot (1973) discussed the *semi-linear* approach developed in terms of the Cartesian definition of finite strain and a local rotation field.

2.3.2 Semi-linear Problems

Apart from the classification of linear and non-linear models given by Bemer et al. (2001), the difficulties in choosing appropriate models for practitioners such as hydrologists and soil scientists are obvious. Based on the findings of van der Knaap (1959) and others, Biot (1973) further introduced semi-linearity in strain-stress relations. Biot (1973) stated that the mechanisms responsible for non-linearity in the strain-stress relationship of porous materials remain the same for isotropic and nonisotropic stress, which is explained as follows:

"The volume of the solid constituting the porous matrix should depend linearly on the fluid pressures and stresses within a wide practical range of values. On the other hand, the strain due to the effective stresses involves modifications of local geometries in the pores caused by changes in contact areas, crack closures, etc. These modifications are essentially non-linear within a range of average strain, which remains small."

This mixed property is termed as semi-linearity and further details on this were given by Biot (1973).

3. Approaches to Viscoelastic Materials with Linear Elasticity

The approach with linear viscoelasticity based on memory functions is also known as hereditary linear elasticity; this concept was developed by Volterra (1909, 1913), who extensively investigated elasticity and related mathematical and mechanical problems. The concept of hereditary linear elasticity has been further pursued by a number of mathematicians, predominantly in Italy. Both linear and non-linear creep and relaxation of viscoelastic materials and mechanics were detailed by Findley et al. (1976); moreover, Wineman (2009) provided an extensive review of non-linear viscoelastic problems. The fractional models of linear elasticity are particularly detailed in a monograph by Mainardi (2010) and also by Mainardi and Spada (2011).

Based on Volterra's pioneering works (Volterra 1907, 1909, 1913) and with more recent notation in integro-differential equations for stress-strain relationships (Mainardi 2010, Koeller 2010, Mainardi and Spada 2011), the fundamental methods for quantifying materials with viscoelasticity are briefly outlined below.

3.1 Creep, Creep Compliance, Relaxation and Relaxation Modulus

The two fundamental concepts for quantifying the interrelationship between stress and strain for viscoelastic materials are *creep* and *relaxation*.

Creep is the increase in strain under loading (or stress), whereas *relaxation* is the decrease in stress when displacement is held constant (Neidigk et al. 2009).

Creep compliance: In a creep test, when a step of constant stress is applied, that is,

$$\sigma = \sigma_0 H(t) \qquad (11.8)$$

where $H(t)$ is the Heaviside step function (given in Chapter 2), and the time-dependent strain $\varepsilon(t)$ is measured, and for a linear material, the stress-strain relationship is represented by (Findley et al. 1976):

$$\varepsilon(t) = \sigma_0 J(t) \qquad (11.9)$$

or

$$J(t) = \frac{e(t)}{\sigma_0} \qquad (11.10)$$

the quantity $J(t)$ is called *creep compliance* (creep strain per unit applied stress), which is a material property.

Relaxation modulus: In a relaxation test, when a step of constant strain is applied, that is,

$$\varepsilon(t) = \varepsilon_0 H(t) \tag{11.11}$$

where $H(t)$ is the Heaviside step function, the stress-strain relationship is represented by:

$$\sigma(t) = \varepsilon_0 G(t) \tag{11.12}$$

or

$$G(t) = \frac{\sigma(t)}{\varepsilon_0} \tag{11.13}$$

where $G(t)$ denotes *relaxation modulus*.

Both *creep compliance J(t)* and *relaxation modulus G(t)* are material-specific properties.

For materials with linear viscoelasticity, Mainardi and Spada (2011) showed that strain $\varepsilon(t)$ and stress $\sigma(t)$ are reciprocally related; their relationship can be written as a set of coupled Volterra integral equations of the second kind:

$$\varepsilon(t) = \int_{0^-}^{t} J(t-\tau) d\sigma(\tau) = \sigma(0)J(t) + \int_0^t J(t-\tau)\frac{d\sigma(\tau)}{d\tau} d\tau \tag{11.14}$$

and

$$\sigma(t) = \int_{0^-}^{t} G(t-\tau) d\varepsilon(\tau) = \varepsilon(0)J(t) + \int_0^t G(t-\tau)\frac{d\varepsilon(\tau)}{d\tau} d\tau \tag{11.15}$$

where $\sigma(t)$ and $\varepsilon(t)$ are the time-dependent functions for *stress* and *strain* respectively, $\sigma(0)$ and $\varepsilon(0)$ are the initial values of $\sigma(t)$ and $\varepsilon(t)$ respectively, $J(t)$ denotes *creep compliance*, $G(t)$ represents *relaxation modulus*, t is time and τ is the integral variable denoting delay.

3.2 Classic Models of Linear Viscoelasticity

Models for quantitative relationship between stress and strain can be developed from either Eq. (11.14) or (11.15). The following widely-used models for elasticity in mechanics (Mainardi and Spada 2011) have options for either creep compliance or relaxation modulus.

[1] Hooke's model

This is the simplest linear model:

$$\sigma(t) = m\varepsilon(t), \qquad \begin{cases} J(t) = \dfrac{1}{m} \\ G(t) = m \end{cases} \tag{11.16}$$

where m denotes relaxation modulus.

[2] Newton's model

This model is of the following form:

$$\sigma(t) = b_1 \frac{d\varepsilon(t)}{dt}, \qquad \begin{cases} J(t) = \dfrac{t}{b_1} \\ G(t) = b_1\delta(t) \end{cases} \tag{11.17}$$

where $\delta(t)$ is the Dirac delta function.

[3] The Kelvin-Voigt model

The Kelvin-Voigt model is a linear combination of Hooke's model and Newton's model.

$$\sigma(t) = m\varepsilon(t) + b_1 \frac{d\varepsilon(t)}{dt} \tag{11.18}$$

with

$$\begin{cases} J(t) = J_1[1 - \exp(-t/\tau_3)], & J_1 = \dfrac{1}{m}, \tau_3 = \dfrac{b_1}{m}, \\ G(t) = G_1 + G_2\delta(t), & G_1 = m, G_2 = b_1 \end{cases} \tag{11.19}$$

[4] Maxwell's model

In addition to the derivative with respect to strain, Maxwell's model also introduces the first-order derivative with respect to stress:

$$\sigma(t) + a_1 \frac{d\sigma(t)}{dt} = b_1 \frac{d\varepsilon(t)}{dt} \tag{11.20}$$

with

$$\begin{cases} J(t) = J_g + J_2 t; & J_g = \dfrac{a_1}{b_1}, J_2 = \dfrac{1}{b_1} \\ G(t) = G_3 \exp(-t/\tau_4); & G_3 = \dfrac{b_1}{a_1}, \tau_4 = a_1 \end{cases} \tag{11.21}$$

[5] Zener's model

This model is written as:

$$\sigma(t) + a_1 \frac{d\sigma(t)}{dt} = m\varepsilon(t) + b_1 \frac{d\varepsilon(t)}{dt} \tag{11.22}$$

with

$$\begin{cases} J(t) = J_g + J_3[1 - \exp(-t/\tau_e)]; & J_g = \dfrac{a_1}{b_1}, J_3 = \dfrac{1}{m} - \dfrac{a_1}{b_1}, \tau_e = \dfrac{b_1}{m} \\ G(t) = G_4 + G_5 \exp(-t/\tau_5); & G_4 = m, G_5 = \dfrac{b_1}{a_1} - m, \tau_5 = a_1 \end{cases} \tag{11.23}$$

Zener's model is more generic and embraces the previous four models.

[6] The anti-Zener model

Herein, derivatives of both stress and strain, along with the second-order derivative of strain, are used:

$$\sigma(t) + a_1 \frac{d\sigma(t)}{dt} = b_1 \frac{d\varepsilon(t)}{dt} + b_2 \frac{d^2\varepsilon(t)}{dt^2} \tag{11.24}$$

with

$$\begin{cases} J(t) = J_4 t + J_5[1 - \exp(-t/\tau_6)]; & J_4 = \frac{1}{b_1}, J_5 = \frac{a_1}{b_1} - \frac{b_2}{b_1^2}, \tau_6 = \frac{b_2}{b_1} \\ \\ G(t) = G_6\delta(t) + G_7 \exp(-t/\tau_7); & G_6 = \frac{b_2}{a_1}, G_7 = \frac{b_1}{a_1} - \frac{b_2}{a_1^2}, \tau_7 = a_1 \end{cases} \tag{11.25}$$

[7] The Burgers model

The Burgers model is more generic than most of the above models, with second-order derivatives for both stress and strain:

$$\sigma(t) + a_1 \frac{d\sigma(t)}{dt} + a_2 \frac{d^2\sigma(t)}{dt^2} = b_1 \frac{d\varepsilon(t)}{dt} + b_2 \frac{d^2\varepsilon(t)}{dt^2} \tag{11.26}$$

with

$$\begin{cases} J(t) = J_6 + J_7 t + J_8[1 - \exp(-t/\tau_8)]; \\ G(t) = G_8 \exp(-t/\tau_9) + G_9 \exp(-t/\tau_{10}); \end{cases} \tag{11.27}$$

4. Fractional Calculus-based Models for Linear Viscoelasticity and Poroviscoelasticity

Investigations into deformation of materials, in terms of elasticity, have been attempted for about 380 years since Galilei conceived this approach in 1638. Physicists, mathematicians and experimentalists across many disciplines have contributed to these topics. However, fractional calculus, pioneered in 1695, was not applied to the study of elasticity and deformation until the first half of the 20th century.

As briefly reviewed in Chapter 3, fractional calculus slowly evolved since 1695 till around 1974 when it was re-launched. The slow evolution of fractional calculus over more than 250 years (from 1695 to 1974) has mainly seen mathematicians and physicists applying fractional calculus to investigate deformation and elasticity. This section briefly summarizes models for linear elasticity, viscoelasticity and poroviscoelasticity, based on fractional calculus.

4.1 Generic Models for Viscoelasticity

From the brief survey of historical developments in research on elasticity and related topics in section 2.2, it can be seen that the early developments in viscoelastic materials with fractional calculus include the contributions of Gemant (1935), Blair and Coppen (1943), Blair (1944), Blair and Reiner (1951), Gerasimov (1948), Rabotnov (1948), Caputo (1966, 1969) and Caputo and Mainardi (1971a, b).

The classical models of linear viscoelasticity in section 3.2 can be written in the following more generic forms:

4.1.1 Rabotnov's Model

Rabotnov (1948) proposed a fractional model for the stress-strain relationship, based on the Abel operator. Zhuravkov and Romanova (2016) re-wrote Rabotnov's model as follows:

$$\sum_{i=0}^{n} a_i \left({}_0D_t^{i\alpha}\sigma(t) \right) = \sum_{k=0}^{m} b_j \left({}_0D_t^{k\alpha}\varepsilon(t) \right) \tag{11.28}$$

For inelastic materials with integer derivatives, Rabotnov's (1948) model is simplified as the following multi-term classic ordinary differential equation (ODE):

$$\sum_{i=0}^{n} a_i \frac{d^i\sigma(t)}{dt^i} = \sum_{k=0}^{m} b_k \frac{d^k\varepsilon(t)}{dt^k} \tag{11.29}$$

which generalizes the classical models in section 3.2. Rabotnov also investigated cases with different values of n and m in his model.

4.1.2 The Mainardi-Spada Model

A generic model based on fractional ODEs was presented by Mainardi and Spada (2011):

$$\sigma(t)+\sum_{k=1}^{p} a_k \frac{d^{\beta_k}\sigma(t)}{dt^{\beta_k}} = m+\sum_{k=1}^{p} b_k \frac{d^{\beta_k}\varepsilon(t)}{dt^{\beta_k}}, \beta_k = k+v-1 \tag{11.30}$$

with

$$\begin{cases} J(t) = J_{f_1} + \sum_{n}^{p} J_n[1 - E_v(-t/\tau_{e,f})^v] + J_{f_2}\dfrac{t^v}{\Gamma(1+v)} \\[3mm] G(t) = G_{f_1} + \sum_{n}^{p} G_n E_v(-t/\tau_{e,g})^v + G_{g_2}\dfrac{t^{-v}}{\Gamma(1-v)} \end{cases} \tag{11.31}$$

where E_v is the Mittag-Leffler function. The Mainardi-Spada model in Eq. (11.30) has specific properties which clearly define the relationship between creep compliance $J(t)$ and relaxation modulus $G(t)$ through Eq. (11.31).

Equation (11.30) generalizes all the classical models in section 3.2 by setting different values of orders β_k with different terms in the summation. It is also a fractional version of the Rabotnov model. Mainardi and Spada (2011) detailed the generalization of these fractional versions of the models. Zhuravkov and Romanova (2016) provided an extensive survey and comment on models based on fractional calculus used in mechanical problems, and devoted extensive discussions on creep and relaxation processes.

4.2 Special Forms of Fractional Models for Viscoelasticity and Poroviscoelasticity

Here, we discuss the Caputo-Mainardi fractional model, which generalizes other classical models with physical parameters clearly defined.

As a special form of Eq. (11.30), the Caputo-Mainardi model (Caputo and Mainardi 1971a) is the fractional generalization of Zener's model:

$$\sigma(t) + \frac{1}{\beta}\frac{d^\lambda \sigma(t)}{dt^\lambda} = m\varepsilon(t) + \frac{1}{\alpha}\frac{d^\lambda \varepsilon(t)}{dt^\lambda} \tag{11.32}$$

where $0 < \lambda \leq 1$ is the order of fractional derivatives. Eq. (11.32) can be written as:

$$\sigma(t) + t_\sigma \frac{d^\lambda \sigma(t)}{dt^\lambda} = m\varepsilon(t) + t_\varepsilon \frac{d^\lambda \varepsilon(t)}{dt^\lambda} \tag{11.33}$$

with t_σ and t_ε as retardation time and relaxation time respectively.

For $\beta = \alpha$, this model describes purely elastic materials under the condition that $\sigma(t) = 0$ and $\varepsilon(t) = 0$ at $t = 0$, that is, under the zero initial condition. For $\frac{\beta - \alpha}{\alpha} \ll 1$, the model describes 'nearly elastic' materials (Caputo and Mainardi 1971a).

The Caputo-Mainardi model in Eq. (11.33) generalizes most of the fractional versions of the models discussed in section 3.2. The fractional generalization of the classical models in section 3.2 is achieved through the substitutions: $\frac{d^\lambda \varepsilon(t)}{dt^\lambda}$ for $\frac{d\varepsilon(t)}{dt}$ and $\frac{d^\lambda \sigma(t)}{dt^\lambda}$ for $\frac{d\sigma(t)}{dt}$ in Eq. (11.33). Debnath (2003b) discussed the models generalized by the Caputo-Mainardi model, which include the following models:

The classic Stokes law: $a = b$;
Fractional-order Newton model or fractional Blair model: $a = m = 0$, $b = E$;
Fractional Voigt model: $a = 0$;
Fractional Maxwell model: $m = 0$, and
The Bagley-Torvik model (Bagley and Torvik 1983), etc.

Experimental findings by others (Bagley and Torvik 1983, Rogers 1983, Koeller 1984) confirmed that the Caputo-Mainardi model in Eq. (11.33) is appropriate for most real materials.

Schiessel et al. (1995) presented a number of similar fractional viscoelastic models, with no reference to the generic fractional models reported by Caputo and Mainardi (1971a) about a quarter of a century earlier.

4.3 Physical Interpretation of Fractional Models with Memory

The explanations of classical models for elasticity are understood in the context of mechanics of elastic materials. The physical interpretation of fractional models for elasticity is another issue to be clarified.

Tarasov (2013) provided an interpretation of the evolution of a dynamic relationship between two quantities, such as the strain-stress relationship, in terms of fractional integrals. His explanation can be applied to models for linear elasticity as well. The following integral equation is similar to the strain-stress relation in Eq. (11.14), with a zero initial condition:

$$\varepsilon(t) = \int_0^t J(t-\tau) \frac{d\sigma(\tau)}{d\tau} d\tau \qquad (11.34)$$

which describes the evolution of a dynamic relationship between strain and stress as a result of the power-law memory of the form:

$$J(t-\tau) = J_0(t-\tau)^{\beta-1} \qquad (11.35)$$

where J_0 is a parameter. When Eq. (11.35) is used in Eq. (11.34), the following integral equation results:

$$\varepsilon(t) = J_0 \int_0^t (t-\tau)^{\beta-1} \frac{d\sigma(\tau)}{d\tau} d\tau, \qquad 0 < \beta < 1 \qquad (11.36)$$

which is comparable to Eq. (11.14) and is, in fact, equivalent to the Caputo left-sided fractional derivatives given by (Kilbas et al. 2006):

$$\varepsilon(t) = \frac{\lambda}{\Gamma(\beta)} \int_0^t (t-\tau)^{\beta-1} \frac{d\sigma(\tau)}{d\tau} d\tau, \qquad 0 < \beta < 1 \qquad (11.37)$$

where $\lambda = \Gamma(\beta)J_0$.

Equations (11.36) and (11.37) have very clear physical implications: the power-law memory function as a kernel in the integral equation explains the delayed strain response to stress. The identity $\lambda = \Gamma(\beta)J_0$ also provides a means to determine the value of β.

As demonstrated for other systems, the Laplace transform of Eq. (11.34) yields a simple operation which can be used to derive the parameters (such as λ, J_0 and β), based on observed data for $\varepsilon(t)$ and $\sigma(t)$. The Laplace transform of Eq. (11.34) subject to a zero initial condition yields:

$$\tilde{\varepsilon}(s) = \tilde{J}(s)s\tilde{\sigma}(s) \qquad (11.38)$$

or

$$\tilde{J}(s) = \frac{\tilde{\varepsilon}(s)}{s\tilde{\sigma}(s)} \qquad (11.39)$$

where $\tilde{\varepsilon}(s)$, $\tilde{J}(s)$ and $\tilde{\sigma}(s)$ are the Laplace transforms of $\varepsilon(t)$, $J(t)$ and $\sigma(t)$ respectively, and s is the Laplace transform variable.

Equation (11.38) results from the identity of Laplace transform $\mathscr{L} \frac{d\sigma(\tau)}{d\tau} = s\sigma(s) - \sigma(0)$ subject to the zero initial condition $\sigma(0) = 0$. Similarly, Eqs. (11.36) and (11.37) can also be solved using the Laplace transform method.

The above analysis is for continuous systems. A parallel analysis can be done for discrete systems with memory, such as the approach by Tarasov (2013), which can be applied to anomalous elasticity in discrete times.

As with other models based on fPDEs, the order of fractional derivatives can be a function of different variables and/or parameters (Colombaro et al. 2018), and other ideas and methods applied to fPDEs for other processes can be equally employed to investigate elastic, plastic and poroviscoelastic materials, such as non-Newtonian time-varying viscosity (Pandey and Hom 2016) and viscoelastic materials under external forces or fields such an electric field (Shiga 1998).

5. Summary

This chapter briefly overviews the concepts of elasticity and their extension to poroviscoelasticity and related applications in the description of porous media and in soil mechanics or geomechanics. The inception of thinking in elasticity of materials started in the 17th century, and has evolved over time until now when fDEs and fPDEs are used for more detailed investigation of poroviscoelasticity and simplified properties.

One issue is the invisible boundary formed between soil physics and soil mechanics, which impedes the communication in methodologies between traditional soil physics, groundwater hydrology and soil mechanics since the 1930s (Philip 1974). To overcome this problem, one option is to consider how to incorporate the stress-strain and energy relations in Eq. (11.7) into the general equation of conservation of mass in Eq. (3.49).

Another issue discussed in this chapter approaches to viscoelastic materials with linear elasticity which has been dominant in literature, which do not consider the non-linear potential energy associated with the stress-strain relation. The choice of linear and non-linear methods depends on the scale of the problem in question: linear models are more appropriate at small scales while non-linear models are more suitable at large scales. With this in mind, physical processes in the field addressed by soil science and groundwater hydrology are more important at large scales, so the non-linear models are clearly more appropriate. This situation requires either the modification of the existing models or development of new models.

Bibliography

Ababou, R. and L.W. Gelhar. 1990. Self-similar randomness and spectral conditioning: Analysis of scale effects in subsurface hydrology. pp. 393–428. *In*: J.H. Cushman [ed.]. Dynamics of Fluids in Hierarchical Porous Media. Academic Press, London, England.

Abdellaoui, B., I. Peral and M. Walias. 2015. Some existence and regularity results for porous media and fast diffusion equations with a gradient term. *Trans. Amer. Math. Soc.* 367: 4757–4791.

Abdel-Rehim, E.A. and R. Gorenflo. 2008. Simulation of the continuous time random walk of the space-fractional diffusion equations. *J. Compt. Appl. Math.* 222: 274–283.

Abel, N.H. 1823. Solution de quelques probléms à l'aide d'intégrales définites, Oeuvres Complétes. 1: 16–18. Grondahl Christiania, Norway, 1881. First appeared in *Mag. Naturvidenkaberne*.

Abramowitz, M. and I. Stegun. 1965. Handbook of Mathematical Functions. Dover Publ. New York, USA.

Adomian, G. 1983. Stochastic Systems. Acad. Press, New York, USA.

Agarwal, R.P., D. O'Regan and S. Staněk. 2010. Positive solutions for Dirichlet problems of singular nonlinear fractional differential equations. *J. Math. Anal. Appl.* 371: 57–68.

Agnese, C., G. Baiamonte and C. Corrao. 2007. Overland flow generation on hillslopes of complex topography: analytical solutions. *Hydrol. Proc.* 21: 1308–1317.

Al-Bassam, M.A. 1965. Some existence theorems on differential equations of generalized order. *J. Reine Angew. Math.* 218(1): 7–78.

Al-Bassam, M.A. and Yu. F. Luchko. 1995. On generalized fractional calculus and its applications to the solution of integro-differential equations. *J. Fract. Calc.* 7: 69–88.

Alexander, S. and R. Orbach. 1982. Density of states on fractals: <fractons>. *J. Physique. Lett.* 43(17): 625–631.

Alkahtani, B.S.T. and A. Atangana. 2016. Controlling the wave movement on the surface of shallow water with the Caputo-Fabrizio derivative with fractional order. *Chaos, Solitons & Fractals* 89: 539–546.

Andries, E., S. Umarov and S. Steinberg. 2006. Monte Carlo random walk simulations based on distributed order differential equations with application to cell biology. *Fract. Calc. Appl. Anal.* 9(4): 351–369.

Atanacković, T.M., L. Oparnica and S. Pilipović. 2009a. Distributional framework for solving fractional differential equations. *Integral Transforms & Special Functions* 20(3-4): 215–222.

Atanacković, T.M., S. Pilipović and D. Zurica. 2009b. Time distributed-order diffusion-wave equation. II. Applications of Laplace and Fourier transformations. *Proc. Royal Soc. A.* 465: 1893–1917.

Atanacković, T.M., S. Pilipović, B. Stanković and D. Zurica. 2014. Fractional Calculus with Applications in Mechanics. ISTE/Wiley, London, England.

Atangana, A. and N. Bildik. 2013. The use of fractional order derivative to predict the groundwater flow. *Math. Prob. Eng.* 2013(543026): 1–9.

Atangana, A. and A. Kilicman. 2013. Analytical solutions of the space-time fractional derivative of advection-dispersion equation. *Math. Prob. Eng.* 853127: 1–9.

Atangana, A. and A. Secer. 2013. A note on fractional order derivatives and table of fractional derivatives of some special functions. *Abstract Appl. Anal.* 2013: 279681.

Atangana, A. 2014. Drawdown in prolate spheroidal-spherical coordinates obtained via Green's function and perturbation methods. *Commun. Nonlinear Sci. Numer. Simulat.* 19(5): 1259–1269.

Atangana, A. and P.D. Vermeulen. 2014. Analytical solutions of a space-time fractional derivative of groundwater flow equation. *Abstract & Appl. Anal.* (381753): 1–11.

Atangana, A. 2016. Derivative with two fractional orders: A new avenue of investigation toward revolution in fractional calculus. *Euro. Phys. J. Plus.* 131(10): 373.

Atangana, A. and B.S.T. Alkahtani. 2016. New model of groundwater flowing within a confine aquifer: application of Caputo-Fabrizio derivative. *Arabian J. Geosci.* 9(1): 1–6.

Atangana, A. and D. Baleanu. 2016. New fractional derivatives with non-local and non-singular kernel: Theory and application to heat transfer model. *Thermal Sci.* 20(2): 763–769, arXiv preprint arXiv: 1602.03408, 2016-arxiv.org.

Atangana, A. 2018. Fractional Operators with Constant and Variable Order with Application to Geo-Hydrology. Acad. Press, London, England.

Athreya, K.B., D. McDonald and P. Ney. 1978. Limit theorems for semi-Markov processes and renewal theory for Markov chains. *Annals Probab.* 6(5): 788–797.

Atkinson, K. 1974. An existence theorem for Abel integral equations. *SIAM J. Math. Anal.* 5(5): 729–736.

Bachu, S. 1995. Flow of variable-density formation water in deep sloping aquifers: review of methods of representation with case studies. *J. Hydrol.* 164: 19–38.

Bachu, S. and K. Michael. 2002. Flow of variable-density formation water in deep sloping aquifers: minimizing the error in representation and analysis when using hydraulic-head distributions. *J. Hydrol.* 259: 49–65.

Baeumer, B., D.A. Benson and M.M. Meerschaert. 2005. Advection and dispersion in time and space. *Phys. A.* 350: 245–262.

Baeumer, B. and M.M. Meerschaert. 2007. Fractional diffusion with two time scales. *Phys. A.* 373: 237–251.

Bagley, R.L. and P.J. Torvik. 1983. A theoretical basis for the application of fractional calculus to viscoelasticity. *J. Rheol.* 27(3): 201–210.

Bagley, R.L. and P.J. Torvik. 1986. On the fractional calculus model of viscoelastic behavior. *J. Rheol.* 30: 133–155.

Bagley, R.L. 1991. The thermorheologically complex material. *Int. J. Eng. Sci.* 29(7): 797–806.

Bagley, R.L. and P.J. Torvik. 2000. On the existence of the order domain and the solution pf distributed order equations. *Intl. J. Appl. Math.* 2: 865–882, 965–987.

Balakrishnan, V. 1960. Fractional powers of closed operators and semi-groups generated by them. *Pac. J. Math.* 10: 419–437.

Balescu, R. 1995. Anomalous transport in turbulent plasmas and continuous time random walks. *Phys. Rev. E.* 51(5): 4807–4822.

Baleanu, D., G.-C. Wu and J.-S. Duan. 2014. Some analytical techniques in fractional calculus: Realities and challenges. pp. 35–62. *In*: J.A.T. Machado, D. Baleanu and A.C. Luo [ed.]. Discontinuity and Complexity in Nonlinear Systems. Springer, Switzerland.

Barenblatt, G.I., I. Zheltov and I. Kochina. 1960. Basic concepts in the theory of seepage of homogeneous liquids in fissured rocks. *J. Appl. Math. Mech.* 24: 852–864.

Barenblatt, G.I., V.M. Entov and V.M. Ryzhik. 1990. Theory of Fluid Flows through Natural Rocks. Kluwer, Dordrecht, The Netherlands.

Barenblatt, G.I., M. Bertsch, A.E. Chertock and V.M. Prostokishin. 2000. Self-similar intermediate asymptotics for a degenerate parabolic filtration-absorption equation. *Proc. Nat. Acad. Sci.* 97(18): 9844–9848.

Barkai, E. 2002. CTRW pathways to the fractional diffusion equation. *Chem. Phys.* 284: 13–27.

Barkai, E. and Y.-C. Cheng. 2003. Aging continuous time random walks. *J. Chem. Phys.* 118(14): 6167–6178.

Barnes, E.W. 1908. A new development of the theory of the hypergeometric functions. *Proc. London Math. Soc.* 2(6): 141–177.

Barnes, E.W. 1910. A transformation of generalized hypergeometric series. *Quart. J. Math.* 41: 136–140.

Baron, T. 1952. Generalized graphical method for the design of fixed bed catalytic reactors. *Chem. Eng. Progress* 48(3): 118–124.

Barrett, J.H. 1954. Differential equations of non-integer order. *Can. J. Math.* 6: 529–541.

Barry, D.A. and G. Sposito. 1989. Analytical solution of a convection-dispersion model with time-dependent transport coefficient. *Water Resour. Res.* 25(12): 2407–2416.

Barry, D.A. and G.C. Sander. 1991. Exact solutions for water infiltration with an arbitrary surface flux or nonlinear solute adsorption. *Water Resour. Res.* 27: 2667–2680.

Barry, D.A., I.G. Lisle, L. Li, H. Prommer, J.-Y. Parlange, G.C. Sander et al. 2001. Similitude applied to centrifugal scaling of unsaturated flow. *Water Resour. Res.* 37(10): 2471–2479.

Baveye, P., C.W. Boast and J.W. Giráldez. 1989. Use of referential coordinates in deforming soils. *Soil Sci. Soc. Am. J.* 53: 1338–1343.

Bear, J. 1972. Dynamics of Fluids in Porous Media. Dover Publ., New York, USA.

Bear, J. 1979. Hydraulics of Groundwater. McGraw-Hill, New York, USA.

Bear, J. and A. Verruijt. 1992. Modeling Groundwater Flow and Pollution. D. Reidel Publ. Co. Dordrecht, The Netherlands.

Bear, J., A.H.-D. Chen, S. Sorek, D. Ouazar and I. Herrera. 1999. Seawater Intrusion in Coastal Aquifers: Concepts, Methods and Practices. Kluwer Acad. Publ. Dordrecht, The Netherlands.

Becker-Kern, P., M.M. Meerschaert and H.P. Scheffler. 2004. Limit theorem for continuous-time random walks with two time scales. *J. Appl. Prob.* 41(2): 455–466.

Bemer, E., M. Boutéca, O. Vincké, H. Hotei and O. Ozanam. 2001. Poromechanics: From linear to nonlinear poroelasticity and poroviscoelasticity. *Oil & Gas Sci. Tech.—Rev. IFP* 56(6): 531–544.

Benano-Melly, L.B., J.-P. Caltagirone, B. Faissat, F. Montel and P. Costeseque. 2001. Modeling Soret coefficient measurement experiments in porous media considering thermal and solutal convection. *Internl. J. Heat Mass Transfer* 44: 1285–1297.

Benson, D.A. 1998. A Fractional Advection-dispersion Equation: Development and Application. Ph.D. Thesis, University of Nevada, Reno, Nevada, USA.

Benson, D.A., S.W. Wheatcraft and M.M. Meerschaert. 2000a. Application of a fractional advection-dispersion equation. *Water Resour. Res.* 36(6): 1403–1412.

Benson, D.A., S.W. Wheatcraft and M.M. Meerschaert. 2000b. The fractional order governing equation of Lévy motion. *Water Resour. Res.* 36(6): 1413–1423.

Benson, D.A., R. Schumer, M.M. Meerschaert and S.W. Wheatcraft. 2001. Fractional dispersion, Lévy motion, and the MADE tracer tests. *Transport in Porous Media* 42: 211–240.

Benson, D.A., C. Tadjeran, M.M. Meerschaert, I. Farnham and G. Pohll. 2004. Radial fractional-order dispersion through fractured rock. *Water Resour. Res.* 40: W12416.

Benson, D.A. and M.M. Meerschaert. 2009. A simple and efficient random walk solution of multi-rate mobile-immobile mass transport equations. *Adv. Water Resour.* 32: 532–539.

Benson, D.A., M.M. Meerschaert and J. Revielle. 2013. Fractional calculus in hydrologic modeling: A numerical perspective. *Adv. Water Resour.* 51: 479–497.

Berkowitz, B. and H. Scher. 1995. On characterization of anomalous dispersion in porous and fractured media. *Water Resour. Res.* 31(6): 1461–1466.

Berkowitz, B., J. Klafter, R. Metzler and H. Scher. 2002. Physical pictures of transport in heterogeneous media: Advection-dispersion, random-walk, and fractional derivative formulations. *Water Resour. Res.* 38(10): 1191, doi:10.1029/2001WR001030.

Berkowitz, B., A. Cortis, M. Dentz and H. Scher. 2006. Modelling non-Fickian transport in geological formation as a continuous time random walk. *Rev. Geophys.* 44: RG2003, 1–49.

Bernoulli, Jacques. 1705. Véritable hypothèse de la résistance des solides, avec la démonstration de la courbure des corps qui font ressort. *Mémoires de mathématique et de physique de l'Académie royale des sciences.* https://hal.archives-ouvertes.fr/ads-00108004/document.

Bhalekar, S. and V. Daftardar-Geiji. 2013. Corrigendum. *Appl. Math. Comput.* 219: 8413–8415.

Bhattarai, S.P., N. Su and D.J. Midmore. 2005. Oxygation unlocks yield potentials of crops in oxygen limited soil environments. *Adv. Agric.* 88: 313–377.

Bhattarai, S.P., L. Pendergast and D.J. Midmore. 2006. Root aeration improves yield and water use efficiency of tomato in heavy clay and saline soils. *Sci. Horticulture* 108(3): 278–288.

Binley, A., S.S. Hubbard, J.A. Huisman, A. Revil, D.A. Robinson, K. Singha et al. 2015. The emergence of hydrogeophysics for improved understanding of subsurface processes over multiple scales. *Water Resour. Res.* 51: 3837–3866.

Biot, M.A. 1935. Le problème de la Consolidation des matières argile uses sous une charge. *Ann. Soc. Sci.* Bruxelles, B55: 110–113.

Biot, M.A. 1939. Non-linear theory of elasticity and the linearized case for a body under initial stress. *Philos. Mag.* 27: 468–489.

Biot, M.A. 1941. General theory of three-dimensional consolidation. *J. Appl. Phys.* 12: 155–164.

Biot, M.A. 1956a. The theory of propagation of elastic waves in a fluid-saturated porous solid. I. Low-frequency range. *J. Appl. Phys.* 26: 182–185.

Biot, M.A. 1956b. The theory of propagation of elastic waves in a fluid-saturated porous solid. II. High-frequency range. *J. Acoustical Soc. Amer.* 28: 179–191.

Biot, M.A. 1956c. Thermoelasticity and irreversible thermodynamics. *J. Appl. Phys.* 27: 240–253.

Biot, M.A. 1973. Nonlinear and semilinear rheology of porous media. *J. Geophys. Res.* 78(23): 4924–4937.

Bird, N.R.A., E. Perrier and M. Rieu. 2000. The water retention function for a model of soil structure with pore and solid fractal distributions. *Eur. J. Soil Sci.* 51: 55–63.

Bird, R.B., W.E. Stewart and E.N. Lightfoot. 1960. Transport Phenomena. Wiley, New York, USA.

Biswas, A. 1967. Hydrologic engineering prior to 600 B.C. *J. Hydraul. Div., Proc. Amer. Soc. Civil Engrs.* 93(HY5): 115–135.

Blair, G.W.S. and F.M.V. Coppen. 1943. The estimation of firmness in soft materials. *Amer. J. Psychol.* 56(2): 234–246.

Blair, G.W.S. 1944. Analytical and integrative aspects of the stress-strain-time problem. *J. Sci. Instrum.* 21: 80–84.

Blair, G.W.S. 1947. The role of psychophysics in rheology. *J. Colloid Sci.* 2(1): 21–32.

Blair, G.W.S. and N. Reiner. 1951. The rheological law underlying the Nutting equation. *Appl. Sci. Res.* 2: 225–234.

Blake, F.C. 1923. The resistance of packing to fluid flow. *Trans. Amer. Inst. Chem. Eng.* 14: 415–421.

Bland, D.R. 1960. The Theory of Linear Viscoelasticity. Pergamon Press, LCCCN-59-14489, Oxford, England.

Blöschl, G. and M. Sivapalan. 1995. Scale issues in hydrological modelling: a review. *Hydrol. Proc.* 9: 251–290.

Bôcher, M. 1909. Introduction to the Study of Integral Equations. *Cambridge Tract.* No. 10, Cambridge.

Bochner, S. 1949. Diffusion equation and stochastics processes. *Proc. Nat. Acad. Sci.* 35: 368–370.

Bolt, G.H. 1979. Soil Chemistry, Physicochemical Models. Elsevier, New York, USA.

Bond, W.J. and P.J. Wierenga. 1990. Immobile water during solute transport in unsaturated sand columns. *Water Resour. Res.* 26(10): 2475–2481.

Bonnet, M. 1982. Methodologie de Modeles de Simulation en Hydrologie. Document 34. Bureau de Recherches Geologique et Miniers, Orleans, France.

Bouchaud, J.-P. and A. Georges. 1990. Anomalous diffusion in disordered media: Statistical mechanics, models, and physical applications. *Phys. Reports.* 4&5: 127–293.

Bourne, D.E. and P.C. Kendall. 1977. Vector Analysis and Cartesian Tensors. 2nd ed. Chapman & Hall, London.

Boussinesq, M.J. 1904. Recherches théoriques sur l'écoulement des nappes d'eau infiltrées dans le sol et sur débit de sources. *J. Math. Pure Appl.* 10: 5–78.

Bouwer, H. 1989. The Bouwer and Rice slug test—An update. *Groundwater* 27(3): 304–309.

Boyce, W.E. and R.C. DoPrima. 1997. Elementary Differential Equations and Boundary Value Problems. 6th ed., Wiley, New York, USA.

Bras, R.L. 1990. Hydrology: An Introduction to Hydrologic Science. Addison-Wesley, Reading, Massachusetts, USA.

Bredehoeff, J.D. 1967. Response of well-aquifer systems to earth tides. *J. Geophys. Res.* 72(12): 3075–3087.

Briciu, A.-E. 2018. Diurnal, semidiurnal, and fortnightly tidal components in orthotidal proglacial rivers. *Environ. Monit. Assess.* 190(160): 1–18.

Briciu, A.-E., D. Mihaila, D.I. Oprea, P.-I. Bistricean and L.G. Lazurca. 2018. Orthotidal signal in the electrical conductivity of an inland river. *Environ. Monit. Assess.* 190(280): 1–15.

Bridge, B.J. and N. Collis-George. 1973. An experimental study of vertical infiltration into a structurally unstable swelling soil, with particular reference to the infiltration throttle. *Aust. J. Soil Res.* 11: 121–132.

Broadbridge, P. 1987. Integrable flow equations that incorporate spatial heterogeneity. *Transport in Porous Media* 2: 129–144.

Broadbridge, P. and I. White. 1987. Time-to-ponding: Comparison of analytic, quasi-analytic, and approximate predictions. *Water Resour. Res.* 23(12): 2302–2310.

Broadbridge, P. and I. White. 1988a. Constant rate rainfall infiltration: A versatile nonlinear model. 1. Analytical solution. *Water Resour. Res.* 24(1): 145–154.

Broadbridge, P. and I. White. 1988b. Constant rate rainfall infiltration: A versatile nonlinear model. 2. Applications of solutions. *Water Resour. Res.* 24(1): 155–162.

Broadbridge, P. 1988. Integrable forms of the one-dimensional flow equation for unsaturated heterogeneous porous media. *J. Math. Phys.* 29(3): 622–627.

Broadbridge, P. 1990. Infiltration in saturated swelling soils and slurries: Exact solutions for constant supply rate. *Soil Sci.* 149(1): 13–22.

Broadbridge, P. and C. Rogers. 1990. Exact solutions for vertical drainage and redistribution in soils. *J. Eng. Math.* 24: 25–43.

Broadbridge, P. and C. Rogers. 1993. On a nonlinear reaction diffusion boundary-value problem: Application of a Lie-Backlund symmetry. *J. Austral. Math. Soc.* (B) 34: 318–332.

Broadbridge, P., M.P. Edwards and J.E. Kearton. 1996. Closed-form solutions for unsaturated flow variable flux boundary conditions. *Adv. Wafer Resour.* 19(4): 207–213.

Bronshtein, I.N. and K.A. Semendyayev. 1979. Handbook of Mathematics. Verlag Harri Deutsch, Van Nostrand Reinhold Co., New York, USA.

Bronshtein, I.N. and K.A. Semendyayev. 1985. Handbook of Mathematics. Verlag Harri Deutsch, New York, USA.

Brooks, R.H. and A.T. Corey. 1964. Hydraulic Properties of Porous Media. Hydrol. Pap. 3, Colorado State Univ., Fort Collins, Colorado, USA.

Brooks, R.H. and A.T. Corey. 1966. Properties of porous media affecting fluid flow. *J. Irrig. Drain. Div., ASCE* 92(IR2): 61–88.

Buckingham, W. 1921. On plastic flow through capillary tubes. *Proc. Am. Soc. Test. Mater.* 21: 1154–1161.

Buckley, S.E. and M.C. Leverett. 1942. Mechanics of fluid displacement in sands. *Trans. AIME* 142: 107–116.

Buckwar, E. and Y. Luchko. 1998. Invariance of a partial differential equation of fractional order under the Lie group scaling transformations. *J. Math. Anal. Appl.* 227: 81–97.

Burbey, T.J. 2001. Stress-strain analyses for aquifer-system characterization. *Ground Water.* 39(1): 128–136.

Burdine, N.T. 1953. Relative permeability calculations from pore-size distribution data. *Petr. Trans. Amer. Inst. Mining Metall. Eng.* 198: 71–77.

Burnett, D. 1935. The distribution of velocities in a slightly non-uniform gas. *Proc. Lond. Math. Soc.* 39: 385–430.

Burnett, D. 1936. The distribution of molecular velocities and the mean motion in a non-uniform gas. *Proc. Lond. Math. Soc.* 40: 382–435.

Butera, S. and M. Di Paola. 2014. A physically based connection between fractional calculus and fractal geometry. *Ann. Phys.* 350: 146–158.

Caceres, M.O. 1986. Coupled generalized master equations for Brownian anisotropically scattered. *Phys. Rev. A.* 33(1): 647–651.

Cajori, F. 1923. The history of notations of the calculus. *Annals Math.* 2nd Ser. 25(1): 1–46.

Cajori, F. 1928. The early history of partial differential equations and of partial differentiation and integration. *Amer. Math. Month.* 35(9): 459–467.

Campbell, S.Y. and J.-Y. Parlange. 1984. Overland flow on converging and diverging surfaces— Assessment of numerical schemes. *J. Hydrol.* 70: 265–275.

Campbell, S.Y., J.-Y. Parlange and C.W. Rose. 1984. Overland flow on converging and diverging surfaces—Kinematic model and similarity solutions. *J. Hydrol.* 67: 367–374.

Camporese, M., S. Ferraris, M. Putti, P. Salandin and P. Teatini. 2006. Hydrological modeling in swelling/ shrinking peat soils. *Water Resour. Res.* 42: W06420.

Caputo, M. 1966. Linear models of dissipation whose Q is almost frequency independent. *Annali Geofis.* 19: 383–393.

Caputo, M. 1967. Linear models of dissipation whose Q is almost frequency independent–II. *Geophys. J. R. Astr. Soc.* 13(5): 529–539.

Caputo, M. 1969. Elasticit`a e Dissipazione. Zanichelli, Bologna (in Italian), Italy.

Caputo, M. and F. Mainardi. 1971a. A new dissipation model based on memory mechanism. *Pure & Appl. Geophys.* (PAGEOPH) 91(1): 134–147. [Reprinted in 2007. *Fract. Calc. Appl. Anal.* 10(3): 309–324].

Caputo, M. and F. Mainardi. 1971b. Linear models of dissipation in anelastic solids. *Riv. Nuovo Cimento.* 1(1): 161–198.

Caputo, M. 1995. Mean fractional order derivatives. Differential equations and filters. *Annals Univ. Ferrara–Sez*, VII – SC. Mat., XLI: 73–84.

Caputo, M. 1999. Diffusion of fluids in porous media with memory. *Geothermics* 29: 113–130.

Caputo, M. 2000. Models of flux in porous media with memory. *Water Resour. Res.* 36(1): 693–705.

Caputo, M. 2001. Distributed order differential equations modelling dielectric induction and diffusion. *Frac. Calc. Appl. Anal.* 4(4): 421–442.

Caputo, M. and M. Fabrizio. 2015. A new definition of fractional derivative without singular kernel. *Progr. Fract. Differ. Appl.* 1(2): 73–85.

Carman, P.C. 1937. Fluid flow through granular beds. *Trans. Inst. Chem. Eng.* 15: S32–S48.

Carman, P.C. 1939. Permeability of saturated sands, soils and clays. *J. Agric. Sci.* 29(02): 262–273.

Carpinteri, A., P. Cornetti and A. Sapora. 2011. A fractional calculus approach to nonlocal elasticity. *Eur. Phys. J. Special Topics* 193: 193–204.

Carslaw, H.S. and J.C. Jaeger. 1947. Conduction of Heat in Solids. Oxford University Press, Oxford, England.

Carslaw, H.S. and J.C. Jaeger. 1959. Conduction of Heat in Solids. 2nd ed., Oxford University Press, Oxford, England.

Cartledge, P. [ed.]. 1998. Cambridge Illustrated History of Ancient Greece. Cambridge Univ. Press, England.

Caserta, A., R. Garra and E. Salusti. 2016. Application of the fractional conservation of mass to gas flow diffusivity equation in heterogeneous porous media. arXiv: 1611.01695v1 [physics-geo-ph] 5 Nov. 2016.

Cauchy, A.-L. 1823. Recherches sur l'équilibre et le mouvement intérieur des corps solides ou fluides, élastiques ou non élastiques. *Bull. Soc. Philomath.* 9–13 = Oeuvres 2(2): 300–304.

Cauchy, A.-L. 1827a. De la pression ou tension dans un corps solide. *Ex. de math.* 2: 42–56 = *Oeuvres* 7(2): 60–78.

Cauchy, A.-L. 1827b. Sur la condensation et la dilatation des corps solides. *Ex. de math.* 2: 60–69 = *Oeuvres* 7(2): 82–83.

Cauchy, A.-L. 1828. Sur quelques théorèmes relatifs à la condensation ou à la dilatation des corps. *Ex. de math.* 3: 237–244 = *Oeuvres* 8(2): 278–287.

Chak, A.M. 1967. A generalization of the Mittag-Leffler function. *Mat. vesnik* (*Bulletin Math.*). 4(19): 257–262.

Chandrasekhar, S. 1943. Stochastic problems in physics and astronomy. *Rev. Modern Phys.* 15(1): 1–89.

Chang, Y.-C. and H.-D. Yeh. 2007. Analytical solution for groundwater flow in an anisotropic sloping aquifer with arbitrarily located multiwells. *J. Hydrol.* 347: 143–152.

Chaves, A.S. 1998. A fractional diffusion equation to describe Levy flights. *Phys. Lett.* 239: 13–16.

Chechkin, A., R. Gorenflo and I.M. Sokolov. 2002. Retarding subdiffusion and acceleration superdiffusion governed by distributed-order fractional diffusion equation. *Phys. Rev. E.* 66: 046129, 1–7.

Chechkin, A.V., R. Gorenflo, I.M. Sokolov and V.Yu. Gonchar. 2003. Distributed order time fractional diffusion equation. *Frac. Calc. Appl. Anal.* 6(3): 259–279.

Chechkin, A., R. Gorenflo and I.M. Sokolov. 2005. Fractional diffusion in inhomogeneous media. *J. Phys. A: Math. Gen.* 38: L679–L684.

Chechkin, A., M. Hofmann and I. Sokolov. 2009. Continuous-time random walk with correlated waiting times. *Phys. Rev. E.* 80: 031112, 1–8.

Chechkin, A.V., H. Kantz and R. Metzler. 2017. Ageing effects in ultraslow continuous time random walks. *Eur. Phys. J. B.* 90: 205.

Chen, J., F. Liu, V. Anh, S. Shen, Q. Liu and C. Liao. 2012. The analytical solution and numerical solution of the fractional diffusion-wave equation with sampling. *Appl. Math. Comput.* 219: 1737–1748.

Chen, Z.X. 1988. Some invariant solutions to two-phase displacement problems including capillary effect. *Soc. Petroleum Eng. J.* 5: 691–700.

Childs, E.C. 1969. An Introduction to the Physical Basis of Soil Water Phenomena. Wiley, London.

Chow, V.T., D.R. Maidment and L.W. Mays. 1988. Applied Hydrology. McGraw-Hill, New York, USA.

Christakos, G. 2012. Random Field Models in Earth Sciences. Dover Publ., Inc., New York.

Chrysikopoulos, C.V., P.V. Roberts and P.K. Kitanidis. 1990. One-dimensional solute transport in porous media with partial well-to-well recirculation: Application to field experiments. *Water Resour. Res.* 26(6): 1189–1195.

Chukbar, K.V. 1995. Stochastic transport and fractional derivatives. *J. Exp. Theor. Phys.* 81(5): 1025–1029.

Cihan, A., E. Perfect and J.S. Tyner. 2007. Water retention models for scale-variant and scale-invariant drainage of mass prefractal porous media. *Vadose Zone J.* 6: 786–792.

Cinlar, E. 1969. On semi-Markov processes on arbitrary spaces. *Proc. Camb. Phil. Soc.* 66: 381–392.

Clebsch, A. 1883. Théorie de l'élasticité des corps solides, traduite par MM. Barré de Saint-Venant et Flamant, avec des Notes étendues de M. de Saint-Venant. Paris.

Coats, K.H. and B.D. Smith. 1964. Dead-end pore volume and dispersion in porous media. *Soc. Pet. Eng. J.* 4: 73–78.

Coimbra, C.F.M. 2003. Mechanics with variable-order differential operators. *Ann. Phys.* 12(11-12): 692–703.

Colombaro, I., R. Garra, A. Giusti and F. Mainardi. 2018. Scott Blair models with time-varying viscosity. *Appl. Math. Lett.* 86: 57–63.

Comolli, A. and M. Dentz. 2017. Anomalous dispersion in correlated porous media: a coupled continuous time random walk approach. *Eur. Phys. J. B.* 90: 166.

Compte, A. 1997. Continuous time random walks on moving fluids. *Phys. Rev. E.* 55(6): 6821–6831.

Compte, A., R. Metzler and J. Camacho. 1997. Biased continuous time random walks between parallel plates. *Phys. Rev. E.* 56(2): 1445–1454.

Cooper, Jr., H.H., J.D. Bredehoeft and I.S. Papadopulos. 1967. Response of a finite-diameter well to an instantaneous charge of water. *Water Resour. Res.* 3(1): 263–269.

Copson, E.T. 1965. Asymptotic Expansions. Cambridge Univ. Press, Cambridge, England.

Costa, F.S., J.A.P.F. Marao, J.C.A. Soares and E.C. de Oliveira. 2015. Similarity solution to fractional nonlinear space-time diffusion-wave equation. *J. Math. Phys.* 56: 033507.

Coulomb, C.A. 1776. Essai sur une application des règles de *Maximu et Minimis* á quelques Problèmes de Statique, relatifs á l'Architecture›, *Mém .... par divers savam.* 350–354.

Cox, D.R. 1967. Renewal Theory. Methuen, London.

Crank, J. 1975. The Mathematics of Diffusion. 2nd ed. Clarendon Press, Oxford.

Craven, T. and G. Csordas. 2006. The Fox-Wright functions and Laguerre multiplier sequences. *J. Math. Anal. Appl.* 314: 109–125.

Crofton, M.W. 1865. Question 1773. *Math. Questions with Their Solutions from the Educated Times.* 4: 71–72.

Culkin, S.L., K. Singha and F.D. Day-Lewis. 2008. Implications of rate-limited mass transfer for aquifer storage and recovery. *Ground Water.* 46(4): 591–605.

Culligan, P.J., J.V. Sinfield, W.E. Maas and D.G. Cory. 2001. Use of NMR relaxation times to differentiate mobile and immobile pore fractions in a wetland soil. *Water Resour. Res.* 37(3): 837–842.

Cushman, J.H. 1990. Dynamics of Fluids in Hierarchical Porous Media. Academic Press, London.

Cushman, J.H. 1991. On diffusion in fractal porous media. *Water Resour. Res.* 27(4): 643–644.

Cushman, J.H. and T.R. Ginn. 1993. Nonlocal dispersion in media with continuously evolving scales of heterogeneity. *Transp. Porous Media.* 13: 123–138.

Cushman, J.H., B.X. Hu and F. Deng. 1995. Nonlocal reactive transport with physical and chemical heterogeneity: Localization errors. *Water Resour. Res.* 31(9): 2219–2237.

Cushman, J.H. and T.R. Ginn. 2000. Fractional advection-dispersion equation: A classic mass balance with convolution-Fickian flux. *Water Resour. Res.* 36(12): 3763–3766.

Cvetkovic, V. 2012. A general memory function for modeling mass transfer in groundwater transport. *Water Resour. Res.* 48: W04528, 1–12.

Dagan, G. 1989. Flow and Transport in Porous Formations. Springer-Verlag, New York.

Darcy, H. 1856. Les fontaines publiques de la ville de Dijon. Dalmont, Paris.

Darrigol, O. 2005. Worlds of Flow. A History of hydrodynamics from the Bernoulis to Prandtl. Oxford Univ. Press, New York.

Daugherty, R.L., J.B. Franzini and E.J. Finnemore. 1989. Fluid Mechanics with Engineering Applications. McGraw-Hill, Singapore.

Davis, H.T. 1924. Fractional operations as applied to a class of Volterra integral equations. *Amer. J. Math.* 46(2): 95–109.

Davis, L.C. 1999. Model of magnetorheological elastomers. *J. Appl. Phys.* 85(6): 3342–3351.

Davis, P.J. 1965. Gamma function and related functions. Chapter 6 in: M. Abramowitz and I.A. Stegun [eds.]. Handbook of Mathematical Functions. Dover, New York.

Day, P.R. 1956. Dispersion of a moving salt-water boundary advancing through saturated sand. *Trans. AGU* 37: 595–601.

de Azevedo, E.N., P.L. de Sousa, R.E. de Sousa and M. Engelsberg. 2006. Concentration-dependent diffusivity and anomalous diffusion: A magnetic resonance imaging study of water ingress in porous zeolite. *Phys. Rev. E.* 73: 011204.

de Jong, J. 1958. Longitudinal and transverse diffusion in granular deposits. *Trans. AGU* 39(1): 67–74.

de Marsily, G. 1986. Quantitative Hydrogeology: Groundwater Hydrology for Engineers. Acad. Press, San Diego.

de Saint-Venant, B. 1871. Theory of unsteady water flow with application to river floods and to propagation of tides in river channels. *C.R. Acad. Sci. Paris* 73: 148–154; 237–240.

De Smedt, F. and P.J. Wierenga. 1979. A generalized solution for solute flow in soils with mobile and immobile water. *Water Resour. Res.* 15(5): 1137–1141.

de Vries, D.A. 1958. Simultaneous transfer of heat and moisture in porous media. *Trans. Amer. Geophys. Union.* 39(5): 909–916.

de Vries, D.A. 1987. The theory of heat and moisture transfer in porous media revisited. *Internl. J. Heat Mass Transfer* 30(7): 1343–1350.

Deans, H.A. 1963. A mathematical model for dispersion in the direction of flow in porous media. *Soc. Petrol. Eng. J.* Mar.: 49–52.

Debler, W.R. 1990. Fluid Mechanics and Fundamentals. Prentice-Hall, Eaglewood Cliffs, New Jersey.

Debnath, L. 2003a. Fractional integral and fractional differential equations in fluid mechanics. *Frac. Calc. Appl. Anal.* 6(2): 119–155.

Debnath, L. 2003b. Recent applications of fractional calculus to science and engineering. *Internl. J. Math. & Math. Sci.* 54: 3413–3442.

Debnath, L. 2004. A brief historical introduction to fractional calculus. *Intl. J. Math. Edu. Sci. Technol.* 35(4): 487–501.

Debnath, L. and D. Bhatta. 2007. Integral Transforms and Their Applications. 2nd ed., Chapman & Hall/ CRC, Boca Raton, Florida.

Deinert, M.R., A. Dather, J.-Y. Parlange and K.B. Cady. 2008. Capillary pressure in a porous medium with distinct pore surface and pore volume fractal dimensions. *Phys. Rev. E.* 77: 021203, 1–3.

Deng, Z.Q., V.P. Singh and L. Bengtsson. 2004. Numerical solution of fractional advection-dispersion equation. *J. Hydraul. Eng.* 130(5): 422–431.

Deng, Z.Q., J.M.P. de Lima and V.P. Singh. 2005. Fractional kinetic model for first flush of stormwater pollutants. *J. Environ. Eng.* 131(2): 232–241.

Deng, Z.Q., J.L.M.P. de Lima, M.I.P. de Lima and V.P. Singh. 2006. A fractional dispersion model for overland solute transport. *Water Resour. Res.* 42: W03416.

Dentz, M. and B. Berkowitz. 2003. Transport behaviour of a passive solute in continuous time random walks and multirate mass transfer. *Water Resour. Res.* 39(5): 1111, 1–20.

Dentz, M., P.K. Kang and T. Le Borgne. 2015. Continuous time random walks for non-local radial solute transport. *Adv. Water Reosur.* 82: 16–26.

Derrick, W.R. and S.I. Grossman. 1987. Introduction to Differential Equations with Boundary Value Problems. West Publ. Com., St. Paul.

Detournay, E. and A.H.-D. Cheng. 1993. Fundamentals of poroelasticity. pp. 113–171. *In*: C. Fairhurst [ed.]. Comprehensive Rock Engineering: Principles, Practice and Projects. Vol. II, Analysis and Design Method. Pergamon Press, Oxford.

Dhont, J.K.G., W. Wiegand, S. Duhr and D. Braun. 2007. Thermodiffusion of charged colloids: Single-particle diffusion. *Langmuir*. 23: 1674–1683.

Di Giuseppe, E., M. Moroni and M. Caputo. 2010. Flux in porous media with memory: Models and experiments. *Transp. Porous Med.* 83: 479–500.

Diethelm K. and A.D. Freed. 1999. On the solution of nonlinear fractional-order differential equations used in the modeling of viscoplasticity. pp. 217–224. *In*: F. Keil, W. Mackens, H. Voss and J. Werther [eds.]. Scientific Computing in Chemical Engineering II. Springer, Berlin, Heidelberg.

Dieulin, A. 1980. Propagation de pollution dans un aquifère alluvial: l'effet de parcours. Thesis, Paris School of Mines-Univ. Paris VI.

Ding, X.-L. and Y.-L. Jiang. 2013. Analytical solutions for the multi-term time-space fractional advection-diffusion equations with mixed boundary conditions. *Nonlinear Anal.: Real World Appl.* 12: 1026–1033.

Dirac, P.A.M. 1934. Discussion of the infinite distribution of electrons in the theory of the positron. *Math. Proc. Camb. Philos. Soc.* 30(2): 150–163.

Dirac, P.A.M. 1947. The Principles of Quantum Mechanics. 3rd ed., Oxford at Clarendon Press.

Dixon, R.M. and D.R. Linden. 1972. Soil air pressure and water infiltration under border irrigation. *Soil Sci. Soc. Amer. Proc.* 36: 948–953.

Djida, J.D., I. Area and A. Atangana. 2016. New numerical scheme of Atangana-Baleanu fractional integral: an application to groundwater flow within leaky aquifer. arXiv preprint arXiv: 1610.08681, 2016—arxiv.org.

Djordjevic, V.D. and T.M. Atanackovic. 2008. Similarity solutions to nonlinear heat conduction and Burgers/Korteweg-de Vries fractional equations. *J. Comput. Appl. Math.* 222: 701–714.

Doeblin, W. 1940. Éléments d'une théorie générale des chaines simples constantes de Markoff. *Ann. Sci. École. Norm. Sup.* Paris III. 57: 61–111.

Doetsch, G. 1956. Anleitung zum Praktischen Gebrauch der Laplace-transformation. Oldenbourg, Munich.

Dooge, J.C.I. 1986. Looking for hydrologic laws. *Water Resour. Res.* 22(9): 46S–58S.

DuChateau, P. and D.W. Zachmann. 1986. Schaum's Outline of Theory and Problems of Partial Differential Equations. McGraw-Hill, Inc., New York.

Dufour, L. 1872. Archives de sciences physiques et naturelles, Genève. 45: 9.

Dufour, L. 1873. On the diffusion of gases through porous partitions and the accompanying temperature changes. *Pogg. Ann.* 148: 490.

Duhamel, J.M.C. 1838. Memoir surcalcul des actions moleculaires developpers par les changement de temperature dans les corps solides. *Memoir de l'Istitute de France.* V.

Dullien, F.A.L. 1992. Porous Media: Fluid Transport and Pores Structure. 2nd Ed. Acad. Press, San Diego.

Dupuit, J. 1848. Etudes théoriqueset practiques sur le movement des eaux dans les canaux découverts et à travers les terrains perméables. Dunod, Paris.

Dupuit, J. 1863. Etudes théoriqueset practiques sur le movement des eaux dans les canaux découverts et à travers les terrains perméables. 2nd ed., Dunod, Paris.

Dutka, J. 1985. On the problem of random flights. *Archive for History of Exact Science* 32(3/4): 351–375.

Dzhebashyan, M.M. and A.B. Nersesyan. 1968. Fractional derivatives and the Chauchy problem for differential equations of fractional order (in Russian). *Izv. Acad. Nauk Armyan. SSR, Ser. Mat.* 3(1): 3–29.

Eagleson, P.S. 1970. Dynamic Hydrology. McGraw-Hill, New York, USA.

Einstein, A. 1905. Über die von der molekularkinetischen Theorie der Wärme geforderte Bewegung von in ruhenden Flüssigkeiten suspendierten Teilchen. *Ann. Der Physik.* 322(8): 549–560. (Investigations on the theory of the Brownian movement. *In*: R. Furth [ed.] and A.D. Cowper (transl.). Methuen, London, 1926).

Elkhoury, J.E., E.E. Brodsky and D.C. Agnew. 2006. Seismic waves increases permeability. *Nature* 441: 1135–1138.

Elvin, M. and N. Su. 1998. Action at a distance: The influence of the Yellow River on Hangzhou Bay since AD 1000. pp. 344–407. *In*: M. Elvin and Liu Ts'ui-jung [eds.]. Sediments of Time: Environment and Society in Chinese History. Cambridge Univ. Press, Cambridge, England.

Encyclopaedia Britannica. Earth Tides. https://www.britannica.com/science/Earth-tide.

Enelund, M., L. Mähler, K. Runesson and B.L. Josefson. 1999. Formulation and integration of the standard linear viscoelastic solid with fractional order rate laws. *Int. J. Solids Struct.* 36: 2417–2442.

Erdélyi, A. 1953a. High Transcendental Functions. McGraw-Hill, New York. Vol. 1.

Erdélyi, A. 1953b. High Transcendental Functions. McGraw-Hill, New York. Vol. 2, Chapter X: Orthogonal Polynomials. pp. 153–231.

Erdélyi, A. 1965. Axially symmetrical potentials and fractional integration. *J. SIAM* 13(1): 216–228.

Euler, L. 1744. Methodus inveniendi lineas curvas maximi minimive proprietate gaudentes. Lausanne & Geneve.

Euler, L. 1748. Introduction in Analysis Infinitorum. M-M Bousquet, Lausanne.

Evans, G., J. Blackledge and P. Yardley. 1999. Analytic Methods for Partial Differential Equations. Springer, London.

Ezzat, M.A., A.S. El-Karamany and S.M. Ezzat. 2012. Two-temperature theory in magneto-thermoelasticity with fractional order dual-phase-lag heat transfer. *Nuclear Eng. Design.* 252: 267–277.

Ezzat, M.A., A.S. El-Karamany and A.A. El-Bary. 2015. On thermos-viscoelasticity with variable thermal conductivity and fractional-order eat transfer. *Internl. J. Thermophys.* 36: 1684–1697.

Fedotov, S., D. Han, M. Johnston and V. Allan. 2019. Asymptotic behavior of the solution of the space dependent variable order fractional diffusion equation: ultra-slow anomalous aggregation. arXiv: 1902.03087v1 [cond-mat.stat-mech], 8 Feb.

Feller, W. 1971. An Introduction to Probability Theory and Its Applications. Vol. 2, Wiley, New York.

Fick, A. 1855a. Ueber diffusion. *Pogg. Ann. Phys. Chem.* 170(4 Reihe 94): 59–86.

Fick, A. 1855b. Ueber Diffusion. *Ann. Phys.* 170(1): 59–86.

Findley, W.N., J.S. Lai and K. Onaran. 1976. Creep and Relaxation of Nonlinear Viscoelastic Materials. North-Holland Publ. Co./Dover Publ. Inc., New York.

Fokker, A.D. 1914. Die mittlere energie rotirerende elektrischer Dipole im Strahlungsfeld. *Ann. Phys. Ser.* 4(Leipzig). 43: 810–820.

Friedrich, C. 1991. Relaxation and retardation function of the Maxwell model with fractional derivatives. *Rheol. Acta.* 30: 151–158.

Food and Agriculture Organisation (FAO) of the United Nations. 2015. The Status of World's Soil Resources. Rome.

Fourier, J.B.J. 1807. Théorie des mouvements de la chaleur dans le corps solides. French Acad., Paris.

Fourier, J. 1822. Théorie Analytique del la Chaleur. Firmin Didot, Paris.

Fox, C. 1928. The asymptotic expansion of generalized hypergeometric functions. *Proc. London Math. Soc. Ser.* 2(5): 389–400.

Fox, C. 1961. The G and H functions as symmetrical Fourier kernels. *Trans. Amer. Math. Soc.* 98(3): 395–429.

Friedrich, C. 1991. Relaxation and retardation functions of the Maxwell model with fractional derivatives. *Rheol. Acta.* 30: 151–158.

Gamerdinger, A.P. and D.I. Kaplan. 2000. Application of a continuous-flow centrifugation method for solute transport in disturbed, unsaturated sediments and illustration of mobile-immobile water. *Water Resour. Res.* 36(7): 1747–1755.

Gao, G., S. Feng, H. Zhan, G. Huang and X. Mao. 2009. Evaluation of anomalous solute transport in a large heterogeneous soil column with mobile-immobile model. *J. Hydrol. Eng.* 14(9): 966–974.

Gao, G., H. Zhan, S. Feng, B. Fu, Y. Ma and G. Huang. 2010. A new mobile-immobile model for reactive solute transport with scale-dependent dispersion. *Water Resour. Res.* 46: W08533, 1–16.

García-Colín, L.S., R.M. Velasco and F.J. Uribe. 2008. Beyond the Navier-Stokes equations: Burnett hydrodynamics. *Phys. Reports.* 465: 149–189.

Gardiner, C.W. 1985. Handbook of Stochastic Methods. 2nd ed., Springer, Berlin.

Gardner, W. and J.A. Widtsoe. 1921. Movement of soil moisture. *Soil Sci.* 11: 215–232.

Garnier, P., E. Perrier, R. Angulo-Jaramillo and P. Baveye. 1997. Numerical model of 3-dimensional anisotropic deformation and 1-dimensional water flow in swelling soils. *Soil Sci.* 162(6): 410–420.

Garnier, P., R. Angulo-Jaramillo, D.A. DiCarlo, T.W.J. Bauters, C.G.J. Darnault, T.S. Steenhuis et al. 1998. Dual-energy synchrotron X-ray measurements of rapid soil density and water content changes in swelling soils during infiltration. *Water Resour. Res.* 34(11): 2837–2842.

Garra, R. 2011. Fractional-calculus model for temperature and pressure waves in fluid-saturated porous rocks. *Phys. Rev E.* 84: 036605.

Garra, R. and E. Salusti. 2013. Application of the nonlocal Darcy law to the propagation of nonlinear thermoelastic waves in fluid saturated porous media. *Phys. D.* 250: 52–57.

Garra, R., E. Salusti and R. Droghei. 2015. Memory effects on nonlinear temperature and pressure wave propagation in the boundary between two fluid-saturated porous rocks. *Adv. Math. Phys.* 2015: 532150.

Gasper, G. and M. Rahman. 2004. Basic Hypergeometric Series. Cambridge Univ. Press, Cambridge.

Gaudet, J.P., H. Jegat, G. Vachuad and P.J. Wierenga. 1977. Solute transfer with exchange between mobile and stagnant water, through unsaturated sand. *Soil Sci. Soc. Amer. J.* 41(4): 665–671.

Gauss, C.F. 1813. Disquisitiones generales circa seriem infinitam. *Comm. soc. reg. sci. Gött. rec.* vol. II, reprinted in Werke 3(1876): 123–162.

Geertsma, J. 1966. Problems of rock mechanics in petroleum production engineering. *Proc. 1st Congr. Internl. Soc. Rock. Mech.* 1: 585–594, Lisbon.

Gefen, Y., A. Aharony and S. Alexander. 1983. Anomalous diffusion on percolating clusters. *Phys. Rev. Lett.* 50(1): 77–80.

Geindreau, C. and J.-L. Auriault. 2002. Magnetohydrodynamic flows in porous media. *J. Fluid Mech.* 466: 343–363.

Gel'fand, I.M. and G.E. Shilov. 1958. Generalized Functions. Vol. 1: Generalized Functions and Operations on Them (Russian), Fizmatgiz, Moscow.

Gel'fand, I.M. and G.E. Shilov. 1964. Generalized Functions. Vol. 1. Academic Press, New York and London. Translated from the Russian.

Gelhar, L.W., C. Welty and K.R. Rehfeldt. 1992. A critical review of data on field-scale dispersion in aquifers. *Water Resour. Res.* 28(7): 1955–1974.

Gelhar, L.W. 1993. Stochastic Subsurface Hydrology Prentice-Hall, Englewood. Cliffs, N.J.

Gemant, A. 1935. The conception of a complex viscosity and its application to dielectrics. *Trans. Faraday Soc.* 31: 1582–1590.

Gemant, A. 1936. A method of analyzing experimental results obtained from elasto-viscous bodies. *Phys.* 7: 311–317.

Gemant, A. 1950. Fractional Phenomena. Chem. Publ. Co., Brooklin.

Gerasimov, A.N. 1948. A generalization of linear laws of deformation and its application to internal friction problem (Russian). *Akad. Nauk SSSR. Prikl. Mat. Mekh.* 12: 251–260.

Gerasimov, D.N., V.A. Kondratieva and O.A. Sinkevich. 2010. An anomalous non-self-similar infiltration and fractional diffusion equation. *Phys. D.* 239: 1593–1597.

Gerke, H.H. and M.Th. van Genuchten. 1993. A dual-porosity model for simulating the preferential movement of water and solutes in structured porous media. *Water Resour. Res.* 29(2): 305–319.

Gerolymatou, E., I. Vardoulakis and R. Hilfer. 2006. Modelling infiltration by means of a nonlinear fractional diffusion model. *J. Phys. D: Appl. Phys.* 39: 4104–4110.

Ghanbarian-Alavijeh, B., H. Millán and G. Huang. 2011. A review of fractal, prefractal and pore-solid fractal models for parameterizing the soil water retention curve. *Can. J. Soil Sci.* 91: 1–14.

Ghanbarian-Alavijeh, B. and A. Hunt. 2012. Comment on More generic capillary pressure and relative permeability models from fractal geometry by Kewen Li. *J. Contam. Hydrol.* 140-141: 21–23.

Ghyben, W.B. 1888. Nota in verband met de voorgenomen putboring nabij Amsterdam. *Tijdschr. Kon. Inst. Ing.* 1888–1889, 1889: 8–22.

Giménez, D., E. Perfect, W.J. Rawls and Ya. Pechepsky. 1997. Fractal models for predicting soil hydraulic properties: a review. *Eng. Geol.* 48: 161–183.

Giraldez, J.V. and G. Sposito. 1978. Moisture profiles during steady vertical flows in swelling soils. *Water Resour. Res.* 14: 314–318.

Giraldez, J.V., G. Sposito and D. Delgado. 1983. A general soil volume change equation. *Soil Sci. Soc. Am. J.* 47: 419–422.

Giraldez, J.V. and G. Sposito. 1985. Infiltration in swelling soil. *Water Resour. Res.* 21: 33–44.

Glöckle, W.G. and T.F. Nonnenmacher. 1995. A fractional calculus approach to self-similar protein dynamics. *Biophys. J.* 68: 46–5.

Gnedenko, B.V. and A.N. Kolmogorov. 1954. Limit Distributions for Sums of Independent Random Variables. Addison-Wesley, Reading, Massachusetts.

Gorenflo, R. and S. Vessella. 1980. Abel Integral Equations: Analysis and Applications. Springer-Verlag, Berlin.

Gorenflo, R. 1987. Nonlinear Abel integral equations: Applications, analysis, numerical methods. pp. 243–259. *In*: P.C. Sabatier [ed.]. Inverse Problems: An Interdisciplinary Study. Academic Press, London.

Gorenflo, R. and A.A. Kilbas. 1995. Asymptotic solution of a nonlinear Abel-Volterra integral equation of second kind. *J. Fract. Calc.* 8: 103–117.

Gorenflo, R. and F. Mainardi. 1998a. Fractional calculus and stable probability distributions. *Arch. Mech.* 59(3): 377–388.

Gorenflo, R. and F. Mainardi. 1998b. Random walk models for space-fractional diffusion processes. *Fract. Calc. Appl. Anal.* 1: 167–191.

Gorenflo, R., A.A. Kilbas and S.V. Rogosin. 1998. On the generalized Mittag-Leffler type functions. *Integr. Transf. Special Funct.* 7: 215–224.

Gorenflo, R., Y. Luchko and F. Mainardi. 1999. Analytical properties and applications of the Wright function. *Fract. Calc. Appl. Anal.* 2(4): 383–414.

Gorenflo, R. and F. Mainardi. 2001. Random walk models approximating symmetric space-fractional diffusion processes. Series Operator Theory: Advances and Applications. Vol. 121: Problems and Methods in Mathematical Physics. Birkhäuser, Verlag Basel, Switzerland, pp. 120–145.

Gorenflo, R. and A. Vivoli. 2003. Fully discrete random walks for space–time fractional diffusion equations. *Signal Proc.* 83: 2411–2420.

Gorenflo, R. and F. Mainardi. 2005. Simply and multiply scaled diffusion limits for continuous time random walk. *J. Phys.: Conf. Ser.* 7: 1–16.

Gorenflo, R., F. Mainardi and A. Vivoli. 2007. Continuous-time random walk and parametric subordination in fractional diffusion. *Chaos, Solitons & Fractals* 34: 87–103.

Gorenflo, R. and F. Mainardi. 2009. Some recent advances in theory and simulation of fractional diffusion processes. *J. Comp. Appl. Math.* 229: 400–415.

Gorenflo, R. and F. Mainardi. 2012. Parametric subordination in fractional diffusion processes. pp. 227–261. *In*: S.C. Lim, J. Klafter and R. Metzler [ed.]. Fractional Dynamics, Recent Advances, World Scientific, Singapore.

Gradshteyn, I.S. and I.M. Ryzhik. 1994. Table of Integrals, Series, and Products. Academic, San Diego, USA.

Grebenkov, D.S. and L. Tupikina. 2018. Heterogeneous continuous-time random walks. *Phys. Rev. E.* 97: 012148.

Green, G. 1828. An Essay on the Application of Mathematical Analysis to the Theories of Electricity and Magnetism. Nottingham, England.

Green, W.A. and G.A. Ampt. 1911. Studies of soil physics: Part 1. The flow of air and water through soils. *J. Agric. Sci.* 4: 1–24.

Griffioen, J.W., D.A. Barry and J.-Y. Parlange. 1998. Interpretation of two-region model parameters. *Water Resour. Res.* 34(3): 373–384.

Gupta, V.K., I. Rodriguez-Iturbe and E.F. Wood [eds.]. 1986. Scale Problems in Hydrology. Reidel, Dordrecht.

Gupta, V.K. and O.J. Mesa. 1988. Runoff generation and hydrologic response via channel network geomorphology—Recent progress and open problems. *J. Hydrol.* 102: 3–28.

Gurtin, M.E. and E. Sternberg. 1962. On the linear theory of viscoelasticity. *Arch. Rational Mech. Anal.* 11(1): 291–356.

Gurtin, M.E. 1973. The linear theory of elasticity. *In*: C. Truesdell [ed.]. Linear Theories of Elasticity and Thermoelasticity. Springer, Berlin.

Hadid, S.B. and Y. Luchko. 1996. An operational method for solving fractional differential equations of an arbitrary order. *Panamer. Math. J.* 6(1): 57–73.

Haggerty, R. and S.M. Gorelick. 1995. Multiple-rate mass transfer for modelling diffusion and surface reactions in media with pore-scale heterogeneity. *Water Resour. Res.* 31(10): 2383–2400.

Haggerty, R., S.A. McKenna and L.C. Meigs. 2000. On the late-time behavior of tracer test breakthrough curves. *Water Resour. Res.* 36(12): 3467–3479.

Haggerty, R., C.F. Harvey, C.F. von Schwerin and L.C. Meigs. 2004. What controls the apparent timescale of solute mass transfer in aquifers and soils? A comparison of experimental results. *Water Resour. Res.* 40(1): W01510, 1–13.

Hahn, M. and S. Umarov. 2011. Fractional Fokker-Planck-Kolmogorov type equations and their associated stochastic differential equations. *Fract. Calculus & Appl. Anal.* 14: 56–79.

Haines, W.B. 1923. The volume changes associated with variations of water content in soils. *J. Agric. Sci.* 13: 296–311.

Haines, W.B. 1927. Studies in the physical properties of soils: IV. A further contribution to the theory of capillary phenomena in soil. *J. Agric. Res.* 17: 264.

Haines, W.B. 1930. Studies in the physical properties of soils: V. The hysteresis effect in capillary properties and the modes of moisture distribution associated therewith. *J. Agric. Sci.* 20: 7.

Hanneken, J.W., B.N.N. Achar, R. Puzio and D.M. Vaught. 2009. Properties of the Mittag-Leffler function for negative alpha. *Phys. Scripta.* T136, 014037: 1–5.

Hantush, M.S. 1962a. Hydraulics of gravity wells in sloping sands. *J. Hydraul. Divis.* 88(4): 1–15.

Hantush, M.S. 1962b. Flow of ground water in sands of nonuniform thickness. Part 2. Approximate theory. *J. Geophys. Res.* 67(2): 711–720.

Hantush, M.S. 1962c. Flow of ground water in sands of nonuniform thickness. Part 3. Flow to wells. *J. Geophys. Res.* 67(4): 1527–1534.

Hantush, M.S. 1964a. Depletion of storage, leakage, and river flow by gravity wells insloping sands. *J. Geophys. Res.* 69(12): 2551–2560.

Hantush, M.S. 1964b. Drawdown around wells of variable discharge. *J. Geophys. Res.* 69(20): 4221–4235.

Hantush, M.S. 1964c. Hydraulics of wells. pp. 281–432. *In*: Ven Te Chow [ed.]. *Adv. Hydrosci.* Vol. 1, Acad. Press, New York, USA.

Hantush, M.S. 1967. Flow to wells in aquifers separated by a semipervious layer. *J. Geophys. Res.* 72(6): 1709–1720.

Haubold, H., A.M. Mathai and R.K. Saxena. 2011. Mittag-Leffler functions and their applications. *J. Appl. Math.* ID 298628: 1–51.

Havlin, S. and D. Ben-Avraham. 1987. Diffusion in disordered media. *Adv. Phys.* 36(6): 695–798.

Hayek, M. 2016. An exact explicit solution for one-dimensional, transient, nonlinear Richards' equation for modeling infiltration with specific hydraulic functions. *J. Hydrol.* 535: 662–670.

He, J.-H. 1998. Approximate analytical solution for seepage flow with fractional derivatives in porous media. *Comput. Methods Appl. Mech. Eng.* 167: 57–68.

Heller, S.R. 1964. Hydroelasticity. *Adv. Hydrosci.* 1: 94–160.

Heller, V. 2011. Scale effects in physical hydraulic engineering models. *J. Hydraul. Res.* 49(3): 293–306.

Henderson, N., J.C. Brêttas and W.F. Sacco. 2010. A three-parameter Kozeny-Carman generalized equation for fractal porous media. *Chem. Eng. Sci.* 65: 4432–4442.

Herrick, M.G., D.A. Benson, M.M. Meerschaert and K.R. McCall. 2002. Hydraulic conductivity, velocity, and the order of the fractional dispersion derivative in a highly heterogeneous system. *Water Resour. Res.* 38(11): 1127 (1–13).

Herrmann, R. 2011. Fractional Calculus: An Introduction for Physicists. World Scientific, Singapore.

Herzberg, A. 1901. Die Wasserversorgung einiger Nordseebaden Zeit. Gasbeleuchtung Wasserversorgung. 44: 818–819, 842–844.

Hewlett, J.D. 1961. Watershed management. US For Service Southeast Forest Exp. Stn. Report. 61–66.

Hewlett, J.D. and A.R. Hibbert. 1967. Factors affecting the response of small watersheds to precipitation in humid areas. pp. 275–290. *In*: W.E. Sopper and H.W. Lull [eds.]. Forest Hydrology. Pergamon Press, New York, USA.

Heymans, N. and I. Podlubny. 2006. Physical interpretation of initial conditions for fractional differential equations with Riemann-Liouville fractional derivatives. *Rheol. Acta.* 45: 765–771.

Hilfer, R. 1995. Exact solutions for a class of fractal time random walks. *Fractals* 3(1): 211–216.

Hilfer, R. 1999. On fractional diffusion and its relation with the continuous time random walks. *Lecture Notes in Physics* 519: 77–82, Springer, Berlin.

Hilfer, R. 2000. Applications of Fractional Calculus in Physics. World Scientific, Singapore.

Hilfer, R. 2008. Threefold introduction to fractional derivatives. pp. 17–73. *In*: R. Klages, G. Radons and I.M. Sokolov [eds.]. Anomalous Transport. Wiley-VCH Verlag Weinheim.

Hilfer, R. and Y. Luchko. 2019. Desiderata for fractional derivatives and integrals. Mathematics. 7(149): 1–5, doi:10.3390/math7020149.

Hillel, D. 1998. Environmental Soil Physics. Acad. Press, Amsterdam.

Horton, R.E. 1933. The role of infiltration in the hydrologic cycle. *Trans. Am. Geophys. Union.* 14: 446–460.

Horton, R.E. 1938. The investigation and application of runoff plot experiments with reference to soil erosion problems. *Proc. Soil Sci. Soc. Amer.* 3: 340–349.

Horton, R.E. 1939. Analysis of runoff-flat experiments with varying infiltration capacity. *Trans. Amer. Geophys. Union.* Part IV: 693–711.

Hsieh, P.A. 1996. Deformation-induced changed in hydraulic head during ground-water withdrawal. *Ground Water.* 34(6): 1082–1089.

Hubbert, M.K. 1940. The theory of ground-water motion. *J. Geol.* 48(8): Part 1 (Nov.–Dec., 1940), 785–944.

Humbert, P. and P. Delerue. 1953. Sur une extention à deux variables de la fonction de Mittag-Leffler. *C.R. Acad. Sci. Paris* 237: 1059–1060.

Irmay, S. 1958. On the theoretical derivation of Darcy and Forchheimer formulas. *Trans. Am. Geophys. Union.* 39(4): 702–707.

Jackson, C.R. 1992. Hillslope infiltration and lateral downslope unsaturated flow. *Water Resour. Res.* 28(9): 2533–2539.

Jacob, N. and H.-G. Leopold. 1993. Pseudo differential operators with variable order of differentiation generating Feller semigroups. *Integr. Equat. Oper. Theory* 17: 544–553.

Jeon, J-.H., E. Barkai and R. Metzler. 2013. Noisy continuous time random walks. *J. Chem. Phys.* 139: 121916.

Jiang, H., F. Liu, I. Turner and K. Burrage. 2012a. Analytical solutions for the multi-term time-space Caputo-Riesz fractional advection-diffusion equations on a finite domain. *J. Math. Anal Appl.* 389: 1117–1127.

Jiang, H., F. Liu, I. Turner and K. Burrage. 2012b. Analytical solutions for the multi-term time-fractional diffusion-wave/diffusion equations in a finite domain. *Computer Math. Appl.* 64: 3377–3388.

Jurlewicz, A., K. Weron and M. Teuerle. 2008. Generalized Mittag–Leffler relaxation: clustering jump continuous-time random walk approach. *Phys. Rev. E.* 78: 011103.

Jurlewicz, A., M.M. Meerschaert and H.P. Scheffler. 2011. Cluster continuous time random walks. *Studia Math.* 205: 13–30.

Jury, W.A., D. Russo and G. Sposito. 1987. The spatial variability of water and solute transport properties in unsaturated soil. *Higardia.* 55(4): 32–56.

Kalbus, E., F. Reinstorf and M. Schirmer. 2006. Measuring methods for groundwater, surface water and their interactions: a review. *Hydrol. Earth Syst. Sci. Discuss.* 3: 1809–1850.

Karalis, K. 1992. Mechanics of Swelling. NATO ASI, Vol. H64: 3–31, Springer-Verlag, Berlin.

Karapetyants, N.K., A.A. Kilbas and M. Saigo. 1996. On the solution of nonlinear Volterra convolution equation with power nonlinearity. *J. Integr. Eqs. & Appl.* 8(4): 429–445.

Karapetiants, N.K., A.A. Kilbas, M. Saigo and G. Samko. 2001. Upper and lower bounds for solutions of nonlinear Volterra convolution integral equations with power nonlinearity. *J. Integr. Eqs. and Appl.* 12(4): 421–448.

Karlinger, M.R. and B.M. Troutman. 1985. An assessment of the instantaneous unit hydrograph derived from the theory of topologically random channel networks. *Water Resour. Res.* 21(11): 1693–1702.

Katchalsky, A. and P.F. Curran. 1965. Nonequilibrium Thermodynamics in Biophysics. Harvard Univ. Press, Cambridge, Mass., USA.

Katz, M. and D. Tall. 2013. A Cauchy-Dirac delta function. *Found. Sci.* 18: 107–123.

Kausel, E. 2010. Early history of soil-structure interaction. *Soil Dyn. Earthquake Eng.* 30: 822–832.

Keller, T., M. Lamande, S. Peth, M. Berli, J.-Y. Delenne and W. Baumgarten. 2013. An interdisciplinary approach towards improved understanding of soil. *Soil & Tillage Res.* 128: 61–81.

Kevorkian, J. 1990. Partial Differential Equations: Analytical Solution Techniques. Wadsworth & Books/ Cole, Pacific Cove, California.

Khuzhayorov, B.Kh., Zh.M. Makhmudov and Sh.Kh. Zikiryaev. 2010. Substance transfer in a porous medium structured with mobile and immobile liquids. *J. Eng. Phys. Thermophys.* 83(2): 263–270.

Kibler, D.F. and D.A. Woolhiser. 1970. The kinematic cascade as a hydrologic model. Colorado State University, Fort Collins, Colo., Hydrol. Pap. No. 39, 27 pp.

Kikuchi, K. and A. Negoro. 1995. Pseudo differential operators and Sobolev spaces of variable order of differentiation. *Rep. Fac. Liberal Arts, Shizuoka Univ. Sci.* 31: 19–27.

Kilbas, A.A. and M. Saigo. 1994. On asymptotic solutions of nonlinear and linear Abel-Volterra integral equations. I. *Surikaisekikenkyusho Kokyuroku* (数理解析研究所講究録). 881: 91–111.

Kilbas, A.A. and M. Saigo. 1995. On solution of integral equation of Abel-Volterra type. *Diff. Integr. Eqs.* 8(5): 993–1011.

Kilbas, A.A., M. Saigo and R. Gorenflo. 1995. On asymptotic solutions of nonlinear Abel-Volterra in equations with quasipolynomial free term. *J. Fract. Calc.* 8: 75–93.

Kilbas, A.A. and M. Saigo. 1996. On solution in closed form of nonlinear integral and differential equations of fractional order. *Surikaisekikenkyusho Kokyuroku* (数理解析研究所講究錄). 963: 39–50.

Kilbas, A.A. and M. Saigo. 1999a. On Solution of nonlinear Abel-Volterra integral equation. *J. Math. Anal. Appl.* 229: 41–60.

Kilbas, A.A. and M. Saigo. 1999b. On the H-function. *J. Appl. Math. Stoch. Anal.* 12(2): 191–204.

Kilbas, A.A. and J.J. Trujillo. 2001. Differential equations of fractional order: Methods, results and problems–I. *Appl. Anal.* 78(1): 153–192.

Kilbas, A.A. and J.J. Trujillo. 2002. Differential equations of fractional order: Methods, results and problems. II. *Appl. Anal.* 81(2): 435–493.

Kilbas, A.A., M. Saigo and J.J. Trujillo. 2002a. On the generalized Wright function. *Fract. Calcul. Appl. Anal.* 5(4): 437–460.

Kilbas, A.A., M. Saigo and R.K. Saxena. 2002b. Solution of Volterra integro-differential equations with generalized Mittag-Leffler function in the kernels. *J. Integr. Eqs. & Appl.* 14(4): 377–396.

Kilbas, A.A., M. Saigo and R.K. Saxena. 2004. Generalised Mittag-Leffler function and generalised fractional calculus operators. *Integral Transforms & Special Func.* 15(1): 31–49.

Kilbas, A.A. 2005. Fractional calculus of the generalized Wright function. *Fract. Calc. Appl. Anal.* 8(2): 113–126.

Kilbas, A.A. and S.A. Marzan. 2005. Nonlinear differential equations with the Caputo fractional derivative in the space of continuously differentiable functions. *Diff. Eqs.* 41(1): 84–89.

Kilbas, A.A., H.M. Srivastava and J.J. Trujillo. 2006. Theory and Applications of Fractional Differential Equations. Elsevier, Amsterdam.

Kilbas, A.A., Yu.F. Luchko, H. Martínez and J.J. Trujillo. 2010. Fractional Fourier transform in the framework of fractional calculus operators. *Integral Transforms & Special Func.* 21(10): 779–795.

Kilbas, A.A., A.A. Koroleva and S.V. Rogosin. 2013. Milti-parametric Mittag-Leffler functions. *Fract. Calc. Appl. Anal.* 16(2): 378–404.

Kirby, M. 1988. Hillslope runoff processes and models. *J. Hydrol.* 100: 315–339.

Kirchhoff, G. 1882. Zur Theorie der Lichtstrahlen. Sitz.-Ber. kgl. Preuß. Akad. Wiss. 22: 641–669.

Kiryakova, V. 1997. All the special functions are fractional differintegrals of elementary functions. *J. Phys. A: Math. Gen.* 30: 5085–5103.

Kiryakova, V. 2010a. The special functions of fractional calculus as generalized fractional calculus operators of some basic functions. *Comput. Math. Appl.* 59: 1128–1141.

Kiryakova, V. 2010b. The multi-index Mittag-Leffler functions as an important class of special functions of fractional calculus. *Comput. Math. Appl.* 59: 1885–1895.

Kiryakova, V.S. 2011. Fractional calculus, special functions and integral transformations: What is the relation? Proc. 40th Jubilee Spring Conf. Union Bulgarian Math., Borovetz, Bulgaria, April 5–9, 2011.

Klafter, J., A. Blumen and M.F. Shlesinger. 1987. Stochastic pathway to anomalous diffusion. *Phys. Rev. A.* 35(7): 3081–3085.

Klonne, F. 1880. Die periodischen schwankungen des wasserspigels in den inundieten kohlenschachten von Dux in der period von 8 April bis 15 September 1979. *Sitzber. Kais. Akad. Wiss.*

Klute, A. 1952. Numerical method for solving the flow equation for water in porous materials. *Soil Sci.* 73: 105–116.

Knight, J.H. 1983. Infiltration function from exact and approximate solutions of Richards' equations. pp. 24–33. *In*: Advances in Infiltration, Proc. Nat. Conf. Adv. Infiltration, Chicago, Dec. 12–13, 1983.

Knobel, R. 2000. An Introduction to the Mathematical Theory of Waves. *Amer. Math. Soc.*, Rhode Island, USA.

Kochubei, A.N. and Y. Luchko. 2019. Basic FC operators and their properties. pp. 23–46. *In*: A. Kochubei and Y. Luchko [eds.]. Handbook of Fractional Calculus with Applications. Vol. 1: Basic Theory. De Gruyter GmbH, Berlin, Germany.

Koeller, R.C. 1984. Applications of fractional calculus to the theory of viscoelasticity. *J. Appl. Mech.* 51(2): 299–307.

Koeller, R.C. 2010. A theory relating creep and relaxation for linear materials with memory. *J. Appl. Mech.* 77: 031008.

Komatsu, T. 1995. On stable-like processes. *In*: S. Watanabe, M. Fukushima, Yu.V. Prohorov and A.N. Shiryaev [eds.]. Probability Theory and Mathematical Statistics. Proc. 7th Japan–Russia Symp., Tokyo, Japan.

Konhauser, J.D.E. 1967. Biorthogonal polynomials suggested by the Laguerre polynomials. *Pacific J. Math.* 21(2): 303–314.

Konno, N. 2010. Quantum walks and elliptic integrals. *Math. Struct. in Comp. Sci.* 20: 1091–1098.

Kostiakov, A.N. 1932. On the dynamics of the coefficient of water-percolation in soils and on the necessity for studying it from a dynamic point of view for purposes of amelioration. Trans. 6th Com. *Internl. Soc. Soil Sci.* Russian Part A: 17–21.

Kozeny, J. 1927. Ueber kapillare Leitung des Wassers im Boden. *Sitzungsber Akad. Wiss., Wien.* 136(2a): 271–306.

Kozeny, J. 1932. Kulturtechnik. 35: 478.

Kreft, A. and A. Zuber. 1978. On the physical meaning of the dispersion equation and its solutions for different initial and boundary conditions. *Chem. Eng. Sci.* 33: 1471–1480.

Kreft, A. and A. Zuber. 1986. Comment on Flux-averaged and volume-averaged concentrations in continuum approaches to solute transport by J.C. Parker and M.Th. Genuchten. *Water Resour. Res.* 22(7): 1157–1158.

Krepysheva, N., L.D. Pietro and M.C. Neel. 2007. Enhanced tracer diffusion in porous media with an impermeable boundary. pp. 171–184. *In*: J. Sabatier, O.P. Agrawal and J.A.T. Machado [eds.]. Advances in Fractional Calculus: Theoretical Developments and Applications in Physics and Engineering. Springer, Dordrecht.

Küntz, M. and P. Lavallee. 2001. Experimental evidence and theoretical analysis of anomalous diffusion during water infiltration into porous building materials. *J. Phys. D: Appl. Phys.* 34: 2547–2554.

Kutner, R. and J. Masoliver. 2017. The continuous time random walk, still trendy: fifty-year history, state of art and outlook. *Eur. Phys. J. B.* 90: 50. DOI: 10.1140/epjb/e2016-70578-3.

Lai, J., S. Mao, J. Qiu, H. Fan, Qian Zhan, Z. Hu. et al. 2016. Investigation progresses and applications of fractional derivative model in geotechnical engineering. *Math. Prob. Eng.* 2016: Article 9183296.

Lallemand-Barres, A. and P. Peaudecerf. 1978. Recherche des relations entre la valeur de la dispersivité macroscopique d'un milieu aquifère, ses autres caractéristiques et les conditions de mesure: Etude bibliographique. *Bull. Bur. Rech. Geol. Min.* Sec. 3, No. 4: 277–284.

Lamé, G. and Clapeyron. 1828. Mémoire sur l'équilibre intérieur des corps solides homogénes. Mémoire .... par divers savans. IV. 1833. The date of the memoir is at least as early as 1828 (Love 1892, 6).

Lamé, G. 1841. Mémoire sur les surfaces isostatiques dans les corps solides homogènes en équilibre d'élasticité. *J. Math. Pures Appl.* 6: 37–60.

Lamé, G. 1859. Lecons sur les Coordinnées Curvilignes et leurs diverses applications. Mallet-Bachelier, Paris.

Lavoie, J.L., T.J. Osler and R. Tremblay. 1976. Fractional derivatives and special functions. *SIAM Rev.* 18(2): 240–268.

Lehnigk, S.H. 1993. The Generalized Feller Equation and Related Topics. Longman, Essex, England.

Leith, J.R. 2003. Fractal scaling of fractional diffusion processes. *Signal Proc.* 83: 2397–2409.

Leibenzon, L.S. 1929. Gas Movement in a Porous Medium (in Russian). *Neft. Khoz.* 10: 497–519.

Lenormand, R. 1992. Use of fractional derivatives for fluid flow in heterogeneous porous media. Proc. ECMOR III, 3rd Eur. Conf. Math. Oil Recovery, 17–19 June 1992: 159–167, Delft Univ. Press. doi: 10.3997/2214-4609.201411072.

Leopold, H.G. 1989. On Besov spaces of variable order of differentiation. *Z. Anal. Anwendungen.* 8: 69–82.

Leopold, H.G. 1991. On function spaces of variable order of differentiation. *Forum Math.* 3: 69–82.

Leverett, M.C. 1939. Flow of oil-water mixture through unconsolidated sands. *Trans. AIME* 32: 149–171.

Leverett, M.C. 1941. Capillary behaviour in porous solids. *Trans. AIME* 142: 152–169.

Le Vot, F., E. Abad and S.B. Yuste. 2017. Continuous-time random-walk model for anomalous diffusion in expanding media. *Phys. Rev. E.* 96: 032117.

Le Vot, F. and S.B. Yuste. 2018. Continuous-time random walks and Fokker-Planck equation in expanding media. *Phys. Rev. E.* 98: 042117.

Levy, P. 1954. Processus semi-markoviens *Proc. Int. Congr. Math.* (Amsterdam). 3: 416–426.

Lewis, M.R. and W.E. Milne. 1938. Analysis of border irrigation. *Agric. Eng.* 19: 267–272.

Li, C. and A. Chen. 2018. Numerical methods for fractional partial differential equations. *Internl. J. Computer Math.* 95(6-7): 1048–1099.

Li, L., D.A. Barry, P.J. Culligan-Hensley and K. Bajracharya. 1994. Mass transfer in soils with local stratification of hydraulic conductivity. *Water Resour. Res.* 30(11): 2891–2900.

Li, X., X. Ling and P. Li. 2009. A new stochastic order based upon Laplace transform with applications. *J. Stat. Plann. Infer.* 139: 2624–2630.

Li, Z., Y. Liu and M. Yamamoto. 2015. Initial-boundary value problems for multi-term time-fractional diffusion equations with positive constant coefficients. *Appl. Math. Comput.* 257: 381–397.

Lighthill, M.H. and G.B. Whitham. 1955. On kinematic waves. I. Flood movement in long rivers. *Proc. Royal Soc. A.* 229: 282–316.

Liingaard, M., A. Augustesen and P.V. Lade. 2004. Characterization of models for time-dependent behaviour of soils. *Intl. J. Geomech.* 4(3): 157–177.

Linz, P. 1985. Analytical and Numerical Methods for Volterra equations. *SIAM Studies in Appl. Math.*, Philadelphia.

Liu, F., V. Anh and I. Turner. 2002. Numerical solution of the fractional-order advection-dispersion equation. pp. 159–164. *In*: S. Wang and N. Fowkes [eds.]. Proc. BAIL2002, Univ. Western Australia, Perth, Australia.

Liu, F., V.V. Anh, I. Turner and P. Zhuang. 2003. Time fractional advection-dispersion equation. *J. Appl. Math. Computing* 13: 233–245.

Liu, F., M.M. Meerschaert, R.J. McGough, P. Zhuang and Q. Liu. 2013. Numerical methods for solving the multi-term time-fractional wave-diffusion equation. *Frac. Calc. Appl. Anal.* 16(1): 9–25.

Lizama, C. 2012. Solutions of two-term time fractional order differential equations with nonlocal initial conditions. *Electronic J. Qual. Theory Differential Eq.* 82: 1–9.

Lockington, D.A. and J.-Y. Parlange. 2003. Anomalous water absorption in porous materials. *J. Phys. D: Appl. Phys.* 36: 760–767.

Logvinova, K. and M.C. Neel. 2007. Solute spreading in heterogeneous aggregated porous media. pp. 185–196. *In*: J. Sabatier, O.P. Agrawal and J.A.T. Machado [eds.]. Advances in Fractional Calculus: Theoretical Developments and Applications in Physics and Engineering. Springer, Dordrecht.

Lorenzo, C.F. and T.T. Hartley. 1998. Initialization, Conceptualization, and Application in the Generalized Fractional Calculus. NASA/TP-1998-208415, Lewis Research Center.

Lorenzo, C.F. and T.T. Hartley. 2000. Initialized fractional calculus. *Int. J. Appl. Mech.* 3(3): 249–265.

Lorenzo, C.F. and T.T. Hartley. 2002. Variable order and distributed order fractional operators. *Nonlinear Dynamics* 29: 57–98.

Love, A.E.H. 1892. A Treatise on the Mathematical Theory of Elasticity. Vol. 1: Cambridge Univ. Press, https://hal.archives-ouvertes.fr/hal-01307751.

Lubich, C. 1986. A stability analysis of convolution quadraturea for Abel-Volterra equations. *IMA J. Numer. Anal.* 6(1): 87–101.

Luchko, Y. 2011. Initial-boundary-value problems for the generalized multi-term time-fractional diffusion equation. *J. Math. Anal. Appl.* 374: 538–548.

Luchko, Yu. and R. Gorenflo. 1998. Scale-invariant solutions of a partial differential equation of fractional order. *Fract. Calc. Appl. Anal.* 1: 63–78.

Luchko, Yu. and R. Gorenflo. 1999. An operational method for solving fractional differential equations with the Caputo derivatives. *Acta Math. Vietn.* 24(2): 207–233.

Luchko, Yu.F. and S.B. Yakubovich. 1993. Operational calculus for the generalized fractional differential operator and applications. *Math. Balkanica, New Ser.* 7: 119–130.

Luchko, Yu., F. Mainardi and Y. Povstenko. 2013. Propagation speed of the maximum of the fractional diffusion-wave equation. *Computers Math. Appl.* 66: 774–784.

Ludu, A. 2016. Differential equations of time dependent order. *AIP Conf. Proc.* 1773: 020005, doi: 10.1063/1.4964959.

Ludwig, C. 1856. Diffusion zwischen ungleich erwärmten Orten gleich zusammengesetzter Lösungen. *Sitzungsber Akad. Wiss. Wien Math-Naturwiss Kl.* 20: 539.

Machado, J.A. and V. Kiryakova. 2017. The chronicles of fractional calculus. *Fract. Calc. Appl. Anal.* 20(2): 307–336.

Machado, J.J., V. Kiryakova and F. Mainardi. 2011. Recent history of fractional calculus. *Commun. Nonlnear Sci. Numer. Simul.* 16: 1140–1153.

Mahdavi, A. 2015. Transient-state analytical solution for groundwater recharge in anisotropic sloping aquifer. *Water Resour. Manage.* 29: 3735–3748.

Maidment, D.R. 1993. Hydrology. Chapter 1. *In*: D.R. Maidment [ed.-in-chief]. Handbook of Hydrology. McGraw-Hill, New York.

Mainardi, F. 1996. Fractional relaxation-oscillation and fractional diffusion-wave phenomena. *Chaos, Solitons & Fractals* 7(9): 1461–1477.

Mainardi, F. 1997. Fractional calculus: Some basic problems in continuum and statistical mechanics. pp. 291–348. *In*: A. Carpinteri and F. Mainardi [eds.]. Fractals and Fractional Calculus in Continuum Mechanics. Springer-Verlag, New York.

Mainardi, F., Y. Luchko and G. Pagnini. 2001. The fundamental solution of the space-time fractional diffusion equation. *Frac. Calc. Appl. Anal.* 4(2): 153–192.

Mainardi, F., G. Pagnini and R.K. Saxena. 2005. Fox H functions in fractional diffusion. *J. Comput. Appl. Math.* 178: 321–331.

Mainardi, F., G. Pagnini and R. Gorenflo. 2007. Some aspects of fractional diffusion equations of single and distributed order. *Appl. Math. Comput.* 187: 295–305.

Mainardi, F., G. Pagnini and R. Gorenflo. 2008. Time-fractional diffusion of distributed order. *J. Vibration & Control* 14(9-10): 1267–1290.

Mainardi, F. 2010. Fractional Calculus and Waves in Linear Viscoelasticity. Imperial Coll. Press, London.

Mainardi, F. and G. Spada. 2011. Creep, relaxation and viscosity properties for basic fractional models in rheology. *Eur. Phys. J. Special Topics* 193: 133–160.

Mainardi, F. 2012. An historical perspective on fractional calculus in linear viscoelasticity. *Fract. Calc. Appl. Anal.* 15(4): 712–717.

Mainardi, F. and R. Garrappa. 2015. On complete monotonicity of the Prabhakar function and non-Debye relaxation in dielectrics. *J. Compt. Phys.* 293: 70–80.

Mandelbot, B.B. 1983. The Fractal Geometry of Nature. W.H. Freeman & Co., New York.

Marshall, T.J., J.W. Holmes and C.W. Rose. 1996. Soil Physics. Cambridge Univ. Press, New York.

Martín, M.A., C. García-Gutiérrez and M. Reyes. 2009. Modeling multifractal features of soil particle size distributions with Kolmogorov fragmentation algorithms. *Vadose Zone J.* 8(1): 202–208.

Martinez, F., J. San, Y.A. Pachepsky and W. Rawls. 2007. Fractional advective-dispersive equation as a model of solute transport in porous media. pp. 199–212. *In*: J. Sabatier, O.P. Agrawal and J.A.T. Machado [eds.]. Advances in Fractional Calculus: Theoretical Developments and Applications in Physics and Engineering. Springer, Dordrecht.

Masoliver, K. and K. Lindenberg. 2017. Continuous time persistent random walk: a review and some generalizations. *Eur. Phys. J. B.* 90: 107, 2–13.

Mathai, A.M. and R.K. Saxena. 1978. The H-Function with Applications in Statistics and Other Disciplines. Wiley, New York.

Maxwell, J.C. 1868. On reciprocal diagrams in space, and their relation to Airy's function of stress. *Proc. Lond. Math. Soc.* 2(1): 58–60.

Mbagwu, J.S.C. 1995. Testing the goodness of fit of infiltration models for highly permeable soils under different tropical soil management systems. *Soil & Tillage Res.* 34: 199–205.

McTigue, D.F. 1986. Thermoelastic response of fluid-saturated porous rock. *J. Geophys. Res.* 91(B9): 9533–9542.

McWhorter, D.B. 1971. Infiltration affected by flow of air. Hydrol. Paper 49, Colorado State University, Fort Collins, Colorado.

McWhorter, D.B. and D.K. Sunada. 1990. Exact integral solutions for two-phase flow. *Water Resour. Res.* 26(3): 399–413.

McWhorter, D.B. and D.K. Sunada. 1992. Reply. *Water Resour. Res.* 28(5): 1479.

Meerschaert, M. and H.-P. Scheffler. 2001. Limit distributions for sums of independent random vectors. Heavy Tails in Theory and Practice. Wiley, New York.

Meerschaert, M.M., D.A. Benson and B. Bäumer. 1999. Multidimensional advection and fractional dispersion. *Phys. Rev. E.* 59(5): 5026–5028.

Meerschaert, M.M., D.A. Benson, H.-P. Scheffler and P. Becker-Kern. 2002. Governing equation and solutions of anomalous random walk limits. *Phys. Rev. E.* 66: 060102(R), 1–4.

Meerschaert, M.M. and C. Tadjeran. 2004. Finite difference approximations for fractional advection–dispersion flow equations. *J. Comput. Appl. Math.* 172(1): 65–77.

Meerschaert, M.M. and C. Tadjeran. 2006. Finite difference approximations for two-sided space-fractional partial differential equations. *Appl. Numer. Math.* 56(1): 80–90.

Meerschaert, M.M., Y. Zhang and B. Baeumer. 2008. Tempered anomalous diffusion in heterogeneous systems. *Geophys. Res. Lett.* 35: L17403.

Meerschaert, M.M., Y. Zhang and B. Baeumer. 2010. Particle tracking for fractional diffusion with two time scales. *Computer & Math. Appl.* 59: 1078–1086.

Meerschaert, M.M. 2012. Chapter 11: Fractional calculus, anomalous diffusion, and probability. pp. 265–284. *In*: R. Metzler, S.C. Lim and J. Klafter [eds.]. Fractional Calculus, Anomalous Diffusion, and Probability. World Scientific, Singapore.

Mehdinejadiani, B., H. Jafari and D. Baleanu. 2013. Derivation of a fractional Boussinesq equation for modelling unconfined groundwater flow. *Eur. Phys. J. Special Topics* 222: 1805–1812.

Meinzer, O.E. 1928. Compressibility and elasticity of artesian aquifers. *Econ. Geol.* 23: 263–291.

Menziani, M., S. Pugnaghi and S. Vicenzi. 2007. Analytical solutions of the linearized Richards equation for discrete arbitrary initial and boundary conditions. *J. Hydrol.* 332(1): 214–225.

Meshkov, S.I. 1974. Viscoelastic Properties of Metals. Metallurgia, Moscow.

Metzler, R., W.G. Glockle and T.F. Nonnenmacher. 1994. Fractional model equation for anomalous diffusion. *Physica A.* 211(1): 13–24.

Metzler, R., J. Klafter and I.M. Sokolov. 1998. Anomalous transport in external fields: Continuous time random walks and fractional diffusion equations extended. *Phys. Rev. E.* 58(2): 16121–1633.

Metzler, R. and A. Compte. 2000. Generalized diffusion-advection schemes and dispersive sedimentation: A fractional approach. *J. Phys. Chem.* 104: 3838–3865.

Metzler, R. and J. Klafter. 2000a. The random walk's guide to anomalous diffusion: A fractional dynamics approach. *Phys. Report.* 339: 1–77.

Metzler, R. and J. Klafter. 2000b. Accelerating Brownian motion: A fractional dynamics approach to fast diffusion. *Europhys. Lett.* 51(5): 492–498.

Metzler, R. and T.F. Nonnenmacher. 2002. Space- and time-fractional diffusion and wave equations, fractional Fokker-Planck equations, and physical motivation. *Chem. Phys.* 284: 67–90.

Metzler, R. and J. Klafter. 2004. The restaurant at the end of random walk: recent development in the description of anomalous transport by fractional dynamics. *J. Phys. A: Math. Gen.* 37: R161–R208.

Michels, L., Y. Méheust, M.A.S. Altoé, E.C. dos Santos, H. Hemmen and R. Droppa, Jr. et al. 2019. Water vapor diffusive transport in a smectite clay: Cationic control of normal versus anomalous diffusion. *Phys Rev. E.* 99: 013102.

Millán, H. and M. González-Posada. 2005. Modelling soil water retention scaling. Comparison of a classical fractal model with a piecewise approach. *Geoderma.* 125: 25–38.

Miller, C.T., G. Christakos, P.T. Imhoff, J.F. McBride and J.A. Pedit. 1998. Multiphase flow and transport modelling in heterogeneous porous media: challenges and approaches. *Adv. Water Resour.* 21(2): 77–120.

Miller, E.E. and R.D. Miller. 1955a. Theory of capillary flow: I. Practical implications. *Proc. Soil Sci. Soc. Am.* 19: 267–271.

Miller, E.E. and R.D. Miller. 1955b. Theory of capillary flow: II. Experimental information. *Proc. Soil Sci. Soc. Am.* 19: 271–275.

Miller, E.E. and R.D. Miller. 1956. Physical theory of capillary flow phenomena. *J. Appl. Phys.* 27(4): 324–332.

Miller, K.S. and B. Ross. 1993. An Introduction to the Fractional Calculus and Fractional Differential Equations. Wiley, New York.

Miller, K.S. and S.G. Samko. 2001. Completely monotonic functions. *Integr. Transf. Special Funct.* 12(4): 389–402.

Miller, K., A. Prudnikov, B. Sachdeva and S. Samko. 1994. Obituary. *Integr. Transf. Special Funct.* 2(1): 81–82.

Mittag-Leffler, G.M. 1903. Sur la nouvelle fonction E_α. *C.R. Acad. Sci. Paris* 137: 554–558.

Moench, A.F. 1995. Convergent radial dispersion in a double-porosity aquifer with fracture skin: Analytical solution and application to a field experiment in fractured chalk. *Water Resour. Res.* 31(8): 1823–1835.

Mogul'skii, A.A. 1976. Large deviations for trajectories of multi-dimensional random walks. *Theory Prob. Appl.* 21(2): 300–15 (translated by W.M. Vasilaky from the Russian).

Momani, S. and Z.M. Odibat. 2007. Fractional Green function for linear time-fractional inhomogeneous partial differential equations in fluid mechanics. *J. Appl. Math. & Computing* 24(1-2): 167–178.

Monin, A.S. and A.M. Yaglom. 1971. Statistical Fluid Mechanics. Vol. I, MIT Press, Cambridge, Mass.

Montroll, E.W. and G.H. Weiss. 1965. Random walks on lattices, II. *J. Math. Phys.* 6: 167–181.

Montroll, E.W. and H. Scher. 1973. Random walks on lattices. IV. Continuous time random walk and influence of absorbing boundaries. *J. Stat. Phys.* 9(2): 101–135.

Morales-Casique, E., S.P. Neuman and A. Guadagnini. 2006. A non-local and localized analyses of non-reactive solute transport in bounded randomly heterogeneous porous media: theoretical framework. *Adv. Water Resour.* 29(8): 1238–1255.

Morel-Seytoux, H.J. 1973. Two-phase flows in porous media. *Adv. Hydrosci.* 9: 119–202.

Mortimer, R.G. and H. Eyring. 1980. Elementary transition state theory of the Soret and Dufour effects. *Proc. Natl. Acad. Sci.* 77(4): 1728–1731.

Mualem, Y. 1976. A new model for predicting the hydraulic conductivity of unsaturated porous media. *Water Resour. Res.* 12: 513–522.

Muralidhar, R. and D. Ramkrishna. 1993. Diffusion in pore fractals. *Trans. Porous Media.* 13: 79–95.

Muskat, M. and M. Meres. 1936. The flow of heterogeneous fluids. *Phys.* 7: 346–363.

Muskat, M. 1937. The Flow of Homogeneous Fluids through Porous Media. McGraw-Hill.

Myint-U, Tyn and L. Debnath. 1987. Partial Differential Equations for Scientists and Engineers. 3rd ed., Prentice Hall, Englewood Cliffs. New Jersey.

Narasimhan, T.N. 1998. Hydraulic characterization of aquifers, reservoir rocks, and soils. A history of ideas. *Water Resour. Res.* 34(1): 33–46.

Narasimhan, T.N. 1999. Fourier's heat conduction equation: History, influence, and connections. *Rev. Geophys.* 37: 151–172.

Narasimhan, T.N. 2009. The dichotomous history of diffusion. *Phys. Today.* 62(7): 48–53.

Natale, G. and E. Salusti. 1996. Transient solutions for temperature and pressure waves in fluid-saturated porous rocks. *Geophys. J. Internl.* 124: 649–656.

Nash, J.E. 1957. The form of the instantaneous unit hydrograph. *IASH Publ.* 45(3-4): 114–121.

Navier, C.-L.M.-H. 1820. Mémoire sur la flexion des plans élastiques. *Bibliothèque de l'Ecole Nationale des Ponts et Chaussées.*

Navier, C.-L.M.-H. 1821. Sur les lois des mouvements des fluides, en ayant égard à l'adhésion des molécules. *Ann. Chimie.* 19: 244–260.

Navier, C.-L.-M.-H. 1827. Mémoire sur les lois de l'équilibre et du mouvement des corps solides élastiques (1821). *Mém. Acad. Sci. Inst. France* 7(2): 375–393.

Neidigk, S., C. Salas, E. Soliman, D. Mercer and M.M.R. Taha. 2009. Creep and relaxation of osteoporotic bones. Proc. SEM Ann. Conf. 1–4 June 2009, Albuquerque, New Mexico.

Negoro, A. 1994. Stable-like process: construction of the transition density and the behavior of sample paths near t = 0. *Osaka J. Math.* 31: 189–214.

Neumann, F.E. 1833. Die thermischen ... Axen des Krystallsystems des Gypses. Pogg. *Ann.* xxii.

Neumann, F.E. 1885. Vorlesungen über die theorie der elasticität der festen Körper und des lichtäthers.

Nielsen, D.R. and J.W. Biggar. 1961. Miscible displacement in soils. 1. Experimental information. *Soil Sci. Soc. Amer. Proc.* 25(10): 1–5.

Nielsen, D.R., M.Th. van Genuchten and J.M. Biggar. 1986. Water flow and solute transport processes in the unsaturated zone. *Water Resour. Res.* 22(9): 89S–108S.

Nielsen, D.R., K. Kutilek, O. Wendroth and J.W. Hopmans. 1997. Selected research opportunities in soil physics. *Sci. Agricola.* 54: 51–77.

Nigmatullin, R.R. 1984. To the theoretical explanation of the universal response. *Physica B.* 123: 739–745.

Nigmantullin, R.R. 1986. The realization of the generalized transfer equation in a medium with fractal geometry. *Phys. Status Solidi B.* 133(1): 425–430.

Nigmatullin, R.Sh. and V.A. Belavin. 1964. Electrolite fractional-differentiating and integrating two-pole network. *Trudy (Trans.) Kazan Aviation Inst.* (Radiotechn. & Electronics). 82: 58–66 (In the Russian).

Noyes, R.M. and R.J. Field. 1974. Oscillatory chemical reactions. *Ann. Rev. Phys. Chem.* 25: 95–119.

Nutting, P.G. 1921a. A study of elastic viscous deformation. *Proc. Amer. Soc. Test Mater.* 21: 1162–1171.

Nutting, P.G. 1921b. A new general law of deformation. *J. Franklin Inst.* 191: 679–685.

Nutting, P.G. 1930. Physical analysis of oil sands. *Bull. Amer. Assoc. Petr. Geol.* 14: 1337–1349.

Oberhettinger, F. and L. Badii. 1973. Tables of Laplace Transforms. Springer-Verlag, Berlin.

Ochoa-Tapia, J.A., F.J. Valdes-Parada and J. Alvarez-Ramirez. 2007. A fractional-order Darcy law. *Physica A.* 374: 1–14.

Odzijewicz, T., A.B. Malinowska, and D.F.M. Torres. 2013. Fractional variational calculus of variable order. pp. 291–301. *In*: A. Almeida, L. Castro and F.-O. Speck [eds.]. Advances in Harmonic Analysis and Operator Theory. The Stefan Samko Anniversary Volume.

Oldham, K.B. and J. Spanier. 1970. The replacement of Fick's law by a formulation involving semidifferentiation. *J. Electroanal. Chem.* 26: 331–341.

Oldham, K.B. and J. Spanier. 1974. The Fractional Calculus. Academic Press, New York.

O'Loughlin, E.M. 1986. Prediction of surface saturation zones in natural catchments by topographic analysis. *Water Resour. Res.* 22(5): 794–804.

Olsen, J.S., J. Mortense and A.S. Telyakovskiy. 2016. A two-sided factional conservation of mass equation. *Adv. Water Resour.* 91: 117–121.

Onsager, L. 1931a. Reciprocal relations in irreversible processes. I. *Phys. Rev.* 15(37): 405–426.

Onsager, L. 1931b. Reciprocal relations in irreversible processes. II. *Phys. Rev.* 15(38): 2265–2279.

Ortigueira, M.D. and J.J. Trujillo. 2012. A unified approach to fractional derivatives. *Commun. Nonlinear Sci. Numer. Simulat.* 17: 5151–5157.

O'Shaunghnessy, B. and I. Procaccia. 1985. Analytical solutions of diffusion in fractal objects. *Phys. Rev. Lett.* 54(5): 455–458.

O'Shaunghnessy, L. 1918. Problem #433. *Amer. Math. Month.* 25(4): 172–173.

Osler, T.J. 1971. Taylor's series generalized for fractional derivatives and applications. *SIAM J. Math. Anal.* 2(1): 37–48.

Özdemir, N. and D. Karadeniz. 2008. Fractional diffusion-wave problem in cylindrical coordinates. *Phys. Lett. A.* 372: 5968–5972.

Pachepsky, Y., D.A. Benson and W. Rawls. 2000. Simulating scale-dependent solute transport in soils with the fractional advection-dispersion equations. *Soil Sci. Soc. Am. J.* 64: 1234–1243.

Pachepsky, Y., D. Timlin and W. Rawls. 2003. Generalised Richards' equation to simulate water transport in unsaturated soils. *J. Hydrol.* 272: 3–13.

Pachepsky, Y.A. and W.J. Rawls. 2003. Soil structure and pedotransfer functions. *Eur. J. Soil Sci.* 54: 443–451.

Paladin, G. and A. Vulpiani. 1987. Anomalous scaling laws in multifractal objects. *Phys. Reports (Rev. Sec. Phys. Lett.).* 156(4): 147–225.

Pandey, V. and S. Holm. 2016. Linking the fractional derivatives and the Lomnitz creep law to non-Newtonian time-varying viscosity. *Phys. Rev. E.* 94: 032606.

Paneva-Konovska, J. 2013. On the multi-index (3m-parametric) Mittag-Leffler functions, fractional calculus relations and series convergence. *Cent. Eur. J. Phys.* 11(10): 1164–1177.

Park, H.W., J. Choe and J.M. Kang. 2010. Pressure behavior of transport in fractal porous media using a fractional calculus approach. *Energy Sources* 22: 881–890.

Parker, J.C. and M.Th. van Genuchten. 1984. Flux-averaged and volume-averaged concentrations in continuum approaches to solute transport. *Water Resour. Res.* 20(7): 866–872.

Parkin, G.W., D.E. Elrich and R.G. Kachanoski. 1992. Cumulative storage of water under constant flux infiltration: analytic solution. *Water Resour. Res.* 28: 2811–2818.

Parlange, J.-Y. and R.E. Smith. 1976. Ponding time for variable rainfall rates. *Can. J. Soil Sci.* 56: 121–123.

Parlange, J.-Y., J.L. Starr, M.Th. van Genuchten, D.A. Barry and J.C. Parker. 1992. Exit condition for miscible displacement experiments. *Soil Sci.* 153(3): 165–171.

Passioura, J.B. and I.R. Cowan. 1968. On solving the non-linear diffusion equation for the radial flow of water to roots. *Agric. Meteorol.* 5: 129–134.

Passioura, J.B. 1971. Hydrodynamic dispersion in aggregated media. *Soil Sci.* 11: 339–344.

Pearson, K. 1905. The problem of the random walk. *Nature* 72(1867): 294.

Perfect, E. and B.D. Kay. 1995. Applications of fractals in soil and tillage research: a review. *Soil & Tillage Res.* 36: 1–20.

Perfect, E. 1999. Estimating mass fractal dimensions from water retention curves. *Geoderma.* 88: 221–231.

Perfect, E., Y. Pachepsky and M.A. Martín. 2009. Fractal and multifractal models applied to porous media. *Vadose Zone J.* 8(1): 174–176.

Perrier, E., M. Rieu, G. Sposito and G. de Marsily. 1996. Models of the water retention curve for soils with a fractal pore size distribution. *Water Resour. Res.* 32(10): 3025–3031.

Pfaff, J.F. 1797. Pbservationes analyticae ad L. Euler Institiones Calculi Integralis. vol. IV, Supplem. II et IV, Historia de 1793, *Nova acta acad. sci. Petropolitanae.* 11: 38–57.

Philip, J.R. 1954a. Some recent advances in hydrologic physics. *J. Inst. Engrs, Australia* 26: 255–259.

Philip, J.R. 1954b. An infiltration equation with physical significance. *Soil Sci.* 77(2): 153–157.

Philip, J.R. 1957a. The theory of infiltration: 1. The infiltration equation and its solution. *Soil Sci.* 83: 345–357.

Philip, J.R. 1957b. The theory of infiltration: 4. Sorptivity and algebraic infiltration equations. *Soil Sci.* 84: 257–264.

Philip, J.R. 1957c. Evaporation, and moisture and heat fields in the soil. *J. Meteorol.* 14: 354–366.

Philip, J.R. and D.A. de Vries. 1957. Moisture movement in porous materials under temperature gradients. *Trans. Amer. Geophys. Union.* 38(2): 222–232.

Philip, J.R. 1960a. A very general class of exact solutions in concentration-dependent diffusion. *Nature* 185(4708): 233.

Philip, J.R. 1960b. General method of exact solution of the concentration-dependent diffusion equation. *Aust. J. Phys.* 13(1): 1–12.

Philip, J.R. and D.A. Farrell. 1964. General solution of the infiltration advance problem in irrigation hydraulics. *J. Geophys. Res.* 69: 621–631.

Philip, J.R. 1967. Sorption and infiltration in heterogeneous media. *Aust. J. Soil Res.* 5: 1–10.

Philip, J.R. 1969a. Theory of infiltration. *Adv. Hydrosci.* 5: 215–296.

Philip, J.R. 1969b. Hydrostatics and hydrodynamics in swelling soils. *Water Resour. Res.* 5(5): 1070–1077.

Philip, J.R. 1969c. Moisture equilibrium in the vertical in swelling soils. I. Basic theory. *Aust. J. Soil Res.* 7: 99–120.

Philip, J.R. 1969d. Moisture equilibrium in the vertical in swelling soils. II. Applications. *Aust. J. Soil Res.* 7: 121–141.

Philip, J.R. and D.E. Smiles. 1969. Kinetics of sorption and volume changes in three-component systems. *Aust. J. Soil Res.* 7: 1–19.

Philip, J.R. 1970a. Flow in porous media. *Ann. Rev. Fluid Mech.* 2: 177-2-4.

Philip, J.R. 1970b. Reply. *Water Resour. Res.* 6(4): 1248–1251.

Philip, J.R. 1970c. Hydrostatics in swelling soils and soil suspensions: Unification of concepts. *Soil Sci.* 109(5): 294–298.

Philip, J.R. 1972. Hydrology of swelling soils. pp. 95–107. *In*: T. Talsma and J.R. Philip [eds.]. Salinity and Water Use. *Proc. Symp. Aust. Acad. Sci.*, Canberra, 1971, Macmillan, London.

Philip, J.R. 1973. On solving the unsaturated flow equation: 1. The flux-concentration relation. *Soil Sci.* 116(5): 328–335.

Philip, J.R. 1974. Fifty years progress in soils physics. *Geoderma.* 12: 265–280.

Philip, J.R. and J.H. Knight. 1974. On solving the unsaturated flow equation: 3. New quasi-analytical technique. *Soil Sci.* 117: 1–13.

Philip, J.R. 1980. Field heterogeneity: Some basic issues. *Water Resour. Res.* 16(2): 443–448.

Philip, J.R. 1985. Reply to Comments on Steady infiltration from spherical cavities. *Soil Sci. Soc. Am. J.* 49: 788–789.

Philip, J.R. 1986. Issues in flow and transport in heterogeneous porous media. *Transport Porous Media* 1: 319–338.

Philip, J.R. 1990. Inverse solution for one-dimensional infiltration, and the ratio of A/K_1. *Water Resour. Res.* 26(9): 2023–2027.

Philip, J.R. 1991a. Hillslope infiltration: Planar slopes. *Water Resour. Res.* 27(1): 109–117.

Philip, J.R. 1991b. Hillslope infiltration: Divergent and convergent slopes. *Water Resour. Res.* 27(6): 1035–1040.

Philip, J.R. 1991c. Infiltration and downslope unsaturated flows in concave and convex topographies. *Water Resour. Res.* 27(6): 1041–1048.

Philip, J.R. 1992a. Flow and volume change in soils and other porous media and in tissues. pp. 3–31. *In*: T.K. Karalis [ed.]. Mechanics of Swelling, NATO ASI, Vol. H64, Springer-Verlag, Berlin.

Philip, J.R. 1992b. Exact solutions for redistribution by nonlinear convection-diffusion. *J. Austral. Math. Soc. Ser. B.* 33: 363–383.

Philip, J.R. 1994. Some exact solutions of convection-diffusion and diffusion equations. *Water Resour. Res.* 30(12): 3545–3551.

Philip, J.R. 1996. Mathematical physics of infiltration on flat and sloping topography. pp. 327–349. *In*: M.F. Wheeler [ed.]. Environmental Studies. The IMA Volumes in Mathematics and Its Applications, Vol. 79. Springer, New York.

Philip, J.R. and J.H. Knight. 1997. Steady infiltration flows with sloping boundaries. *Water Resour. Res.* 33(8): 1833–1841.

Pickens, J.F., R.E. Jackson, K.J. Inch and W.F. Merritt. 1981. Measurement of distribution coefficients using radial injection dual-tracer test. *Water Resour. Res.* 17(3): 529–544.

Pirson, S.J. 1953. Performance of fractured oil reservoirs. *Bull. Amer. Assoc. Petrol. Geologists* 37(2): 232–244.

Pitcher, E. and W.E. Sewell. 1938. Existence theorems for solutions of differential equations of non-integer order. *Bull. Amer. Math. Soc.* 44(2): 100–107 (A correction in 44(12): 888).

Planck, M. 1917. Uber einen Satz der statistischen Dynamik und seine Erweiterung in der Quantentheorie. *Sitzungsber. Preuß. Akad. Wiss., phys.-math. Kl.*, 10.5: 324–341.

Platten, J.K. and P. Costesèque. 2004. Charles Soret. A short biography. *Eur. Phys. J. E.* 15: 235–239.

Pochhammer, L. 1890. Zur Theorie der Euler'schen Integrale. *Mathematische Annalen.* 35(4): 495–526.

Podlubny, I. 1999. Fractional Differential Equations. Acad. Press, San Diego, California.

Podlubny, I. 2002. Geometric and physical interpretation of fractional integration and fractional differentiation. *Fract. Calc. Appl. Anal.* 5(4): 367–386.

Polubarinova-Kochina, P.Ia. 1952. Teoriia dvizhenia gruntovykh vod (Theory of Motion of Ground Waters), Gostekhizdat.

Polubarinova-Kochina, P.Ya. 1962. Theory of Ground-water Movement. Princeton, NJ, Princeton Univ. Press.

Pólya, G. 1921. Über eine Aufgabe der Wahrscheinlichkeitsrechnung betreffend die Irrfahrt im Straßennetz. *Math. Ann.* 84(1-2): 149–160.

Polyanin, A. and A.V. Manzhirov. 1998. Handbook of Integral Equations. CRC Press, Boca Raton, Florida.

Poole, E.G.C. 1936. Introduction to the Theory of Linear Differential Equations. Oxford Univ. Press, Oxford.

Post, E.L. 1918. A solution of $\dfrac{\partial^{1/2} y}{\partial x^{1/2}} - \dfrac{y}{x} = 0$ for Problem #433 by O'Shaughnessy. *Amer. Math. Month.* 25(4): 172–173.

Post, E.L. 1919. Discussion of the solution $(d/dx)^{1/2} y = y/x$. *Amer. Math. Monthly* 26: 37–39.

Postelnicu, A. 2004. Influence of a magnetic field on heat and mass transfer by natural convection from vertical surfaces in porous media considering Soret and Dufour effects. *Internl. J. Heat Mass Transfer* 47: 1467–1472.

Prabhakar, T.R. 1971. A singular integral equation associated with a generalized Mittag-Leffler function in the kernel. *Yokohoma Math. J.* 19: 7–15.

Prajapati, J.C., R.K. Jana, R.K. Saxena and A.K. Shukla. 2013. Some results on the generalized Mittag-Leffler function operator. *J. Inequal. Appl.* 33: 1–6.

Pratt, W.E. and D.W. Johnson. 1926. Local subsidence of the Goose Creek oil field. *J. Geol.* 34: 577–590.

Prevedello, C.L., J.M.T. Loyola, K. Reichardt and D.R. Nielsen. 2009. New analytic solution related to the richards, philip, and green–ampt equations for infiltration. *Vadose Zone J.* 8: 127–135.

Pyke, R. 1961. Markov renewal processes: Definition and preliminary properties. *Annals Math. Stat.* 32(4): 1231–1242.

Qi, H. and J. Liu. 2010. Time-fractional radial diffusion in hollow geometries. *Meccanica.* 45: 577–583.

Raats, P.A.C. 1965. Development of equations describing transport of mass and momentum in porous media, with special reference to soils. PhD Thesis, Univ. Illinois, Urbana-Champaign.

Raats, P.A.C. and A. Klute. 1968. Transport in soils: the balance of momentum. *Soil Sci. Soc. Amer. Proc.* 32: 452–456.

Raats, P.A.C. 2001. Developments in soil-water physics since the mid 1960s. *Geoderma.* 100: 355–387.

Raats, P.A.C. and M.Th. van Genuchten. 2006. Milestones in soil physics. *Soil Sci.* 171(1): S21–S28.

Rabotnov, Yu.N. 1948. Equilibrium of an elastic medium with after-effect (in Russian). *Prikladnaya Mat. i Mekhanika (J. Appl. Math. Mech.).* 12(1): 53–62. Reprinted as: Rabotnov, Yu.N. 2014. *Fract. Cal. Appl. Anal.* 17(3): 684–696.

Rador, T. and S. Taneri. 2006. Random walks with shrinking steps: First-passage characteristics. *Phys. Rev. E.* 73: 036118.

Raghavan, R. 2012. Fractional derivatives: Application to transient flow. *J. Petrol. Sci. Eng.* 80: 7–13.

Raghavan, R. and C. Chen. 2019. The Theis solution for subdiffusive flow in rocks. *Oil & Gas Sci. Tech. Rev.* 74(6): 1–10.

Rammal, R. and G. Toulouse. 1983. Random walks on fractal structures and percolation clusters. *J. Physique Lett.* 44(1): 13–22.

Rangarajan, G. and M. Ding. 2000. Anomalous diffusion and the first passage time problem. *Phys. Rev. E.* 62(1): 120–135.

Rasmussen, W.O. 1994. Infiltration-advance equation for radial spreading. *Water Resour. Res.* 30(4): 929–937.

Rasmussen, T.C. and L.A. Crawford. 1996. Identifying and removing barometric pressure effects in confined and unconfined aquifers. *Ground Water.* 35(3): 502–511.

Rawls, W.L., D.L. Brakensiek and K.E. Saxton. 1982. Estimation of soil water properties. *Trans. ASAE* 25: 1316–1320 and 1328.

Rawls, W.L. and D.L. Brakensiek. 1995. Utilizing fractal principles for predicting soil hydraulic properties. *J. Soil Water Cons.* 50(5): 463–465.

Razminia, K., A. Razminia and J.J. Trujilo. 2015a. Analysis of radial composite systems based on fractal theory and fractional calculus. *Signal Process* 107: 378–388.

Razminia, K., A. Razminia and D.M.F. Torre. 2015b. Pressure responses of a vertically hydraulic fractured well in a reservoir with fractal structure. *Appl. Math. Comput.* 257: 374–380.

Razminia, K., A. Razminia and D. Baleanu. 2019. Fractal-fractional modelling of partially penetrating wells. *Chaos, Solitons & Fractals* 119: 135–142.

Renard, P., J. Gómez-Hernández and S. Ezzedine 2005. Ch. 147: Characterization of Porous and Fractured Media. Groundwater. *Encyclopedia of Hydrol. Sci.,* Wiley.

Ricciuti, C. and B. Toaldo. 2017. Semi-Markov models and motion in heterogeneous media. *J. Stat. Phys.* 169: 340–361.

Richards, L.A. 1931. Capillary conduction of liquids in porous mediums. *Phys.* 1(1): 318–333.

Richardson, L.F. 1926. Atmospheric diffusion shown on a distance-neighbour graph. *Proc. Roy. Soc. Ser. A.* 110(756): 709–737.

Riesz, M. 1938. Intégrales de Riemann-Liouville et potentiels. *Acta Szeged.* 9: 1–42.

Rieu, M. and G. Sposito. 1991. Fractal fragmentation, soil porosity, and soil water properties: I. Theory. *Soil Sci. Soc. Am. J.* 55: 1231–1238.

Risken, H. 1996. The Fokker-Planck Equation. 2nd ed., Springer, Berlin.

Roberts-Austen, W.C. 1896. On the diffusion of metals. *Philos. Trans. R. Soc. London* 187: 383–413.

Rodriguez-Iturbe, I. and J.B. Valdés. 1979. The geomorphologic structure of hydrologic response. *Water Resour. Res.* 15(6): 1409–1420.

Rogers, L. 1983. Operators and fractional derivatives for viscoelastic constitutive equations. *J. Rheol.* 27: 351–372.

Rogosin, S. and F. Mainardi. 2014. George William Scott Blair—The pioneer of fractional calculus in rheology. ArXiv: 1404.3295: 10.1685/journal.caim.481.

Rojstaczer, S. 1988. Determination of fluid flow properties from the response of water levels in wells to atmospheric loading. *Water Resour. Res.* 24(11): 1927–1938.

Rojstaczer, S. and F. Riley. 1990. Response of the water level in a well to earth tides and atmospheric loading under unconfined conditions. *Water Resour. Res.* 26(8): 1803–1817.

Romkens, M.J.M. and S.N. Prasad. 2006. Rain infiltration into swelling/shrinking/cracking soils. *Agric. Water Management* 86(1-2): 196–205.

Ross, B. [ed.]. 1975. Fractional calculus and its applications. Proc. The Internl. Conf. Fract. Calc. & Its Appl. Univ. New Havean, Connecticut, June 1974. Springer-Verlag, New York, USA.

Rubinshtein, L.I. 1948. K voprosu rasprostraneniia tepla v geterogennykh sredakh (On the problem of the process of propagation of heat in heterogeneous media). *Izv. A&ad. Nauk SSSR, Ser. Geogr.* No. 1.

Saichev, A. and G. Zaslavsky. 1997. Fractional kinetic equations: solutions and applications. *Chaos.* 7(4): 753–764.

Sachkov, Yu.L. 2008. Maxwell strata in the Euler elastic problem. *J. Dyn. Control Sys.* 14(2): 169–234.

Saffman, P.G. 1959. A theory of dispersion in a porous medium. *J. Fluid Mech.* 6(3): 321–349.

Saigo, M. and A.A. Kilbas. 1994. On asymptotic solutions of nonlinear and linear Abel-Volterra integral equations. II. *Surikaisekikenkyusho Kokyuroku* (数理解析研究所講究録). 881: 112–129.

Samko, S.G. and B. Ross. 1993. Integration and differentiation to a variable fractional order. *Integral Transforms Spec. Funct.* 1(4): 277–300.

Samko, S.G., A.A. Kilbas and O.I. Marichev. 1993. Fractional Integrals and Derivatives: Theory and Applications. Gordon & Breach, Amsterdam. [Engl. Transl. from Russian, Integrals and Derivatives of Fractional Order and Some of Their Applications. Nauka i Tekhnika, Minsk 1987].

Samko, S.G. 1995. Fractional integration and differentiation of variable order. *Anal. Math.* 21: 213–236.

Samko, S.G. 2013. Fractional integration and differentiation of variable order: an overview. *Nonlinear Dyn.* 71: 653–662.

Sander, G.C., J.-Y. Parlange and W.L. Hogarth. 1988. Air and water flow: II. Gravitational flow with an arbitrary flux boundary condition. *J. Hydrol.* 99(5): 225–234.

Sander, G.C., J.-Y. Parlange, I.G. Lisle and S.W. Weeks. 2005. Exact solutions to radially symmetric two-phase flow for an arbitrary diffusivity. *Adv. Water Resour.* 28(10): 1112–1121.

Sandev, T., A.V. Chechkin, N. Korabel, H. Kantz, I.M. Sokolov and R. Metzler. 2015. Distributed-order diffusion equations and multifractality: Models and solutions. *Phys. Rev. E.* 92: 042117.

Sardin, M., D. Schweich, G.J. Leij and M.Th. van Genuchten. 1991. Modeling the nonequilibrium transport of linearly interacting solutes in porous media: A review. *Water Resour. Res.* 27(9): 2287–2307.

Saxena, R.K., S.L. Kalla and R. Saxena. 2011. Multivariate analogue of generalized Mittag-Leffler function. *Integral Transf. Sp. Funct.* 22(7): 533–548.

Saxena, R.K., J.P. Chauhan, R.K. Jana and A.K. Shukla. 2015. Further results on the generalized Mittag-Leffler function operator. *J. Inequal. Appl.* 2015(75): 1–12.

Scalas, E., R. Gorenflo and F. Mainardi. 2004. Uncoupled continuous-time random walks: Solution and limiting behavior of the master equation. *Phys. Rev. E.* 69: 011107.

Scheidegger, A.E. 1954. Statistical hydrodynamics in porous media. *J. Appl. Phys.* 25(8): 994–1001.

Scheidegger, A.E. 1961. General theory of dispersion in porous media. *J. Geophys. Res.* 66(10): 3273–3278.

Schiessel, H., R. Metzler, A. Blumen and T.F. Nonnenmacher. 1995. Generalized viscoelastic models: their fractional equations with solutions. *J. Phys. A: Math. Gen.* 28: 6567–6584.

Schneider, W.R. and W. Wyss. 1989. Fractional diffusion and wave equations. *J. Math. Phys.* 30(1): 134–144.

Schofield, R.K. and G.W.S. Blair. 1930. The influence of the proximity of a solid wall on the consistency of viscous and plastic materials. *J. Phys. Chem.* 34: 248–262; III. 35: 1212–1215 (1931); IV. 39: 973–981 (1935).

Schofield, R.K. and G.W.S. Blair. 1933. The relationship between viscosity, elasticity and plastic strength of a soft materials as illustrated by some mechanical properties of flour dough. III. *Proc. Roy. Soc. A.* 141: 72–85.

Schulz, J.H.P., E. Barkai and R. Metzler. 2014. Aging renewal theory and application to random walks. *Phys. Rev. X.* 4: 011028, 1–24.

Schumer, R., D.A. Benson, M.M. Meerschaert and B. Baeumer. 2003a. Multiscaling fractional advection-dispersion equations and their solutions. *Water Resour. Res.* 39(1): 1022, 1–11.

Schumer, R., D.A. Benson, M.M. Meerschaert and B. Baeumer. 2003b. Fractal mobile-immobile solute transport. *Water Resour. Res.* 39(10): 1296, 1–12.

Schumer, R., M.M. Meerschaert and B. Baeumer. 2009. Fractional advection-dispersion equation for modeling transport at the Earth surface. *J. Geophys. Res.* 114: F00A07.

Schumer, R., B. Baeumer and M.M. Meerschaert. 2011. External behavior of a coupled continuous time random walk. *Phys. A.* 390: 505–511.

Schwarzenbach, R.P., B.I. Escher, K. Fenner, T.B. Hofstetter, C.A. Johnson, U. von Gunten et al. 2006. The challenge of micropollutants in aquatic systems. *Sci.* 313: 1072–1077.

Seyfried, M.S. and P.S.C. Rao. 1987. Solute transport in undisturbed columns of an aggregated tropical soil: Preferential flow effects. *Soil Sci. Amer. J.* 51: 1434–1444.

Shen, C. and M. Phanikumar. 2009. An efficient space-fractional dispersion approximation for stream solute transport modeling. *Adv. Resour. Res.* 32: 1482–1494.

Sherman, L.K. 1932. Streamflow from rainfall using the unit-graph method. *Eng. New Records.* 108: 501–505.

Shiga, T. 1998. Deformation and viscoelastic behavior of polymer gel in electric fields. *Proc. Japan Acad., Ser. B, Phys. Biol. Sci.* 74: 6–11.

Shih, D.C.-F. 2016. Storage in confined aquifer: Spectral analysis of groundwater in responses to Earth tides and barometric effect. *Hydrol. Proc.* 32: 1927–1935.

Shreve, R.L. 1967. Infinite topologically random channel networks. *J. Geol.* 75: 178–186.

Shukla, A.K. and J.C. Prajapati. 2007. On a generalization of Mittag-Leffler function and its properties. *J. Math. Anal. Appl.* 336: 797–811.

Sierpiński, W. 1916. Sur une courbe cantorienne qui contient une image biunivoque et continue de toute courbe donnée. *C. R. Acad. Sci. Paris* (in French) 162: 629–632.

Singh, P.K., S.K. Mishra and M.K. Jain. 2014. A review of the synthetic unit hydrograph: from the empirical UH to advanced geomorphological methods. *Hydrol. Sci. J.* 59(2): 239–261.

SkØien, J.O., G. Blöschl and A.W. Western. 2003. Characteristic space scales and timescales in hydrology. *Water Resour. Res.* 39(10): 1304, doi:10.1029/2002WR001736.

Smiles, D.E. and M.J. Rosenthal. 1968. The movement of water in swelling materials. *Aust. J. Soil Res.* 6: 237–248.

Smiles, D.E. 1974. Infiltration into a swelling material. *Soil Sci.* 117(3): 140–147.

Smiles, D.E. 2000a. Use of material coordinates in porous media solute and water flow. *Chem. Eng. Sci.* 80: 215–220.

Smiles, D.E. 2000b. Material coordinate and solute movement in consolidating clay. *Chem. Eng. Sci.* 55: 773–781.

Smiles, D.E. 2000c. Hydrology of swelling soils: a review. *Aust. J. Soil Sci.* 38: 510–521.

Smiles, D.E. and P.A.C. Raats. 2005. Hydrology of swelling clay soils. pp. 1011–1026. *In*: M.G. Anderson [ed.]. Encyclopedia of Hydrological Sciences, Wiley, Chichester, England.

Smit, W. and H. de Vries. 1970. Rheological models containing fractional derivatives. *Rheol. Acta.* 9(4): 525–534.

Smith, R.E. and J.-Y. Parlange. 1978. A parameter-efficient hydrologic infiltration model. *Water Resour. Res.* 14(3): 533–538.

Smith, R.E. 1983. Approximate soil water movement by kinetic characteristics. *Soil Sci. Soc. Amer. J.* 47: 3–8.

Smith, R.E. 2002. Infiltration theory for hydrologic applications. *Water Resour. Monogr.* 15, Amer. Geophy. Union, Washington, D.C.

Smith, W.L. 1955. Regenerative stochastic processes. *Proc. Boy. Soc.* (London), Ser. A: 232: 6–31.

Sokolov, I.M., A.V. Chechkin and K. Klafter. 2004. Distributed-order fractional kinetics. *Acta Physica Polonica B.* 35: 1323–1341.

Soret, C. 1879. Etat d'équilibre des dissolutions dont deux parties sont portées à des températures différentes. *Arch. Sci. Phys. Natl. Geneve.* t.II: 48.

Soret, C. 1880. Influence de la température sur la distribution des sels dans leurs solutions. *Compte-Rendu de l'Académie des Sci. Paris* 91: 289–291.

Southwell, R.V. 1913. On the General theory of elastic stability. *Phil. Trans. Roy. Soc. Ser. A.* 213: 187–244.

Spane, F.A. 2002. Considering barometric pressure in groundwater flow investigations. *Water Resour. Res.* 38(6): 1078.

Spiegel, M.R. 1965. Laplace Transforms. McGraw-Hill, New York.

Sposito, G. 1995. Recent advances associated with soil water in the unsaturated zone. *Rev. Geophys, Supplement.* 1059–1065.

Sposito, G. 1998. Scale Dependence and Scale Invariance in Hydrology. Cambridge Univ. Press, Cambridge, England.

Srivastava, H.M. 1968. On an extension of the Mittag-Leffler function. *Yokohama Math. J.* 16(2): 77–88.

Srivastava, H.M. and R.G. Buschman. 1992. Theory and Applications of Convolution Integral Equations. Kluwer Acad. Publ. Dordrecht, The Netherlands.

Srivastava, H.M. and Ž. Tomovski. 2009. Fractional calculus with an integral operator containing a generalized Mittag-Leffler function in the kernel. *Appl. Math. Comput.* 211: 198–210.

Stanković, B. 1970. On the function of E.M. Wright. *Publ. de L'Inst. Math., Nouvelle serie.* 10(24): 113–124.

Starr, J.L. and J.-Y. Parlange. 1976. Solute transport in saturated soil column. *Soil Sci.* 121: 364–372.

Sternberg, Y.M. 1969. Some approximate solutions of radial flow problems. *J. Hydrol.* 7: 158–166.

Stoker, J.J. 1958. Water Waves. Wiley, New York.

Stokes, G.G. 1845. On the theories of the internal friction of fluids in motion, and of the equilibrium and motion of elastic solids. *Trans. Cambridge Phil. Soc.* 8: 287–319.

Stokes, G.G. 1851. On the effect of the internal friction of fluids on the motion of pendulums (1850). *Trans. Cambridge Phil. Soc.* 9: 8–106.

Straka, P. 2018. Variable order fractional Fokker-Planck equations derived from continuous time random walks. *Physica A: Stat. Mech. Appl.* 503: 451–463.

Strutt, J.W. (Lord Rayleigh). 1873. *Proc. Math. Soc. London.* 4: 357.

Strutt, J.W. (Lord Rayleigh). 1894. Theory of Sound. MacMillan, London (1st ed. 1877). I: 78; (2nd, ed. 1894) I: 102.

Su, N. 1993. Hydrological and hydraulic modelling of runoff-producing areas in small rural catchments. Ph.D. Thesis, The Australian National University, Canberra, Australia.

Su, N. 1994. A formula for computation of time-varying recharge of groundwater. *J. Hydrol.* 160: 123–135.

Su, N. 1995. Development of the Fokker-Planck equation and its solutions for modeling transport of conservative and reactive solutes in physically heterogeneous media. *Water Resour. Res.* 31(12): 3025–3032.

Su, N., F. Liu and J. Barringer. 1997. Forecasting floods at variable catchment scales using stochastic systems hydro-geomorphic models and digital topographic data. Proc. Modelling & Simulation (MODSIM97). Hobart, Tasmania, Australia, 8–11 Dec. 1997: 1695–1700 (https://www.mssanz.org.au/MODSIM97/Vol%204/Su5.pdf).

Su, N. 2002. The modified Richards equation and exact solutions for soil-water dynamics on eroding hillslopes. *Water Resour. Res.* 38(6): Article 8(1–6).

Su, N. 2004. Generalisation of various hydrological and environmental transport models using Fokker-Planck equation. *Environ. Model. & Software* 19(4): 345–356.

Su, N., G. Sander, F. Liu, V. Anh and D. Barry. 2005. Similarity solutions for solute transport in fractal porous media using a time- and scale-dependent dispersivity. *Appl. Math. Model.* 29: 852–870.

Su, N. and D. Midmore. 2005. Two-phase flow of water and air during aerated subsurface drip irrigation. *J. Hydrol.* 313(3-4): 158–165.

Su, N. and D. Midmore. 2006. Addendum to Two-phase flow of water and air during aerated subsurface drip irrigation. *J. Hydrol.* 330(3-4): 765.

Su, N. 2007. Radial water infiltration–advance–evaporation processes during irrigation using point source emitters in rigid and swelling soils. *J. Hydrol.* 344: 190–197. A correction to a figure: 2008. *J. Hydrol.* 360: 297.

Su, N. 2009a. Equations of anomalous adsorption onto swelling porous media. *Materials Letters* 63(28): 2483–2485.

Su, N. 2009b. *N*-dimensional fractional Fokker-Planck equation and its solutions for anomalous radial two-phase flow in porous media. *Applied Math. Comput.* 213(2): 506–515.

Su, N. 2010. Theory of infiltration: Infiltration into swelling soils in a material coordinate. *J. Hydrol.* 395: 103–108.

Su, N. 2012. Distributed-order infiltration, absorption and water exchange in swelling soils with mobile and immobile zones. *J. Hydrol.* 468-469: 1–10.

Su, N. 2014. Mass-time and space-time fractional partial differential equations of water movement in soils: theoretical framework and application to infiltration. *J. Hydrol.* 519: 1792–1803.

Su, N., P.N. Nelson and S. Connor. 2015. The distributed-order fractional-wave equation of groundwater flow: Theory and application to pumping and slug tests. *J. Hydrol.* 529: 1263–1273.

Su, N. 2017a. The fractional Boussinesq equation of groundwater flow and its applications. *J. Hydrol.* 547: 403–412.

Su, N. 2017b. Exact and approximate solutions of fractional partial differential equations for water movement in soils. *Hydrol.* 4(8): 1–13.

Su, X. 2009. Boundary value problem for a coupled system of nonlinear fractional differential equations. *Appl. Math. Lett.* 22: 64–69.

Suarez, D.L., J.D. Rhoades, R. Lavado and C.M. Grieve. 1984. Effect of pH on saturated hydraulic conductivity and soil dispersion. *Soil Sci. Soc. Am. J.* 481: 50–55.

Suarez, D.L. 1985. Chemical effects on infiltration. pp. 416–419. *In:* D.G. DeCoursey [ed.]. Proceedings National Resources Modeling Symposium. USDA. ARS, Washington, D.C.

Sun, H., W. Chen and Y. Chen. 2009. Variable-order fractional differential operators in anomalous diffusion modelling. *Physica A.* 388: 4586–4592.

Sun, H., W. Chen, H. Sheng and Y. Chen. 2010. On mean square displacement behaviors of anomalous diffusion with variable and random orders. *Phys. Lett. A.* 374: 906–910.

Sun, H., Y. Chen and W. Chen. 2011a. Random-order fractional differential equation models. *Signal Proc.* 91: 525–530.

Sun, H., W. Chen, H. Wei and Y.Q. Chen. 2011b. A comparative study of constant-order and variable-order fractional models in characterizing memory property of systems. *Eur. Phys. J.* 193: 185–192.

Sun, H., M.M. Meerschaert, Y. Zhang, J. Zhu and W. Chen. 2013. A fractal Richards' equation to capture the non-Boltzmann scaling of water transport in unsaturated media. *Adv. Water Resour.* 52: 292–295.

Sun, H., Y. Zhang, W. Chen and D.M. Reeves. 2014. Use of variable-index fractional-derivative model to capture transient dispersion in heterogeneous media. *J. Contam. Hydrol.* 157: 47–58.

Sun, H., A. Chang, Y. Zhang and W. Chen. 2019. A review on variable-order fractional differential equations: Mathematical foundations, physical models, numerical methods and applications. *Fract. Calc. Appl. Anal.* 22(1): 27–59.

Sun, Y., Y. Xiao, C. Zheng and K.F. Hanif. 2016. Modelling long-term deformation of granular soils incorporating the concept of factional calculus. *Acta Mech. Sin.* 32(1): 112–124.

Talsma, T. and A. van der Lelij. 1976. Infiltration and water movement in an *in situ* swelling soil during prolonged ponding. *Aust. J. Soil Res.* 14: 337–349.

Tamarkin, J.D. 1930. On integrable solutions of Abel's integral equation. *Ann. Math.* 2nd *Ser.* 31(2): 219–229.

Tarasov, V.E. 2005a. Fractional hydrodynamic equations for fractal media. *Ann. Phys.* 318(2): 286–307.

Tarasov, V.E. 2005b. Continuous medium model for fractal media. *Phys. Lett. A.* 336: 167–174.

Tarasov, V.E. 2013. Review of some promising fractional physical models. *Internl. J. Modern Phys. B.* 27(9): 1330005, 1–32.

Tarasov, V.E. 2016. On chain rule for fractional derivatives. *Commun. Nonlinear Sci. Numer. Simulat.* 30: 1–4.

Tarasov, V.E. 2018. No nonlocality. No fractional derivative. *Commun. Nonlinear Sci. Numer. Simulat.* 62: 157–163.

Tatom, F.B. 1995. The relationships between fractional calculus and fractals. *Fractals* 3(1): 217–229.

Tavares, D., R. Almeida and D.F.M. Torres. 2016. Caputo derivatives of fractional variable order: Numerical approximations. *Commun. Nonlinear Sci. Numer. Simulat.* 35: 69–87.

Taylor, G.I. 1953. Dispersion of soluble matter in solvent flowing slowly through a tube. *Proc. Roy. Soc. A.* 219(1137): 186–203.

Taylor, G.I. 1954. Dispersion of matter in turbulent flow through a pipe. *Proc. Roy. Soc. A.* 223(1155): 446–486.

Taylor, R. 2014. Hydrology: When wells run dry. *Nature* 516: 179–180.

Tejedor, V. and R. Metzler. 2010. Anomalous diffusion in correlated continuous time random walks. *J. Phys. A: Math Theor.* 43: 082002, 1–11.

Telyakovskiy, A.S., S. Kurita and M.B. Allen. 2016. Polynomial-based approximate solutions to the Boussinesq equation near a well. *Adv. Water Resour.* 96: 68–73.

Tempany, H.A. 1917. The shrinkage of soils. *J. Agric. Sci.* 8: 312–330.

Temple, G. 1953. Theories and applications of generalized functions. *J. Lond. Math. Soc.* 28: 134–148.

Temple, G. 1963. The theory of week functions. 1. *Proc. Royal Soc. London Ser. A, Math. Phys. Sci.* 276(1365): 149–167.

Terzaghi, K. 1923. Die berechnung der durchlassigkeitsziffer des tones aus dem verlauf der hydrodynamischen spannungserscheinungen. Akademie der Wissenschaften in Wein, Sitzungberichte. Mathematisch-Naturewissenschaftliche Klasse. Part IIa. 132(3-4): 125–138.

Terzaghi, K. 1956. Theoretical Soil Mechanics. 8th Printing, Wiley, New York.

Thurston, R.H. 1895. Materials of Construction. John Wiley, New York.

Thomson, Sir, W. (Lord Kelvin). 1842. On the uniform motion of heat in homogeneous solid bodies and its connection with the mathematical theory of electricity. *Cambridge Math. J.* 3: 71–84.

Thomson, W. (Lord Kelvin). 1855. On the Thermo-elastic and Thermo-magnetic Properties of Matter. *Quarterly J. Math.* I: 57–77.

Thomson, W. (Lord Kelvin). 1855. On the thermo-elastic and thermo-magnetic properties of matter. *Quart. J. Math.* 1(1855–1857) 55–77 = (with notes and additions) Phil. Mag. 5(5) (1878), 4–27 = Pt. VII of On the dynamical theory of heat. Papers 1: 291–316 (218, 262).

Thomson, Sir, W. 1884. Art. Elasticity in Encyclopaedia Britannica and Mathematical and Physical Papers, Vol. III. Also, Lectures on Molecular Dynamic. Baltimore.

Thomson, Sir, W. (Lord Kelvin). 1890. Mathematical and Physical Papers. Volume III. Elasticity, Heat, Electro-Magnetism (Collected papers from May 1841 to 1890). C.J. Clay & Sons, Cambridge Univ. Press, Cambridge, London.

Timashev, S.F., Y.S. Polyakov, P.I. Misurkin and S.G. Lakeev. 2010. Anomalous diffusion as a stochastic component in the dynamics of complex processes. *Phys. Rev. E.: Stat. Nonlin. Soft Matter Phys.* 81: 041128.

Tomovski, Ž., T.K. Pogány and H.M. Srivastava. 2014. Laplace type integral expressions for a certain three-parameter family of generalized Mittag-Leffler functions with applications involving complete monotonicity. *J. Franklin Inst.* 351: 5437–5454.

Troutman, B.M. and M.R. Karlinger. 1984. On the expected width function for topologically random channel networks. *J. Appl. Probab.* 21: 836–849.

Troutman, B.M. and M.R. Karlinger. 1985. Unit hydrograph approximations assuming linear flow through topologically random channel networks. *Water Resour. Res.* 21(5): 743–754.

Troutman, B.M. and M.R. Karlinger. 1986. Averaging properties of channel network using methods in stochastic branching theory. pp. 185–216. *In*: V.K. Gupta, I. Rodriguez-Iturbe and E.F. Wood [eds.]. Scale Problems in Hydrology. Reidel, Dordrecht.

Truesdell, C. 1953. Notes on the history of the general equations of hydrodynamics. *Amer. Math. Month.* 60(7): 445–458.

Truesdell, C. and W. Noll. 2004. The non-linear field theories of mechanics. *In*: S.S. Antman [ed.]. The Non-Linear Field Theories of Mechanics. Springer, Berlin.

Tyler, S.W. and S.W. Wheatcraft. 1990. Fractal processes in soil water retention. *Water Resour. Res.* 26(5): 1047–1054.

Tyler, S.W. and S.W. Wheatcraft. 1992a. Reply. *Water Resour. Res.* 28(2): 603–604.

Tyler, S.W. and S.W. Wheatcraft. 1992b. Fractal scaling of soil particle-size distributions: Analysis and limitations. *Soil Sci. Soc. Amer. J.* 56: 362–369.

Tzou, D.Y. 1995. A unified field approach for heat conduction from macro- to micro-scales. *ASME J. Heat Transfer* 117(1): 8–16.

Uchaikin, V.V. and V.V. Saenko. 2003. Stochastic solution to partial differential equations of fractional orders. *Siberian J. Numer. Math.* 6(2): 197–203.

Umarov, S. and R. Gorenflo. 2005a. Cauchy and nonlocal multi-point problems for distributed order pseudo-differential equations. Part one. *J. Analysis & Its Appl.* 24(3): 449–466.

Umarov, S. and R. Gorenflo. 2005b. On multi-dimensional random walk models approximating symmetrical space-fractional diffusion processes. *Fract. Calc. Appl. Anal.* 8(1): 73–88.

Umarov, S. and E. Saydamatov. 2006. A fractional analog of the Duhamel principle. *Fract. Calc. Appl. Anal.* 9(1): 57–70.

Umarov, S. and S. Steinberg. 2006. Random walk models associated with distributed fractional order differential equations. *IMS Lect. Notes–Monogr. Ser. High Dimen. Prob.* 51: 117–127.

Umarov, S. and S. Steinberg. 2009. Variable order differential equations and diffusion processes with changing modes. arXiv: 0903.2524v1 [math-ph].

United National Environment Programme (UNEP). 2017. Frontiers 2017: Emerging Issues of Environmental Concern. Nairobi, Kenya.

UNESCO. 2018. Nature-based Solutions for Water: The United Nations World Water Development Report 2018.

University of St. Andrews. 2000. Claude Louis Marie Henri Navier. https://www-history.mcs.st-andrews. ac.uk/Biographies/Navier.html.

Unterberger, A. and J. Bokobza. 1965a. Les opérateurs pseudodifferentiels d'ordre variable. *C.R. Acad, Sci. Paris Ser. A.* 261: 2271–2273.

Unterberger, A. and J. Bokobza. 1965b. Sur une generalisation des opérateurs de Calderon-Zygmund et des espaces H^s. *C.R. Acad. Sei. Paris Ser. A.* 260: 3265–3267.

Unterberger, A. 1973. Sobolev spaces of variable order and problems of convexity for partial differential operators with constant coefficients. *Astérisque.* 2-3: 325–341.

Valdes-Parada, F., J.A. Ochoa-Tapia and J. Alvarez-Ramirez. 2007. Effective medium equations for fractional Fick's law in porous media. *Phys. A: Stat. Mech. Its Appl.* 373: 339–353.

Valério, D. and J.S. da Costa. 2011. Variable-order fractional derivatives and their numerical approximations. *Signal Proc.* 91: 470–483.

Valério, D., J.J. Trujillo, M. Rivero, J.A.T. Machado and D. Baleanu. 2013. Fractional calculus: A survey of useful formulas. *Eur. Phys. J. Special Topics* 222: 1827–1846.

Valério, D., J.J. Machado and V. Kiryakova. 2014. Some pioneers of the applications of fractional calculus. *Fract. Calc. Appl. Anal.* 17(2): 552–578.

van der Heijde, P.K.M. 1988. Spatial and temporal scales in groundwater modelling. pp. 195–223. *In*: T. Rosswall, R.G. Woodmansee and P.J. Risser [eds.]. Scales and Global Change. SCOPE 35. Wiley, Chichester, England.

van der Knaap. 1959. Nonlinear behavior of elastic porous media. *Trans. AIME* 216: 179–187.

van der Kamp, G. and J.E. Gale. 1983. Theory of earth tide and barometric effects in porous formations with compressible grains. *Water Resour. Res.* 19(2): 538–544.

van Genuchten, M. and W.J. Alves. 1982. Analytical solutions of the one-dimensional convection-dispersion solute transport equation. US Dept. Agric. Tech. Bull. No. 1661, 151 p.

van Genuchten, M.Th. and P.J. Wierenga. 1976. Mass transfer studies in sorbing porous media, I, Analytical solutions. *Soil Sci. Soc. Am. J.* 40(4): 473–480.

van Genuchten, M.Th. 1980. A closed-form equation for predicting the hydraulic conductivity of unsaturated soils. *Soil Sci. Soc. Am. J.* 44: 892–898.

Vazquez, J.L. 2007. The Porous Media Equation: Mathematical Theory. Oxford Univ. Press, NY.

Veal, D.G. 1966. A Computer Solution of Converging, Subcritical Overland Flow. M.S. Thesis, Cornell Univ., New York.

Verruijt, A. 1969. Elastic storage of aquifers. pp. 331–376. *In*: J.J.M. de Wiest [ed.]. Flow through Porous Media. Acad. Press, New York, USA.

Villermaux, J. and W.P.M. van Swaay. 1969. Modèle reprèsentatif de la distribution des temps de sèjour dans un rèacteur semi-infini à dispersion axiale avec zoness tagnantes. Application à l'ècoullement ruisselant dans des colonnes d›anneaux Raschig. *Chem. Eng. Sci.* 24: 1007–1011.

Višik, M.I. and G.I. Èskin. 1967. Convolution equations of variable order. *Tr. Mosk. Mat. Obs.* 16: 25–50.

Voigt, W. 1882. Allgemeine Formeln für die Bestimmung der Elasticitätsconstanten von Krystallen durch die Beobachtung der Biegung und Drillung von Prismen. *Ann. Phys.* 16(2): 273–321, 398–416.

Voigt, W. 1887. Theoretische Studien über die Elasticitätsverhältnisse der Krystalle. *Abh. Ges. Wiss. Göttingen.* 34: 100.

Voller, V.R. 2011. On a fractional derivative form of the Green-Ampt infiltration model. *Water Resour. Res.* 34(2): 257–262.

Voller, V.R. 2014. Fractional Stefan problems. *Internl. J. Heat Mass Trans.* 74: 269–277.

von Helmholtz, H. 1876. Wied. *Ann.* 3.

Volterra, V. 1907. Sur l'équilibre des corps élastiques multiplement connexes. Annales scientifiques *de l'É.N.S.* 3e série, tome 24: 401–517. http://www.numdam.org/item?id=ASENS_1907_3_24__401_0.

Volterra, V. 1909. Sulle equazioni integro-differenziali della theoria dell'elasticita. *Atti Reale Accad. Naz. Lincei. Rend. Cl. Sci. Fis., Mat. e Natur.* 18: 295–300.

Volterra, V. 1913. Lecons sur les fonctions de Lignes. Gauthier-Villard, Paris.

Volterra, V. 1928. Sur la théorie mathématique des phénomènes des héréditaires. *J. Math. Pure Appl.* 7: 249–298.

Volterra, V. 1929. Alcune osservazioni sui fenomeni ereditarii. Rendiconti *della Reale Accad. Nazionale dei Lincei, Classe Sci. Fis. Mat e Nat.* (Ser. VII) 19: 585–595.

Volterra, V. 1940. Energia nei fenomeni elastici ereditari. *Acta Pontificia Acad. Sci.* 4: 115–128.

Volterra, V. 1959. Theory of Functionals and of Integral and Integrodifferential Equations, Dover, New York [First published in 1930].

Vot, F.L., E. Abad and S.B. Yuste. 2017. Continuous-time random-walk for anomalous diffusion in expanding media. *Phys. Rev. E.* 96: 032117.

Wagner, P. 2004. On the explicit calculation of fundamental solutions. *J. Math. Anal. Appl.* 297: 404–418.

Wang, H. 2000. Theory of Linear Poroelasticity with Applications to Geomechanics and Hydrogeology. Princeton Univ. Press, Princeton, New Jersey, USA.

Wanner, G. 2010. Kepler, Newton and numerical analysis. *Acta Numerica.* 2010: 561–598.

Weeks, S.W., G.C. Sander and J.-Y. Parlange. 2003. n-dimensional first integral and similarity solutions for two-phase flow. *ANZIAM J.* 44: 365–380.

Weiss, R. 1972. Product integration for the Generalized Abel equation. *Math. Comput.* 26(117): 177–190.

Werner, P.W. 1953. On non-artesian groundwater flow. *Geofisica pura e applicate.* 25(1): 37–43.

Werner, P.W. 1957. Some problems in non-artesian ground-water flow. *Trans. Amer. Geophys. Union.* 38(4): 511–518.

Wheatcraft, S.W. and S.W. Tyler. 1988. An explanation of scale-dependent dispersivity in heterogeneous aquifers using concepts of fractal geometry. *Water Resour. Res.* 24(4): 566–578.

Wheatcraft, S.W. and J.H. Cushman. 1991. Hierarchical approaches to transport in heterogeneous porous media. *Rev. Geophys. Supplement.* 263–269.

Wheatcraft, S.W. and M.M. Meerschaert. 2008. Fractional conservation of mass. *Adv. Water Resour.* 31: 1377–1381.

White, I. and M.J. Sully. 1987. Macroscopic and microscopic capillary length and time scales from field infiltration. *Water Resour. Res.* 23(8): 1514–1522.

Wiman, A. 1905. Über den fundamentalsatz in der teorie der funktionen. $E_\alpha(x)$. *Acta Math.* 29(1): 191–201.

Wineman, A. 2009. Nonlinear viscoelastic solids. *Math. & Mech. Solids* 14: 300–366.

Wright, E.M. 1933. On the coefficients of power series having exponential singularities. *J. London Math. Soc.* 1-8: 71–79.

Wright, E.M. 1935. The asymptotic expansion of the generalized hypergeometric function. *J. London Math. Soc.* 10: 287–293.

Wyckoff, R.D. and H.G. Botset. 1936. The flow of gas-liquid mixtures through unconsolidated sands. *Phys.* 7: 325–345.

Wyss, W. 1986. The fractional diffusion equation. *J. Math. Phys.* 27(11): 2782–2785.

Xu, P. and Yu, B. 2008. Developing a new form of permeability and Kozeny–Carman constant for homogeneous porous media by means of fractal geometry. *Adv. Water Resour.* 31: 74–81.

Yakubovich, S. and Yu. Luchko. 1994. The Hypergeometric Approach to Integral Transforms and Convolutions. *Math. Its Appl.* 287. Kluwer Acad. Publ., Dordrecht.

Yalin, M.S. 1971. Theory of Hydraulic Models. McMillan, London.

Yeh, H.-D. and Y.-C. Chang. 2013. Recent advances in modeling of well hydraulics. *Adv. Water Resour.* 51: 27–51.

Yu, B. and J. Li. 2001. Some fractal characters of porous media. *Fractals* 9(3): 365–372.

Yu, R. and H. Zhang. 2006. New function of Mittag-Leffler type and its application in the fractional diffusion-wave equation. *Chaos, Solitons & Fractals* 30: 946–955.

Yuan, F. and Z. Lu. 2005. Analytical solutions for vertical flow in unsaturated, rooted soils with variable surface fluxes. *Vadose Zone J.* 4: 1210–1218.

Zaslavsky, G.M. 1992. Anomalous transport and fractal kinetics. pp. 481–500. *In*: H.K. Moffatt, G.M. Zaslavsky, P. Compte and M. Tabor [eds.]. Topological Aspects of the Dynamics in Fluids and Plasmas. Kluwer, Dordrecht.

Zaslavsky, G.M. 1994a. Renormalization group theory of anomalous transport in systems with Hamiltonian chaos. *Chaos* 4(1): 25–33.

Zaslavsky, G.M. 1994b. Fractional kinetic equation for Hamiltonian chaos. *Phys. D: Nonlinear Phenomena.* 76(1-3): 110–122.

Zaslavsky, G.M. 2002. Chaos, fractional kinetics, and anomalous transport. *Phys. Rep.* 371: 461–580.

Zaslavsky, D. and A.S. Rogowski. 1969. Hydrologic and morphologic implications of anisotropy and infiltration in soil profile development. *Soil Sci. Soc. Am. Proc.* 33: 594–599.

Zayernouri, M. and G.E. Karniadakis. 2015. Fractional spectral collocation methods for linear and nonlinear variable order FPDEs. *J. Comput. Phys.* 293: 312–338.

Zener, C.M. 1938. Internal frictions in solids. II. General theory of thermoelastic internal friction. *Phys. Rev.* 53: 90–99.

Zener, C.M. 1948. Elasticity and Unelasticity of Metals. Univ. Chicago Press, Chicago.

Zhang, Y., B. Baeumer and D.A. Benson. 2006a. Relationship between flux and resident concentrations for anomalous dispersion. *Geophys. Res. Lett.* 33: L18407.

Zhang, Y., D.A. Benson, M.M. Meerschaert and H.P. Scheffler. 2006b. On using random walks to solve the space-fractional advection-dispersion equations. *J. Stat. Phys.* 123(1): 89–110.

Zhang, Y., M.M. Meerschaert and B. Baeumer. 2008. Particle tracking for time-fractional diffusion. *Phys. Rev. E.* 78: 036705.

Zhang, Y., D.A. Benson and D.M. Reeves. 2009. Time and space nonlocalities underlying fractional-derivative models: Distinction and literature review of field applications. *Adv. Water Resour.* 32: 561–581.

Zhang, Y., C. Green and B. Baeumer. 2014. Linking aquifer spatial properties and non-Fickian transport in mobile–immobile like alluvial settings. *J. Hydrol.* 512: 315–331.

Zhang, Y., C.T. Green and G.R. Tick. 2015. Peclet number as affected by molecular diffusion controls transient anomalous transport in alluvial aquifer-aquitard complexes. *J. Contam. Hydrol.* 177-178: 220–238.

Zhang, Y., L. Chen, D.M. Reeves and H.G. Sun. 2016. A fractional-order tempered-stable continuity model to capture surface water runoff. *J. Vibration & Control* 22(8): 1993–2003.

Zhang, Y., H.-G. Sun, B. Lu, R. Garrard and R.M. Neupauer. 2017. Identify source location and release time for pollutants undergoing super-diffusion and decay: Parameter analysis and model evaluation. *Adv. Water Resour.* 107: 517–524.

Zhao, R.-J. and Y.-L. Zhuang. 1963. Regional principles of rainfall-runoff relations. *J. East China Tech. Univ. Water Resour.* (Huadong Shuili Xueyuan Xuebao) 1: 53–67.

Zhao, R.-J. 1992. The Xinanjiang model applied in China. *J. Hydrol.* 135: 371–381.

Zhuang, P., F. Liu, I. Turner and Y.T. Gu. 2014. Finite volume and finite element methods for solving a one-dimensional space-fractional Boussinesq equation. *Appl. Math. Model.* 38: 3860–3870.

Zhuravkov, M.A. and N.S. Romanova. 2016. Review of methods and approaches for mechanical problem solutions based on fractional calculus. *Math. Mech. Solids* 21(5): 595–620.

Zwanzig, R.W. 1961. Statistical mechanics of irreversibility. pp. 106–141. *In*: W.E. Brittin, B.W. Downs and J. Downs [eds.]. Lectures in Theoretical Physics. Vol. III, Interscience, New York.

Zwanzig, R.W. 1964. Incoherent inelastic neutron scattering and self-diffusion. *Phys. Rev.* 133(1A): A50–A51.

Zwillinger, D. 1998. Handbook of Differential Equations. 3rd ed., Academic Press, San Diego.

Index